PRACTICAL PROBLEMS WITH SOLUTIONS IN WASTEWATER ENGINEERING

AUTHORS

Dr. S.N. Kaul obtained his bachelor and master degrees from IIT, Kanpur and was awarded Ph.D. degree from the University of New Castle Upon Tyne, UK. Dr. Kaul joined CSIR-National Environmental Engineering Research Institute [NEERI], Nagpur in 1970 and worked for over three decades. Dr. Kaul superannuated as Acting Director and Director Grade Scientist from NEERI and was also associated with MIT College of Engineering, Pune as its Principal. Dr. Kaul has published over 350 Papers in National and International journals and authored over 30 books on Environmental Science and Engineering, 8 Patents and 150 Conference papers. Dr. Kaul has designed and operated several waste treatment plants in India.

Dr. Kaul has supervised over a dozen of Ph.D. and 40 M.Tech. students. In addition, Dr. Kaul was responsible for organizing several National and International Conferences and has traveled extensively in India and abroad concerning various R&D programs. Dr. Kaul has received several National and International awards and is expert member of various important National committees on Environmental Science and Engineering.

Dr. D.R. Saini obtained his B.Tech. and M.Tech. from HBTI, Kanpur and Ph.D. (Chem. Engg.) from IIT-Delhi. Presently working as Principal Scientist at NCL, Pune. Dr. Saini has co-authored three books, six chapters and published seventy papers in international journals. Dr. Saini has guided 3 Ph.D. and few M.Tech. students.

Er. Prateek Kaul has obtained his B Tech and M Tech from NIT, Nagpur. He has worked in several multinational industries in India and abroad and has number of patents and publications to his credit.

PRACTICAL PROBLEMS WITH SOLUTIONS IN WASTEWATER ENGINEERING

VOLUME 4

Authors

Prof. (Dr.) S.N. Kaul
Dr. D.R. Saini
Er. Prateek Kaul

2016

Daya Publishing House®

A Division of

Astral International (P) Ltd

New Delhi 110 002

Practical Problems with Solutions in Wastewater Engineering (6 Volumes Set)

Volume 1: Page 0001-0398
Volume 2: Page 0399-0798
Volume 3: Page 0799-1198
Volume 4: Page 1199-1598
Volume 5: Page 1599-1998
Volume 6: Page 1999-2376

Cataloging in Publication Data—DK
   Courtesy: D.K. Agencies (P) Ltd. <docinfo@dkagencies.com>

Kaul, S. N., author.
   Practical problems with solutions in wastewater engineering / authors, Prof. (Dr.) S.N. Kaul, Dr. D.R. Saini, Er. Prateek Kaul.
   6 volumes   cm
      Includes bibliographical references (pages   ) and index.
      ISBN 9789351308973 (Vol. 4)
      ISBN 9789351243717 (International Edition)

      1. Sewage—Purification.   I. Saini, D. R., author.   II. Kaul, Prateek, author.   III. Title.
   DDC 628.3   23

Published by	:	**Daya Publishing House**® *A Division of* **Astral International Pvt. Ltd.** – ISO 9001:2008 Certified Company – 4760-61/23, Ansari Road, Darya Ganj New Delhi-110 002 Ph. 011-43549197, 23278134 E-mail: info@astralint.com Website: www.astralint.com
Laser Typesetting	:	**Classic Computer Services**, Delhi - 110 035
Printed at	:	**Thomson Press India Limited**

Dedicated to the Sacred Memory

of

Sri Sri Sarda Maa

and

Swami Vivekananda

PREFACE

Solving problems is one of basic function of engineers. Text books in the field of wastewater engineering devote most of their pages to the development and explanation of the theory. This Book is devoted to the pragmatic idea that performance and understanding of calculations should have more proportion in illustrating the principles. In order to help the reader An overview of wastewater treatment processes has been presented in Chapter-1.

This book is organized wherein a problem is stated and the necessary equations are given in the beginning followed by the solution. The present book stresses on the following topics which are of great importance to the country at large, *viz.*:

❑ Sludges

❑ Nutrient Removal

❑ Natural System of Treatment

❑ Stream Sanitation

❑ Process Economics

❑ Elimination of VOCs

❑ Data Analysis-Statistical Techniques

This book will be useful to students, teachers, practising engineers, and also the researchers in the field of wastewater treatment. Some of the topics bring together information which probably has not been accumulated nor presented in one place before. Ample notes have been given at the end of each problem to assist the reader in comprehending the entire process of the systems calculation, (salient points to ponder). Sufficient references have also been added in the problems.

Towards the end of the book a large number of references have been added to help the user of the book (Bibliography). In addition, two appendices have been given for necessary valuable data, some salient information in Wastewater engineering.

The authors have over three decades of experiences in teaching, research, and application of full scale plants in the field and, therefore a need was felt to pass the experience to all the stake holders. We are sure that the readers will enjoy reading the book as we did while collecting and collating the information for writing this book. The authors would like to thanks Dr. A. Gupta, Director Grade Scientist, CSIR-NEERI, Nagpur for writing a chapter on Statistical Techniques for us in this book.

The trust reposed by the Astral International (P) Ltd. New Delhi is thankfully acknowledged. In addition authors thank all persons who directly or indirectly helped us to bring out this book particularly the students, all our beloved parents, teachers and associates in the CSIR family.

Prof. (Dr.) S.N. Kaul

Dr. D.R. Saini

Er. Prateek Kaul

CONTENTS (VOL. 1-6)

ABBREVIATATION AND ACRONYMS

AACE	American Association of Cost Engineers
Al2(SO4)3.14.3(H2O)	Alum
ASCE	American Society of Civil Engineers
ASME	American Society of Mechanical Engineers
AWWA	American Water Works Association
AE	Aeration Efficiency
BOD	Biochemical oxygen demand
BDOC	Biodegradable Dissolved Organic Carbon
BNR	Biological Nutrient Removal
BPR	Biological Phosphorus Removal
BPT	Best Practicable Technology
BOD/COD	Biological or Chemical oxygen Demand
CBOD	Carbonaceous Biochemical Oxygen Demand at 5 days and 20°C
CFID	Continuous feed and intermittent discharge
$Ca(OH)_2$	Lime
CH_4	Methane
COD	Chemical oxygen demand
DO	Dissolved oxygen
DAF	Dissolved Air Flotation
DBP	Disinfection By-product
DCS	Distributed Control System
DO	Dissolved Oxygen
DOC	Dissolved Organic Carbon
EPEA	Environmental Protection and Enhancement Act
EC	Equipment cost
EC EDR	European Community Environmental Directive Requirements

ENR	Engineering News Record
EPA	Environmental Protection Agency
F/M	Food to micro-organisms ratio
Fe2(SO4)3	Ferric sulfate
FeCl3.6H2O	Ferric chloride
FeSO4.7H2O	Ferrous sulfate
F/M	Food to Microorganism ratio
G	Velocity Gradient
GCDWQ	Guidelines for Canadian Drinking Water Quality
GWUDI	Groundwater under the direct influence of surface water
GAC	Granular activated carbon
GCC	Gulf Cooperation Council
gfd	Gallon per square foot per day
H2S	Hydrogen sulfide
ha	Hectare
HRT	Hydraulic retention time
HPC	Heterotrophic Plate Count
HRT	Hydraulic Retention Time
IX	Ion exchange
IFID	Intermittent feed and intermittent discharge
KOH	Potassium hydroxide
MF	Microfiltration
mgd	Million of gallons per day
MLSS	Mixed liquor suspended solids
MWTP	Municipal waste-water treatment plant
MAC	Maximum Acceptable Concentration
MLSS	Mixed Liquor Suspended Solids
NH3-N	Ammonia nitrogen
N	Nitrogen
Na2S2O5	Sodium metabisulfite
Na2SO3	Sodium sulfite
NaOH	Sodium hydroxide

NF	Nan filtration
NGO	Non-governmental organization
NPDES	National pollutant discharge elimination system
NTS	Natural treatment system
NSF	National Sanitation Foundation
NTU	Nephelometric Turbidity Unit
OTE	Oxygen Transfer Efficiency
OTR	Oxygen Transfer Rate
ORP	Oxidation Reduction Potential
OU	Odour Unit
O&M	Operation and maintenance
OF	Overflow
ORP	Oxidation-reduction potential
P	Phosphorus
P2O7	Polyphosphate
PAC	Powered activated carbon
POTWs	Publicly owned treatment works
PO4-3	Orthophosphate
PVC	Polyvinyl chloride
PLC	Programmable Logic Controllers
QA/QC	Quality Assurance/Quality Control
RBC	Rotating Biological Contactor
RBC	Rotating biological contactor
RI	Rapid infiltration
RO	Reverse osmosis
SBR	Sequencing batch reactor
SCADA	Supervisory control and data acquisition
SO_2	Sulfur dioxide
SR	Slow rate
SRT	Solids retention time
SS	Suspended solids
SAR	Sodium Adsorption Ratio

SBR	Sequencing Batch Reactor
SRT	Sludge Retention Time
TBOD	Total Biochemical Oxygen Demand at 5 days and 20 oC
TOC	Total Organic Carbon
TP	Total Phosphorus
TSS	Total Suspended Solids
TTHM	Total Trihalomethanes
TCC	Total construction costs
TIC	Total indirect cost
TOC	Total organic carbon
TOD	Total oxygen demand
TSS	Total dissolved solids
UF	Ultra filtration
UNEP	United Nations Environment Programme
USAID	United States Agency for International Development
USEPA	United States Environmental Protection Agency
UV	Ultraviolet
UC	Uniformity Coefficient
UV	Ultraviolet
WaTER	Water Treatment Estimation Routine
WEF	Water Environment Federation
WWTP	Waste-water treatment plant

[a]ratio of apparent volumetric mass transfer coefficient in wastewater to clean water

$$\dfrac{10^3\,L}{m^3}$$

$$= 0.7 \times 10^{-6}\ mol$$

The total moles in both phases is :

$$n_B = n_{Ba} + n_{Bw} = 3.2 \times 10^{-6}\ mol$$

From the density of pure benzene, we calculate the volume as:

$$V_B = \dfrac{n_B}{C_B} = \dfrac{n_B M_B}{\rho_B} = \dfrac{(3.2\times10^{-6}\ mol)(78.11g\ mol^{-1})}{0.8787\ g\ cm^{-3}} \times \dfrac{10^3\,\mu L}{cm^3}$$

$$= 0.28\ \mu L.$$

Example 6.40

Flow Through a Circular Pipe

Determine the average velocity and the discharge through an 18 in. diameter pipe placed on a slope of 0.003 ft/ft if the n value for clean uncoated cast iron pipe is assumed to be 0.013 and the depth of flow in the pipe is 10 in.

1. Required expression for flow in a circular conduits (partially full)

$$Cos\left(\dfrac{\theta}{2}\right) = 1 - \dfrac{2\,y_D}{D}$$

$$A = \dfrac{D^2}{4}\left[\dfrac{\pi\theta}{360} - \dfrac{Sin\,\theta}{2}\right]$$

$$WP = \dfrac{\theta}{360}\,(2\,\pi r)$$

where   A   = Cross-sectional area (ft^2)

   D   = Diameter of circular section (ft)

   R   = Radius of circular section (ft)

   WP = Wetted perimeter (ft)

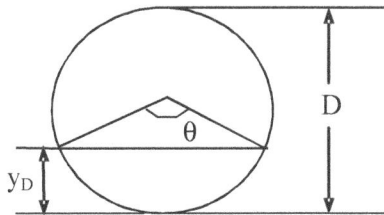

Figure 5 : Cross-section of circular pipe.

Dake has shown that the best performance of a circular section occurs when :

$$4\theta - 6\theta(\text{Cos}\theta) + \text{Sin}(2\theta) = 0$$

or, $\theta = 154°$

or, $y_D = 1.9r$

It means that a circular section discharge most effectively when it is following about 95% full. A circular section passes its maximum flow at $y_D = 1.9$ r.

2. Required flow through a partially flowing circular pipe

  • Value of θ :

$$\cos\left(\frac{\theta}{2}\right) = 1 - \frac{2\, y_D}{D}$$

$$= 1 - \frac{2(10)}{18}$$

$$q = 160°$$

Cross-sectional area (A) :

$$A = \frac{D^2}{4}\left[\frac{\pi\theta}{360} - \frac{\sin\theta}{2}\right]$$

$$= \frac{(18/12)^2}{4}\left[\frac{\pi(160)}{360} - \frac{\sin(160)}{2}\right]$$

$$= 0.72 \text{ ft}^2$$

Wetted perimeter (WP) :

$$WP = \frac{\theta}{360}\,(\pi D)$$

$$= \frac{160}{360}[\pi(18/12)]$$

$$= 2.09 \text{ ft}$$

Hydraulic radius (R) :

$$R = \frac{A}{WP} = \frac{0.72}{2.09} = 0.34 \text{ ft}$$

Average velocity (Manning's equation) :

$$V = \frac{1.49}{n}\, R^{2/3}\, S^{0.5}$$

$$= \frac{1.49}{0.013}(0.34)^{2/3}(0.003)^{0.5}$$

Flow rate (Q) :

$$Q = A\ V$$
$$= (0.72)(3.06) = 2.2\ ft^3/s.$$

Example 6.41

Wastewater Pump Characteristics

The characteristics of a centrifugal pump operating at two different speeds are listed below. Graph these curves and connect the best efficiency points (bep) with a dashed line. Calculate head-discharge values for an operating speed of 1450 revolutions per minute (rpm) and plot the curve. Finally, sketch the pump operating envelope between 60 and 120 percent of the best efficiency points.

Table 18

	Speed = 1750 rpm			*Speed = 1150 rpm*	
Discharge (gpm)	*Head (ft)*	*Efficiency (%)*	*Discharge (gpm)*	*Head (ft)*	*Efficiency (%)*
0	220	-	0	96	-
1500	216	63	1000	93	65
2500	203	81	1500	89	77
3000	192	85	2000	82	83
3300	182	86[a]	2200	77	84[a]
3500	176	85	2500	70	83
4500	120	72	3000	49	71

[a]best efficiency point.

Solution

The characteristics curves for 1750 rpm and 1150 rpm are drawn in Figure. 1, with efficiency values listed at the plotting points. The best efficiency points are connected with a dashed line.

The head-discharge values for an operating speed of 1450 rpm calculated using equation 1 and 2. For example, using the data of 1500 gpm and 216 ft at 1750 rpm, one plotting points is calculated as :

$$Q_2 = Q_1 \frac{N_2}{N_1} = 1500 \frac{1450}{1750} = 1240\ \text{gpm} \tag{1}$$

$$H_2 = H_1 \left(\frac{N_2}{N_1} \right)^2 = 216 \left(\frac{1450}{1750} \right)^2 = 148\ \text{ft} \tag{2}$$

All of the values calculated from the data at 1750 rpm are 1240 gpm at 148 ft, 2040 at 140, 2450 at 132, 2730 at 125 (bep), 2860 at 121, and 3730 gpm fat 82.

The boundaries of the operating envelope at 1750 rpm are calculated as 60 and 120 percent of the discharge at the best operating point of 3300 gpm.

$$0.60 \times 3300 = 2000 \text{ gpm}$$

$$1.20 \times 3300 = 4000 \text{ gpm}$$

At 1150 rpm for a bep of 2200 gpm, the limits are 1300 gpm and 2600 gpm. The pump operating envelope.

Note :

1. **Pump Characteristics :** Normally, a pump can be furnished with impellers of different diameters, within a specified range, for each size casing. Since discharge of a pump changes with impeller diameter and operating speed, manufacturers publish data showing the characteristic curves for impellers of several diameters operating at two or more speeds. For a given impeller diameter operated at different speeds, the discharge is directly proportional to the speed, head is proportional to the square of the speed, and power input varies with the cube of the speed :

$$\frac{Q_1}{Q_2} = \frac{N_1}{N_2} \tag{1}$$

$$\frac{H_1}{H_2} = \frac{N_1^2}{N_2^2} \tag{2}$$

$$\frac{P_{i1}}{P_{i2}} = \frac{N_1^3}{N_2^3} \tag{3}$$

where Q = discharge, gallons per minute (liters per second)

H = head, feet (meters)

P_i = power input, horsepower (kilowatts)

For a pump operating at the same speed, a change in impeller affects discharge, head, and power input approximately as follows :

$$\frac{Q_1}{Q_2} = \frac{D_1}{D_2}$$

$$\frac{H_1}{H_2} = \frac{D_1^2}{D_2^2}$$

$$\frac{P_{i1}}{P_{i2}} = \frac{D_1^3}{D_2^3}$$

where D = impeller diameter, inches (centimeters).

2. **Power and Efficiency :** The power output of a pump is the work done per unit of time in lifting the water to higher elevation. The equation relating discharge and head to power delivered is

$$P_o = \frac{w\,QH}{550 \times 60} = \frac{QH}{3960}$$

where P_o = power output, horsepower

Q = discharge, gallons per minute

H = head, feet

w = unit weight of water = 8.34 lb/gal

550 = foot pounds/second per horsepower

60 = seconds per minute

$$3960 = 550 \times \frac{60}{8.34}$$

In SI metric units, power output of a pump is

P_o = wQH = 0.0098 QH (SI units)

where P_o = power output, kilowatts (kN.m/s)

Q = discharge, liters per second

H = head, meters

w = unit weight of water = 0.0098.

The efficiency of a pump is the ratio of the power output to measured power input. (Power input is some-times called brake horsepower bhp.)

$$E_p = \frac{P_o}{P_i}$$

where E_p = pump efficiency, dimensionless

P_i = power input (brake horsepower), horsepower (kilowatts)

P_o = power output, horsepower (kilowatts)

Centrifugal pump efficiency usually ranges from 60 to 85 percent.

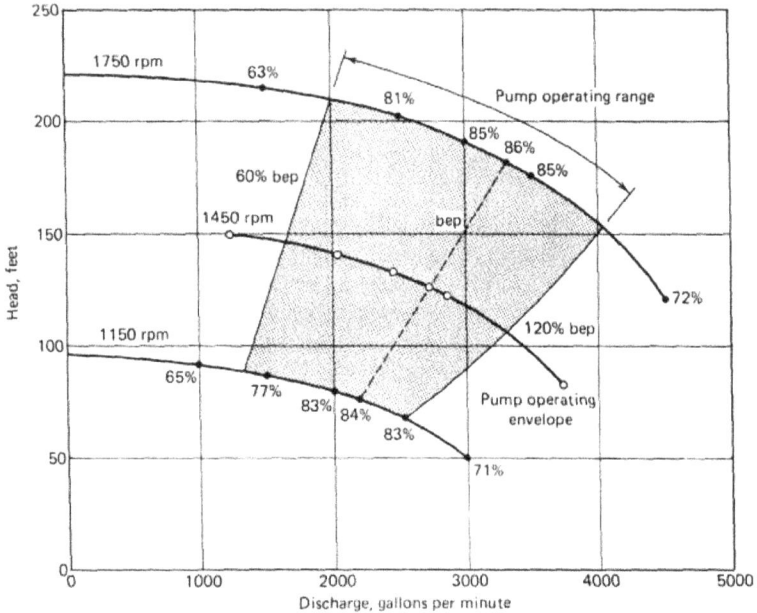

Figure 6 : Characteristic pump curves showing the pump operating envelope.

Example 6.42

Pumps: Net Positive Suction Head

Determine the net positive suction head (NPSH) for a pump. Assume the following data:

Elevation of the discharge point from sump well (h_{st})	: 25 m
Overall head losses of the system ($h_{fs} + h_{fd}$)	: 5 m
[Total head loss in the suction pipe + discharge pipe]	
Vertical distance for the sump well level to the pump center line (h_e)	: 1.5 m
Discharge velocity (V_2)	: 1.25 m/s
Suction friction losses	: 1.0 m
Diameter of discharge pipe	: 0.250 m
Diameter of suction pipe	: 0.225 m

Solution

1. Required expression of net positive suction head [NPSH]

    NPSH is the amount of energy possessed by a fluid at the inlet to the pump. It is the inlet dynamic head that pushes the fluid into the pump impeller blades. If NPSH does not exist at the suction side, cavitation will occur.

It is possible to pump groundwater from any depth, if impeller of the pump are laid out in series [heads are added], and this principle is also used in the design of deep-well pumps.

The pushing pressure absolute and not gage converts to available head or energy at the suction side of the pump. The surface of the sump is below the pump, so this available head must be substracted by h_e. The other substraction are the friction losses (suction side) h_{fs}. For the source tank fluid level above the centreline of the pump impeller (h_s) will be added increasing the available energy.

$$\text{NPSH} = \left[\frac{P_{atm} - P_V}{\gamma}\right] - h_e(+h_s) - h_{fs}$$

Point-1 : Sump well level

Point-2 : Inlet to the pump

Energy equation

$$\frac{V_1^2}{2g} + Z_1 + \frac{P_1}{\gamma} - h_{fs} = \frac{V_2^2}{2g} + Z_2 + \frac{P_2}{\gamma} + 0 + 0 + \left[\frac{P_{atm} - P_V}{\gamma}\right] - h_{fs}$$

$$= \frac{V_2^2}{2g} + h_e + \frac{P_2}{\gamma} \tag{1}$$

[P_2 = Absolute pressure at the inlet to the pump]

$$\frac{P_2}{\gamma} + \frac{V_2^2}{2g} = \frac{P_{2,g} + P_{atm} - P_V}{\gamma} + \frac{V_2^2}{2g}$$

By definition, NASH is :

$$\text{NPSH} = \left[\frac{P_{atm} - P_V}{\gamma}\right] - h_{fs} - h_e(\text{or} + h_s)$$

Therefore,

$$\text{NPSH} = \frac{P_2}{\gamma} + \frac{V_2^2}{2g} = \frac{V_2^2}{2g} + \frac{P_{2,g} + P_{atm} - P_V}{\gamma}$$

$$= \frac{V_2^2}{2g} + \left[h_i + \frac{P_{atm} - P_V}{\gamma}\right]$$

As,

$$\frac{P_{2,g}}{\gamma} + \frac{P_{atm} - P_V}{\gamma} = h_i + \frac{P_{atm} - P_V}{\gamma}$$

where $P_{2,g}$ = Gage pressure at the inlet to the pump

           P_V = Vapour pressure of the fluid

           h_e = Suction lift

           h_s = Suction head

2. Required net positive suction head (NPSH)

$$NPSH = \frac{P_{atm} - P_V}{\gamma} - h_e(+h_s) - h_{fs}$$

$$P_{atm} = 101325 \ N/m^2$$

$$h_e = 1.5 \ m$$

$$h_{fs} = 1.0 \ m$$

$$P_V \ at \ 25°C = 3167 \ N/m^3$$

$$\gamma \ at \ 25°C = \rho \ g = 998(9.81) \ N/m^3$$

Therefore,

$$NPSH = \frac{101325 - 2340}{998(9.81)} - 1.5 - 1.0$$

$$= 10.110 - 1.5 - 1.0$$

$$= 7.6 \ m \ of \ water$$

[No cavitation as it is + ve].

Note :

- *Safety margin regarding NPSH :* Provide a margin of safety in the neighbourhood of 90% of the calculated NPSH. If the computed NPSH is around 7 m, assume it to be 0.9(7 m) = 6.3 m.

Example 6.43

Pumps : Total Developed Head

Determine the following for a centrifugal pump :

- Will cavitation occur?

- Total dynamic heads (inlet and outlet condition of the pump)

- Manometric heads (inlet and outlet).

Assume the following data :

Elevation above the sump   : 25 m

Friction losses and velocity at the discharge side of the pump   : 15 m and 1.5m/s

Suction frictional losses	: 0.75 m/s
Operating drive	: 1200 rpm
Diameter of suction pipe	: 200 mm
Diameter of discharge pipe	: 175 mm
Distance (vertical) from the sump pool of the pump centerline	: 1.5 m

Solution

1. Required cavitation effect

    Applying Bernoulli equation between sump pool level-1 and inlet to the pump 2 :

    $$\frac{V_1^2}{2g} + Z_1 + \frac{P_1}{\gamma} + h_q - h_f + h_p = \frac{V_2^2}{2g} + Z_2 + \frac{P_2}{\gamma}$$

    Cross-section area of discharge pipe $(A_d) = \frac{\pi D^2}{4} = \frac{3.14(0.175)^2}{4}$

    $$= 0.024 \text{ m}^2$$

    Cross-section area of suction pipe $(A_2) = \frac{\pi D^2}{4} = \frac{3.14(0.200)^2}{4}$

    $$= 0.0314 \text{ m}^2$$

    Discharge flow rate $= V_d(A_d)$

    $$= 1.5 \text{ m/s } (0.024 \text{ m}^2)$$

    $$= 0.036 \text{ m}^3/\text{s}$$

    Applying continuity equation :

    $$V_2(A_2) = 0.036 \text{ m}^3/\text{s}$$

    $$V_2 = \frac{0.036 \text{ m}^3/\text{s}}{0.0314 \text{ m}^2} = 1.146 \text{ m/s}$$

    Therefore, Bernoulli equation yields

    $$[0+0+0+0-0.75+0] = \frac{(1.146)^2}{2(9.81)} + 1.5 + \frac{P_2}{\gamma}$$

    or $\quad \frac{P_2}{\gamma} = -[0.070+1.5+0.75] \text{ m}$

    $$= -2.32 \text{ m [pressure used in equation is 0, this value corresponds to the manometric head to the pump]}$$

Vapour pressure at 25°C = 3.167 kN/m² = 0.323 m of water

[1 atmosphere = 10.34 m of water]

Pump cavitation = – (10.34 – 0.323) = – 10.017 m and is much less than – 2.32 m; and no cavitation will occur in the pump.

2. Required inlet and outlet total developed head

Applying Bernoulli equation between sump level 1 and the discharge point 2 :

$$\frac{P_1}{\gamma} + \frac{V_1^2}{2g} - h_{lp} - h_{fs} - h_{fd} + h_p = \frac{P_2}{\gamma} + \frac{V_2^2}{2g} + h_{st}$$

$$[-h_{lp} + h_p] = \text{Total developed head (TDH)} = \left[\frac{P_2}{\gamma} - \frac{P_1}{\gamma}\right] + \left[\frac{V_2^2}{2g} - \frac{V_1^2}{2g}\right]$$

$$+ h_{st} + h_{fs} + h_{fd}$$

where h_{lp} = Head losses in pump

h_p = Pump head

h_{fs} = Frictional losses on suction side

h_{fd} = Frictional losses on discharge side

h_{st} = Elevation above sump level

$$\left[\frac{P_2}{\gamma} - \frac{P_1}{\gamma}\right] = 0; \qquad h_{st} = 25 \text{ m}$$

$$V_1 = 0$$

$$\frac{V_2^2}{2g} = \frac{(1.5)^2}{2(9.81)} = 0.1147 \text{ m}$$

$$h_{st} + h_{fd} = 15 \text{ m}$$

Therefore, TDH = (0 + 0.1147 + 15 + 25) m = 40.1147 m

TDH between inlet and outlet of the pump is:

$$\text{TDH} = \text{TDH}_{vivo} = h_{abo} + h_{vo} - (h_{abi} + h_{vi})$$

$$40.1147 \text{ m} = h_{abo} + \frac{(1.5)^2}{2(9.81)} - \left[(-2.32 + 10.34) + \frac{(1.146)^2}{2(9.81)}\right]$$

$$= h_{abo} + 0.1147 - [8.02 + 0.070]$$

$$h_{abo} = 48.07 \text{ m}$$

Therefore, outlet manometric head = (48.07 – 10.34) m

$$= 37.77 \text{ m of water.}$$

3. Required inlet and outlet dynamic head

Inlet dynamic head = $h_{abi} + h_{vi}$ [i refers to inlet to the pump]

$$= (-2.32 + 10.34) + \frac{(1.146)^2}{2(9.81)}$$

$$= 8.02 + 0.07 = 8.09 \text{ m}$$

Outlet dynamic head = $h_{abo} + h_{vo}$; [0 refers to outlet to the pump)

$$= 48.07 + \frac{(1.5)^2}{2(9.81)}$$

$$= 48.07 + 0.1147 = 48.1847 \text{ m}.$$

Note :

- Required frictional losses

  * Pipes (h_{ft}) = $f\left(\dfrac{L}{D}\right)\dfrac{V^2}{2g}$

  * Fittings, etc. (h_{fm}) = $K\dfrac{V^2}{2g}$

Example 6.44

Pumps : Total Developed Head

Determine the total developed head (TDH) for a specific pump with the following data:

Pressure at the outlet of the pump	:	200 kN/m² (guage)
Discharge flow rate	:	0.20 m³/s
Diameter of the outlet discharge pipe	:	0.4 m
Water temperature	:	25°C
Motor housepower	:	66.666
Motor efficiency	:	90%
Pump loss $(-h_{lp})$		

Solution

1. Required expression for the total developed head (TDH) pump efficiency (h), and head equivalent to brake power input to the pump form a prime mover (h_{brake})

    Total developed head (TDH)

    $$TDH = -h_{lp} + h_p$$

where h_{lp} = Pump loss

           h_p = Head given to pump

$$TDH = \left(\frac{P_{g,o}}{\gamma} + \frac{V_o^2}{2g} \right)$$

where $P_{g,o}$ = Guage pressure at outlet of a pump

           V_o = Discharge velocity at the outlet of a pump

Brake efficiency (η)

$$\eta = \frac{1}{h_{brake}} \left(\frac{P_{g,o}}{\gamma} + \frac{V_o^2}{2g} \right)$$

where h_{brake} = Net head input to the pump

Brake head (h_{brake})

$$h_{brake} = \frac{745.7 h_p}{Q\gamma}$$

where Q = Rate of flow (m^3/s)

           h_p = Net head developed by the pump (N.m/s)

           γ = Specific weight (N/m^3)

2. Required total developed head (TDH)

$$V_o = \frac{Q}{\pi D^2/4} = \frac{0.2(4)}{3.14(0.4)^2} = 1.59 \ m/s$$

$$TDH = \frac{P_{g,o}}{\gamma} + \frac{V_o^2}{2g}$$

$$= \frac{200(1000)}{998(9.81)} + \frac{(1.59)^2}{2(9.81)}; \ [\gamma = \rho g]$$

$$TDH = -h_{lp} + h_p = (20.43 + 0.13) = 20.56 \ m \ of \ water.$$

3. Required head equivalent to brake power input to the pump from a prime mover

$$h_{brake} = \frac{745.7 h_p}{Q\gamma}$$

$$h_p = TDH + h_{lp}$$

$$= 20.56 + 0.20 \ ; \ [h_{lp} = 0.2]$$

$$= 20.76 \ m \ of \ water$$

HP input to the pump = 0.9 × 66.666

$$= 60 \text{ hp}$$

$$h_{brake} = \frac{745.7 \text{ hp}}{Q\gamma}$$

$$= \frac{745.7(60)}{0.2 \text{ m}^3/\text{s} (998 \times 9.81)} = 22.8 \text{ m of water.}$$

4. Required pump efficiency (η)

$$\eta = \frac{1}{h_{brake}} \left[\frac{P_{g,o}}{\gamma} + \frac{V_o^2}{2g} \right]$$

$$= \frac{TDH}{h_{brake}}$$

$$= \frac{20.56 \text{ m of water}}{22.80 \text{ m of water}} = 0.89 (= 89\%).$$

Example 6.45

Pumps : Pump Scaling Laws [Homologous Pump]

Determine the total developed head (H), flow rate (Q), brake power (P_{brake}), and pump efficiency (η) if the pump impeller diameter (D) is changed from 0.25 to 0.30 m. The rotational speed remains the same. Assume the following data:

Head developed (H_1) at D = 0.25 m   : 15 m

Rate of flow (Q_1) at D = 0.25 m   : 0.15 m³/s

Power (brake) (P_{brake})$_1$ at D = 0.25 m   : 50 kw

Pump efficiency (η_1)   : 60%.

Solution

1. Required expressions for H, Q, P$_{brake,}$ and h

$$\left[\frac{Hg}{\omega^2 D^2} \right]_2 = \left[\frac{Hg}{\omega^2 D^2} \right]_1$$

$$H_2 = \left(\frac{D_2}{D_1} \right)^2 \times H_1; \text{ [At same speed (ω)]}$$

$$\left[\frac{Q}{\omega D^3} \right]_2 = \left[\frac{Q}{\omega D^3} \right]_1$$

$$Q_2 = \left(\frac{D_2}{D_1}\right)^3 Q_1$$

$$\left[\frac{P_{brake}}{\rho\omega^3 D^5}\right] = \left[\frac{P_{brake}}{\rho\omega^3 D^5}\right]_1 = C_p\eta; \quad [C_p = 1 \text{ for the same pump}]$$

where C_p = Pumping coefficient

$$[P_{brake}]_2 = \left(\frac{D_2}{D_1}\right)^5 [P_{brake}]_1.$$

2. Required H, Q, P_{brake} and h at D = 0.30 m and same speed for the same pumps

$$H_2 = \left[\frac{D_2}{D_1}\right]^2 H_1$$

$$= \left[\frac{0.30}{0.25}\right]^2 (15) = 21.6 \text{ m of water}$$

$$Q_2 = \left[\frac{0.30}{0.25}\right]^3 (0.15) = 0.259 \text{ m}^3/\text{s}$$

$$[P_{brake}]_2 = \left[\frac{0.30}{0.25}\right]^5 (50) = 124.4 \text{ kw}$$

$$(C_p\eta)_2 = (C_p\eta)_1$$

$$\eta_2 = \eta_1; \quad [C_p \text{ remains same for the pump}]$$

$$\eta_2 = 60\%.$$

Example 6.46

Pumps : Pump Scaling Laws [Homologous Pump]

Determine the resulting head (H), flow rate, (Q), power brake (P_{brake}), and pump efficiency (h) using pump scaling laws for an operating speed of 1200 rpm. Assume the following data :

Before changing the speed to 1200 rpm, the speed of the pump : 900 rpm

Head developed at 900 rpm   : 11m

Flow rate at 11 m head   : 0.2 m³/s

Pump efficiency (η_1)   : 60%

Solution

1. Required expressions of H, Q, P_{brake}, and h for pumps

    Total developed head (TDH) = f $(\rho, \omega, D, Q, \mu, \epsilon, l)$

    where ρ = Mass density

        w = Rotational speed of the impeller

        μ = Dynamic viscosity

        D = Impeller diameter

        Q = Flow rate

        ϵ = Roughness of chamber

        l = Characteristic length

    The dimensionless numbers are (ϵ is small and represent l) :

    Therefore, $\Delta P = f (\rho, \omega, D, Q)$

    Only two dimensionless numbers are [(5 – 3) = 2] :

    $$\pi_1 = \frac{\Delta P}{\rho \omega^2 D^2}$$

    $$\pi_2 = \frac{Q}{\omega D^3}$$

    Final relationship is :

    $$\frac{\Delta P}{\rho \omega^2 D^2} = \phi \left(\frac{Q}{\omega D^3} \right)$$

    $$\frac{\Delta P}{\gamma} = H \text{ (Total developed head)}$$

    $$\frac{Hg}{\omega^2 D^2} = \phi \left(\frac{Q}{\omega D^3} \right); \quad [\gamma = \rho g]$$

    and $\left[\dfrac{P}{\rho \omega^3 D^5} \right]_1 = \left[\dfrac{P}{\rho \omega^3 D^5} \right]_2 = C_p \eta$

    where C_p = Power coefficient

        h = Pump efficiency

        P = Power given to the fluid

    Therefore, $h_{brake} = \dfrac{P_{brake}}{\gamma Q}$

    $(C_p \eta)_2 = (C_p \eta)_1$ or $\eta_2 = \eta_1$ [if C_p remains some]

And, the simplified

scaling laws for a given pump operating at different speeds (w) are:

$$H_2 = \left(\frac{\omega_2}{\omega_1}\right)^2 H_1$$

$$Q_2 = \left(\frac{\omega_2}{\omega_1}\right) Q_1$$

$$P_2 = \left(\frac{\omega_2}{\omega_1}\right)^3 P_1$$

$$[P_{brake}]_2 = \left(\frac{\omega_2}{\omega_1}\right)^3 [P_{brake}]_1$$

Similarly, pumps of constant rotational speed (w), but of different diameter (D), the simplified scaling laws are :

$$H_2 = \left(\frac{D_2}{D_1}\right)^2 H_1$$

$$Q_2 = \left(\frac{D_2}{D_1}\right)^3 Q_1$$

$$P_2 = \left(\frac{D_2}{D_1}\right)^5 P_1$$

$$[P_{brake}]_2 = \left(\frac{D_2}{D_1}\right)^5 [P_{brake}]_1$$

2. Required H, Q, P_{brake} and η at 1200 rpm^2

$$\left[\frac{Hg}{\omega^2 D^2}\right]_2 = \left[\frac{Hg}{\omega^2 D^2}\right]_1$$

$$H_2 = \left(\frac{\omega_2}{\omega_1}\right)^2 H_1$$

$$= \left(\frac{1200}{900}\right)^2 11 = 19.56 \text{ m}$$

$$\left[\frac{Q}{\omega D^3}\right]_2 = \left[\frac{Q}{\omega D^3}\right]_1$$

$$Q_2 = \left(\frac{\omega_2}{\omega_1}\right) Q_1$$

$$= \left(\frac{1200}{900}\right)(0.2) = 0.267 \text{ m}^3/\text{s}$$

$$\left[\frac{P_{brake}}{\rho \omega^3 D^5}\right]_2 = \left[\frac{P_{brake}}{\rho \omega^3 D^5}\right]_1$$

$$[P_{brake}]_2 = \left(\frac{\omega_2}{\omega_1}\right)^3 [P_{brake}]_1$$

$$= \left(\frac{1200}{900}\right)^3 (25); \ [P_{brake}]_1 = 25 \text{ kW}$$

$$= 59.26 \text{ kW}$$

$$\eta_2 = \eta_1$$

$$\eta_2 = 60\%; \ [Cp = \text{power coefficient remains some}].$$

Note :

[A] Axial Pump

Axial pumps may be used for pumping treatment plant effluents and settled wastewater influent.

Axial pumps are dependable with only routine maintenance operation, relatively quiet operation, provide even pumping flow, and do not require large space.

Axial pumps are suitable only at low heads, and cannot pump wastewater containing trash or other solids that may clog it.

[B] Pneumatic Ejector

Mostly used in lift stations, to pump treatment plant screenings, and to pump raw sludge and sludge cake.

Little maintenance required, automatically operated, not easily clogged by solids; closed system prevents escape of sewer gas.

Low efficiency, large volume of air needed, and low volume pumping capability.

[C] Displacement Pump

For pumping primary, activated, digested, chemical sludges, and for chemical feed systems.

Constant capacity against widely varying head, self primary, handless large solids, large discharge head may be provided.

Not suitable for operation requiring smooth steady discharge, low volumes handled; maintenance may be required for lubricating valves and pistons.

Example 6.47

Sludge Pumping Head loss

Use the following equations to determine the piping head loss 1500 ft of 6 in pipe :

$$\frac{H}{L}(\text{Head loss}) = \frac{16Sy}{3WD} + \frac{\eta V}{WD^2}$$

The velocity where sludge becomes turbulent may be bounded by the following equations for lower and upper critical velocities. The Hazen-William equation is used when flow is turbulent

$$V_{lc} \text{ (Lower critical velocity)} = \frac{1000\,\eta + 103\sqrt{94\,\eta^2 + D^2 Sy\,\rho}}{D\rho} \quad (\text{ft/s})$$

$$V_{uc} \text{ (Upper critical velocity)} = \frac{1500\,\eta + 127\sqrt{140\,\eta^2 + D^2 Sy\,\rho}}{D\rho} \quad (\text{ft/s})$$

where D is the pipe diameter (ft), Sy is the shear stress at the yield point where sludge begins to flow (lb/ft^2), η is the coefficient of rigidity (lb/ft^2.s), W is the weight of water (64.4 lb/ft^3), ρ is the density of sludge (= W weight of water, lb/ft^3)

Assume the following data :

Digesting sludge concentration	:	3.5%
Yield shear stress (Sy)	:	0.13 lb/ft^2
Rigidity coefficient (η)	:	0.012 lb/ft^2.s
Velocity in pipe	:	1.5 ft/s
Flow	:	100 gpm

Solution

1. Determination of velocities

$$V_{lc} = \frac{1000 \times 0.012 + 103\sqrt{94(0.012)^2 + (0.5)^2 \times 0.13 \times 64.4}}{0.5 \times 64.4}$$

$$= 5 \text{ ft/s}$$

$$V_{uc} = \frac{(1500 \times 0.012) + 127\sqrt{140(0.012)^2 + (0.5)^2 (0.13)(64.4)}}{0.5 \times 64.4}$$

$$= 6.4 \text{ ft/s}$$

2. Laminar head loss

$$\frac{H}{L} = \frac{16 \times 0.012}{3 \times 64.4 \times 0.5} + \frac{0.13 \times 1.5}{64.4 \times (0.5)^2}$$

= 0.043 ft water/ft

H = 1500 × 0.043 = 64.5 ft, [L = 1500 ft]

3. Use of Hazen–Williams equation

$$\frac{H}{L} = 0.002083 \times L \left(\frac{100}{C}\right)^{1.85} \times \left(\frac{Q^{1.85}}{D^{4.8655}}\right)$$

where H is the head loss (ft), L is the length of pipe (ft), C is the coefficient, Q is the flow (gpm), and D is the pipe diameter (in.)

$$\frac{H}{L} = 0.002083 \times 1500 \left(\frac{100}{100}\right)^{1.85} \times \left(\frac{100^{1.85}}{6^{4.8655}}\right)$$

= 4.6 ft

H turbulent is far less than H laminar.

Example 6.48

Head loss using Sludge Rheology

Estimate the head loss in a 6 in. pipe line, 10000 ft long conveying untreated sludge at an average flow rate of 600 gal/min. Assume the following data:

Yield stress (s_y) :   0.035 lb/ft²

Coefficient of rigidity (η) :   0.03 lb/ft.s

Specific gravity   :   1.02

Solution

1. Required Reynolds number (R_e)

Average velocity (V) in pipe

$$A = \frac{\pi D^2}{4} = \frac{3.14(0.5)^2}{4} = 0.2 \text{ ft}^2$$

$$V = \frac{Q}{A} = \frac{(600 \text{ gal/min})\left(\dfrac{1 \text{ ft}^3}{7.48 \text{ gal}}\right)\left(\dfrac{1 \text{ min}}{60 \text{ s}}\right)}{0.2 \text{ ft}^2} = 6.7 \text{ ft/sec}$$

Reynolds number $(\gamma V \Delta / \eta)$

$$R_e = \frac{\gamma V D}{\eta} \text{ or } \frac{\rho V D}{\eta} \text{ [SI units]}$$

$$= \frac{(62.4 \times 1.02 \text{ lb/ft}^3)(6.7 \text{ ft/s})(0.5 \text{ ft})}{0.03 \text{ lb/ft.s}}$$

$$= 7.107 \times 10^3.$$

2. Required Hedstrom number [H_e]

$$H_e = \frac{D^2 s_y g\gamma}{\eta^2} \quad \text{[US customary units]}$$

$$= \frac{D^2 s_y \rho}{\eta^2} \quad \text{[SI units]}$$

where s_y = Yield stress [lb_f/ft^2 (N/m^2)]

g = Gravity constant (32.2 ft/s^2)

η = Rigidity coefficient [lb/ft.s (kg/m.s)]

ρ = Specific mass of sludge (kg/m^3)

γ = Specific weight of sludge (lb/ft^3)

D = Pipe diameter [ft (m)]

V = Average velocity [ft/s (m/s)]

$$H_e = \frac{(0.5 \text{ ft})^2 (0.035 \text{ lb/ft}^2)(32.2 \text{ ft/s}^2)(62.4 \times 1.02 \text{ lb/ft}^3)}{(0.03 \text{ lb/ft.s})^2}$$

$$= 3.98 \times 10^4$$

Using the graphical plot of friction factor (f) versus R_e at various H_e, f = 0.007.

3. Required Pressure Drop

Pressure drop Dp for turbulent conditions is expressed as:

$$\Delta p = \frac{2f\gamma L V^2}{gD} \quad \text{[US customary units]}$$

$$= \frac{2f\rho L V^2}{D} \quad \text{[SI units]}$$

where Δp = Pressure drop [lb_f/ft^2 (N/m^2)]

L = Pipe length [ft (m)]

f = Friction factor

$$\Delta p = \frac{2(0.007)(62.4 \times 1.02 \text{ lb/ft}^3)(10,000 \text{ ft})(6.7 \text{ ft/s})^2}{(32.2 \text{ ft/s}^2)(0.5 \text{ ft})}$$

$$= 24845 \text{ lb/ft}^2$$

And is equivalent to $\dfrac{24845 \text{ lb/ft}^2}{62.4 \text{ lb/ft}^3}$ = 398 ft of water column.

Example 6.49

Maximum Velocity in a Pipe Carrying Digested Sludge

What will be the maximum velocity at which a digested sludge with 92 percent moisture will flow in a laminar state (R_e = 2000) through a 12 inches diameter pipe?

Solution

From the figure related to variation of yield stress (τ_o) with moisture content of sewage sludges, it is seen that 92% falls below critical moisture content for digested sludge. Therefore, the sludges will be expected to behave as a plastic liquid. Assuming that the digestion can be rated as good,

$$\tau_o \text{ (yield stress) = 0.05 lb–force/ft}^2$$

$$\mu_R = 0.023 \text{ lb–mass/ft–s}$$

[Variation of coefficient of rigidity with moisture content of sewage sludges]

$$\rho = 62.4 \times 1.032 = 64.4 \text{ lb mass/ft}^2$$

[Variation of specific gravity with moisture content of sewage sludges]

The absolute viscosity for a plastic liquid can be calculated as:

$$\mu = \left(\mu_R + \frac{g_c \tau_o D}{6 \overline{V}} \right), \text{ for circular pipes}$$

$$\mu = 0.023 + \frac{32.17 \times 0.05 \times 1}{\overline{V}}$$

$$R_e = \frac{\overline{V} D \rho}{\mu} = \frac{\overline{V}^2 \times 1 \times 64.4}{0.023 \overline{V} + 0.269} = 2000 \text{ and}$$

$$\overline{V} = 3.26 \text{ ft/s}.$$

Example 6.50

Sludge Flow and Sludge Concentration

1. Sludge flow

    1.1 Many wastewater sludges are non-Newtonian fluids with plastic rather than viscous properties. Flow hydraulics is further complicated because most sludges are thixotrophic and plastic properties change during stirring and turbulence. Gases or air released during flow add to the difficulty of identifying probable hydraulic behaviour.

    Fundamental friction losses increase with solids content and decrease with increase in temperature. Laminar or transitional flow persists at relatively high velocities (1.5 to 4.5 ft/s for thick sludges flowing in pipes 5 to 12 inches in diameter). All sludges behave more or less like water at turbulent velocities.

The head loss (h/L) in pipes can be estimated by using Poiseuille's equation for viscous liquid:

$$\frac{h}{L} = \frac{32}{g} \frac{v}{D^2} \left(\frac{\eta}{\rho} + \frac{1}{6} \frac{\tau_y}{\rho} \frac{D}{v} \right)$$

where L and D are respectively the length and diameter of the pipe, v is the velocity of solids with transporting liquid, η and τ_y are respectively their coefficient of rigidity and shearing stress at yield point, and ρ is the mass density. The terms in the bracket are analogues to the kinematic viscosity of a Newtonian liquid. For dilute mixture, the shearing stress at yield point approaches zero and the coefficient of rigidity can be approximated by the dynamic viscosity of the fluid (v).

Thixotropy makes the determination of η/ρ and τ_y/ρ difficult. Common values are $\eta = 0.03$ and $\tau_o = 1$ for thick sludges (sludge solids greater than 2%).

1.2 Determine the head loss for digested, and plain sedimentation sludges. Assume the following data:

Velocity of flow : 1.5 ft/s

Length of pipe : 150 ft

Diameter of pipe : 1 ft

Solids concentration : 10%

η/ρ : 5×10^{-4} ft^2/s

τ_y/ρ : 1.510^{-3} (ft/s)^2

1.3 Head loss in a pipe :

$$\frac{h}{L} = \frac{32}{g} \frac{v}{D^2} \left(\frac{\eta}{\rho} + \frac{1}{6} \frac{\tau_y}{\rho} \frac{D}{v} \right)$$

$$= \frac{32}{32.2} \times \frac{1.5}{1} \left(5 \times 10^{-4} + \frac{1}{6} \times \frac{1.5 \times 10^{-3} \times 1}{1.5} \right)$$

H = 150 × 1.5 ×(6.67 × 10^{-4})

= 0.15 ft.

1.4 Kinematic viscosity of the sludges (v_p) :

$$v_p = \left(\frac{\eta}{\rho} + \frac{1}{6} \frac{\tau_y}{\rho} \frac{D}{v} \right)$$

$$= \left(5\times10^{-4} + \frac{1}{6}\times\frac{1.5\times10^{-3}\times1}{1.5} \right)$$

$$= 6.67 \times 10^{-4} \text{ ft}^2/\text{s}$$

$$R_e = \frac{v\times D}{v_p} = \frac{1.5\times1}{6.67\times10^{-4}} = 2250$$

In turbulent flow, head loss of fairly homogeneous wastewater solids, digested primary solids, and activated sludge is not increased more than 1% for each 1% of solids concentration.

Fresh, plain sedimentation solids are transported at head losses 1.5 to 4 times those of water. Heavier and larger solids settle out when the flow velocities are small and obstruct the flow.

2. Sludge concentration

The geometry of the flux curve (sludge solids at given sludge concentration) can be exploited by deciding on an underflow concentration (C_u), and drawing a tangent to the flux curve to determine the limiting flux (vc_{min}), on the ordinate and the limiting concentration (C_a) at the point of tangency. Using the similar triangles property the minimum surface area can be calculated :

$$\frac{(vc)_{min}}{C_u} = \frac{(vc)_a}{(C_u - C_a)}$$

$$(vc)_{min} = \frac{(v_a C_a) C_u}{(C_u - C_a)}$$

Rate of withdrawal is equal to the minimum flux $(vc)_{min}$

$$\frac{Q}{A} = (vc)_{min} = \frac{(v_a C_a) C_u}{(C_u - C_a)}$$

$$\text{or } A = \frac{Q(C_u - C_a)}{(v_a C_a) C_u}$$

2.1 Determine the requisite minimum surface area for a thickener. Use the following data

Wastewater flow (Q)	:	2 MGD (86 L/s)
Underflow concentration (C_u)	:	4,000 mg/L
Observed concentration (C_a)	:	2000 mg/L, and
Interface velocity (v_a) at the point of tangency to the solid flux point	:	1.5×10^{-2} cm/s

$$\text{Surface area (A)} = \frac{Q(C_u - C_a)}{[(v_a C_a)C_u]}$$

$$= \frac{46 \times 10^3 (4-2)}{1.5 \times 10^{-2} \times 2 \times 4} = 7.67 \times 10^5 \text{ cm}^2$$

$$= 76.7 \text{ m}^2 \ (826 \text{ ft}^2)$$

$$\text{Surface loading} = \frac{2 \times 10^6}{826} = 2425 \text{ gal/ft}^2.$$

Note :

1. In order to avoid speticity in gravitational thickners, the sludge must be introduced at rates between 6000 to 10000 gpd/ft^2

2. At lower loadings, the tank contents may have to be chlorinated

3. Solids surface loading may be as high as 30 lb/ft^2.d and as low as 5 lb/ft^2.d for activated sludge yielding solids concentrations of 10% and 2.5% respectively

4. The resistance of compressible cake has been observed to rise in proportion to the applied pressure

$$\frac{R_1}{R_2} = \left(\frac{\Delta p_1}{\Delta p_2}\right)^s \qquad [R, \text{ m/kg}]$$

where s is the coefficient of compressibility

According to Trubnick and Mueller (Biological Treatment of Sewage and Industrial Wastes, Vol.2, McCabe and Eckenfelder, Jr. (Eds.), Reinhold, N.Y. 1958)

$$\frac{R_1}{R_2} = \frac{[1 + k\Delta p_1]^p}{[1 + k\Delta p_2]^q}$$

where k is the coefficient, and p and q are compressibility coefficients.

5. Coagulant (conditioner) requirements for chemical sludge conditioning are generally expressed as percentage ratio of pure chemical to the weight of the solids fraction on dry basis. Component parts are:

 Liquid–fraction requirement (approximated by the stoichiometry of the idealized chemical reactions).

 Solids–fraction requirement (based on experience).

Genter (Trans. Am. Soc. Civil Engrs. 111, 641, 1946) suggested the following relationship for FeCl$_3$:

$$p_c = \left[\frac{1.08 \times 10^{-4} \, A p}{(1-p)}\right] + 1.6 \, p_v/p_f$$

where A is the alkalinity of the sludge moisture (mg/L as $CaCO_3$), p_c, p, p_v and p_f are the percentage of chemicals ($FeCl_3$), moisture, volatile matter and fixed solids in the sludge (all on dry basis), respectively.

The coagulant requirements can be reduced by digesting the sludge prior to coagulating it for dewatering. The first term of the equation for p_c is greatly magnified by digestion and it can be reduced either by adding lime (as precipitating agent or by washing out the share of the alkalinity with water (of alkalinity by elutriation).

The second term of the equation is obtained from operating results for vacuum filtration of $FeCl_3$ treated wastewater (as it is a function of volatile content of the sludge:

1 mg $CaCO_3$ (100) combines with (2/3 × 162.2/100) = 1.08 of $FeCl_3$ (162.2):

$$2FeCl_3 + 3Ca(HCO_3)_2 = 2Fe(HCO_3) + 3CaCl_2 + 6CO_2).$$

Example 6.51

Design Calculations of Power Requirements for Liquids and Gases

1. Basic equation of power requirements for non-compressible fluids, the integrated form of the total mechanical–energy balance reduces to

$$W_o = \Delta Z + \Delta\left(\frac{V^2}{2\alpha g_c}\right) + \Delta pv + \Sigma F$$

All terms have usual meanings except α : Correction coefficient = 1.0 for turbulent flow and 0.5 for viscous flow.

1.1 Application of the total mechanical–energy balance to non–compressible flow system

Determine the horse power of the motor required to drive the pump.

Assume the following data:

Water temperature	: 61°F (Constant throughout)
Length of pipe and diameter	: 1000 ft, 2.067 in ID
Gate valves and standard tee elbows	: 2 Nos., 3 Nos. and 90° bend
Efficiency of pump	: 40% (Includes the losses at entrance and exit of the pump housing)
Pumping rate	: 50 gal/min

Water is to be pumped from a large reservoir into the top of an over head tank (open to atmosphere), the elevation difference : 70 ft.

1.1.1 Total mechanical energy balance between point–1 (surface of water in reservoir) and point–2 (Just outside of the pipe at discharge point)

$$W_o = (Z_2 - Z_1) + \left(\frac{V_2^2 - V_1^2}{2\alpha g_c \quad 2\alpha g_c} \right) + (p_2 v_2 - p_1 v_1) + \Sigma F$$

Points 1 and 2 are taken where the linear velocity of the fluid is negligible, therefore:

$$\frac{V_2^2}{2\alpha g_c} - \frac{V_1^2}{2\alpha g_c} = 0$$

$p_2 = p_1 = 1$ atm

$v_1 = v_2 =$ Non-compressible liquid

$Z_2 - Z_1 = 70$ lb$_f$-ft/lb mass (assuming $g = g_c$)

1.1.2 Required frictional losses

$$\text{Average velocity} = \frac{\text{Flow}}{\text{Cross-sectional area}}$$

$$= \frac{(50)(144)}{(60)(7.481)(2.067)^2(0.785)}$$

$$= 4.78 \text{ ft/s}$$

Viscosity of water at 61°F = 1.12 Cp = (1.12) (0.000672) lb/ft.s

Density of water at 61°F = 62.3 lb/ft³

$$R_e \text{ in 2 in. pipe} = \frac{(2.067)(4.78)(62.3)}{(12)(1.12)(0.000672)} = 68,000$$

$$\frac{\epsilon}{D} = \frac{(0.00015)\ 12}{2.067} = 0.00087 ,$$

[Commercial steel pipe, E = 0.00015 ft.]

Friction factor (f) = 0.0057, [Plot of Re – Vs. Fannung's friction factor]

Total equivalent length (L$_e$) for fittings and valves

$$= \frac{(2)(7)(2.067)}{12} + \frac{(3)(32)(2.067)}{12}$$

$$= 19 \text{ ft}$$

Friction due to flow through pipe and fittings

$$= 2f \left[\frac{V^2(L + L_e)}{g_c\ D} \right]$$

$$= 2 \times 0.0057 \left[\frac{(4.78)^2 (1000+19)12}{(32.17)(2.067)} \right] = 47.9 \text{ lb}_f - \text{ft/lb m}$$

Friction due to contraction (F_c) and enlargement (F_e)

$$F = F_c + F_e$$

$$= \frac{K_c V_2^2}{2 \alpha g_c} + \frac{(V_1 - V_2)^2}{2 \alpha g_c}, \quad [K_c = 0.5]$$

$$= \frac{(0.5)(4.78)^2}{(2)(1)(32.17)} + \frac{(4.78-0)^2}{2 \alpha g_c} = 0.53 \text{ lb}_f - \text{ft/lb m}$$

$$\Sigma F = 47.9 + 0.53 = 48.4 \frac{\text{lb}_f - \text{ft}}{\text{lb m}}$$

1.1.3   Theoretical mechanical energy balance (W_o)

$$W_o = 70 + 48.4 = 118.4 \text{ lb}_f - \text{ft/lb m}$$

$$1 \text{ hp} = 550 \text{ lb}_f . \text{ft/s}$$

hp of motor required to drive the pump

$$= \frac{(118.4)(50)(62.3)}{(0.40)(60)(7.481)(550)}$$

$$= 3.74 \text{ hp (say 4.0 hp)}$$

2.   Application of total energy balance for the flow of an ideal gas

For gas

$$W_o = \Delta Z + \Delta h + \Delta \left(\frac{V^2}{2 \alpha g_c} \right) - Q$$

h = enthalpy = u + pv lb$_f$.ft/lb.m

2.1   Determine the total energy (lb$_f$.ft/lbm) supplied by compressor located between points 1@ and 2. Assume the following data:

Pipe ID : 2.067 in.

Turbulent conditions, nitrogen gas flowing at constant mass rate through a long, straight, horizontal pipe upstream (point–1) temperature : 70°F and 15 psia and pressure

Average linear velocity   : 70 ft/s

Downstream (point–2) temperature : 140°F and 50 psia and pressure

External heater provided between   : 10 BTU
point–1 and 2 and supplies to the
flowing gas

Heat capacity (Cp) of gas   : 7.0 BTU/lb md–°F

2.1.1 Total energy balance between point 1 and 2 (1 lb of gas flowing)

$$W = (h_2 - h_1) + \frac{V_2^2}{2g_c} - \frac{V_1^2}{2\,g_c} - Q$$

$(h_2 - h_1)$ = Enthalpy change

V_1 = 50 ft/s

$$V_2 = \frac{(50)(460 + 140)(15)}{(460 + 70)(50)} = V_1 \left(\frac{T_2}{T_1} \right)\left(\frac{p_1}{p_2} \right) = 17 \text{ ft/s, [Ideal gas]}$$

1 BTU = 778 lb$_f$.ft

Q = (10) (778) lb$_f$.ft/lb m

2.1.2 Total energy supplied by compressor (W)

N$_2$ as ideal gas

Enthalpy change = $(h_2 - h_1) = \dfrac{Cp}{M}(T_2 - T_1)$

$$= 7\left(\frac{140 - 70}{28} \right)778 = 13,600 \ \frac{\text{lb}_f.\text{ft}}{\text{lb m}}$$

$$W = 13,600 + \frac{(17)^2}{(2)(32.17)} - \frac{(50)^2}{(2)(32.17)} - 7780$$

$$= 5790 \ \frac{\text{lb}_f.\text{ft}}{\text{lb m}}$$

2.2 Application of total energy balance for flow of a non-ideal gas (steam turbine)

Super-heated steam enters a turbine under such conditions that the enthalpy of the entering steam is 1340 BTU/lb. On the same basis, the enthalpy of the steam leaving the turbine is 990 BTU/lb. If the turbine operates under adiabetic conditions and changes in kinetic energy and elevation potential energy are negligible, determine the maximum amount of energy obtainable from the turbine per pound of entering stream.

2.2.1 Energy balance (1 lb of gas flowing)

W = $(h_2 - h_1)$ + KE + PE - Q

= $(h_2 - h_1)$ + 0 + 0 - 0, (Adiabatic)

= 990 - 1340 = - 350 BTU/lb–sleam

Maximum energy obtainable from turbine = 350 BTU/lb–stream.

2.3 Application of the total mechanical energy balance for the flow of a compressible fluid with high pressure

Determine the pressure in the pipe at the exit from the pump, and the mechanical energy as foot–pound force per minute added to the air by the pump (assuming the pump operation is isothermal). Assume

Air flow rate through a straight, horizontal steel pipe	: 15 lb/min, 2.067 in ID
Pipe length (L)	: 3000 ft
Downstream air pressure, and temperature	: 5 psig and 70°F
Air pressure at the entrance of the pump along with the temperature	: 10 psig and 80°F.

2.3.1 Total mechanical energy balance for the system between points 2 and 3

Point–1 : Entrance to pump

Point–2 : Exit from the pump

Point–3 : Downstream end of pipe

General equation :

$$W = \Delta Z + \Delta \left(\frac{V^2}{2\alpha g_c} \right) + \Delta pv + \Sigma F - Q \tag{1}$$

When ΔZ, V, V, are elevation change, Velocity of flow, and specific volume of the gas respectively.

Amount of heat exchanged between surroundings and the system is unknown, the total energy balance cannot be used to solve this problem.

$$\int_2^3 vdp + \int_2^3 \frac{V\,dV}{\alpha g_c} = -\int_2^3 dF = -\int_2^3 \frac{2f V^2\,dL}{g_c\,D} \tag{2}$$

Mass velocity G (lb m/ft².sec) is constant, and

$$V = Gv$$

$$dV = G\,dv$$

Therefore, $$-\int_2^3 \frac{dp}{v} = \int_2^3 \frac{G^2 dv}{\alpha g_c v} + \int_2^3 \frac{2f G^2\,dL}{g_c\,D} \tag{3}$$

Ideal gas equation [specific volume of gas] :

$$v = \frac{RT}{Mp}$$

where M = Molecular weight (29 lb/lb–mol), R = G as constant [1545lb$_f$/ft² (ft³)/(lb mol) (°R)], and T = Temperature (°R)

$$\frac{M}{2\,RT_{avg}}(p_2^2 - p_3^2) = \frac{G^2}{\alpha g_c}\ln\left(\frac{v_3}{v_2}\right) + \frac{2\,f_{avg}\,G^2\,L}{g_c\,D} \qquad (4)$$

T$_{avg}$ represents the average absolute temperature between points 2 and 3, and temperature variations upto 20% from the average value will introduce only a small error in the final result

The error introduced by using f$_{avg}$ (constant) based on average temperature and pressure instead of the exact integrated value is not important unless pressure variations are considerably greater than those involved in this problem

$$T = 535°R$$

$$\mu_{air} = 0.018 \text{ Cp} = (0.018)(0.000672) \text{ lb/ft s}$$

$$G = \frac{(15)(144)(4)}{(60)(2.067)^2(\pi)} = 10.77 \text{ lb/ft}^2.s$$

$$R_e = \frac{DG}{\mu} = \frac{(2.067)(10.77)}{(12)(0.018)(0.000672)} = 153,000$$

$$\alpha = 1$$

$$\frac{\in}{D} = \frac{(0.00015)\,12}{2.067} = 0.00087 \text{ , [Roughness factor]}$$

Therefore, f$_{avg}$ = 0.0052, [from the friction factor plot at ∈/D = 0.00087 and R$_e$ = 153,000]

$$p_3 = (5 + 14.7)(144) = 2840 \text{ lb}_f/\text{ft}^2$$

$$\frac{v_3}{v_2} = \frac{T_3}{T_2}\left(\frac{p_2}{p_3}\right) = \frac{530\,p_2}{(540)(2840)} = \frac{p_2}{2890}$$

Substituting into equation (4):

$$\frac{29}{(2)(1545)(535)}[p_2^2 - (2840)^2] = \frac{(10.77)^2}{32.17}\ln\left(\frac{p_2}{2890}\right)$$

$$+ \frac{(2)(0.0052)\,(10.77)^2\,(3000)(12)}{(32.17)(2.067)}$$

$$p_2 = 6750 \text{ psf} = \frac{6750}{144} \text{ psia} = 47 \text{ psia.}$$

2.3.2 Total mechanical energy added by the pump between points 1 and 3

$$\int_1^3 vdp + \int_1^3 \frac{VdV}{\alpha g_c} = \int_1^3 dW - \int_1^3 dF \tag{5}$$

$$\int_1^3 dF = \int_2^3 dF$$

(By definition except that occurring at the pump frictional loss)

and
$$\int_1^3 vdp + \int_2^3 vdp + \int_1^2 \frac{vdv}{\alpha g_c} + \int_2^3 \frac{Vdv}{\alpha g_c} = W - \int_2^3 dF \tag{6}$$

Subtracting equation (2) from equation (6) :

$$\int_1^2 vdp + \int_1^2 \frac{VdV}{\alpha g_c} = W$$

Although pumps and compressor operate near adiabatic conditions, the pump operation will be assumed as isothermal in this case:

$$\int_1^2 vdp = \frac{RT}{M} \ln\left(\frac{p_2}{p_1}\right)$$

$$v_1 = \frac{(10.77)(359)(540)(14.7)}{(29)(492)(24.7)} = 87 \text{ ft/s} = Gv_1$$

$$v_2 = v_1 \frac{p_2}{p_1} = \frac{(87)(24.7)}{47} = 46 \text{ ft/s}$$

$$W = \frac{RT}{M} \ln\left(\frac{p_2}{p_1}\right) + \frac{V_2^2}{2 g_c} - \frac{V_1^2}{2 g_c}$$

$$= \frac{(1545)(540)}{29} \ln\left(\frac{47}{24.7}\right) + \frac{(46)^2}{(2)(32.17)} - \frac{(87)^2}{(2)(32.17)}$$

$$= 18,400 \text{ lb}_f.\text{ft/lb m}$$

Mechanical energy (total) added to air by pump (under isothermal condition)

$$= (18,400) \text{ lb}_f. \text{ ft (15 lbm/min)}$$

$$= 276 \text{ lb}_f.\text{ft/min}$$

[The total energy supplied to the pump could be determined if isothermal efficiency of the pump (including any end effects caused by the pump housing) were known].

Note :

1. Theoretical power [Isothermal compression [(ideal gas) pv = constant] for any number of stages can be expressed as:

$$\text{Power}(P) = p_1 v_1 \ln\left(\frac{p_2}{p_1}\right)$$

$$h_p = 3.03 \times 10^{-5} \; p_1 Q_1 \ln\left(\frac{p_2}{p_1}\right)$$

   where P is the power requirement (lb_f.ft/lbm), p_1 is the intake pressure (lb_f/ft^2), v_1 is the specific volume of gas at intake conditions (ft^3/lbm), p_2 is the final delivery pressure (lb_f/ft^2), and Q is the gas rate at intake conditions (ft^3/s).

2. Similarly, for ideal gas under going adiabatic (isentropic) compression (pv^k = constant), the following equations apply:

   Single-stage compression

$$\text{Power }(P) = \frac{k}{k-1} p_1 v_1 \left[\left(\frac{p_2}{p_1}\right)^{\frac{k-1}{k}} - 1\right]$$

$$hp = \frac{3.03 \times 10^{-5} \; p_1 Q_1}{k-1} \left[\left(\frac{p_2}{p_1}\right)^{\frac{k-1}{k}} - 1\right]$$

$$p_2 = p_1 \left(\frac{v_1}{v_2}\right)^k = p_1 \left(\frac{T_2}{T_1}\right)^{\frac{k}{k-1}}$$

$$T_2 = T_1 \left(\frac{v_1}{v_2}\right)^{k-1} = T_1 \left(\frac{p_2}{p_1}\right)^{\frac{k-1}{k}}$$

   Multistage compression

   [Assuming equal division of work between cylinders and inter cooling of gas to the original intake temperature]

$$hp = \frac{3.03 \times 10^{-5} \; kN}{k-1} \; p_1 Q_1 \left[\left(\frac{p_2}{p_1}\right)^{\frac{k-1}{kN}} - 1\right]$$

   where k = (Cp/Cv) is the ratio of specific heats, T is the temperature (°R), N is the number of stages of compression and rest terms follow the same nomenclature as indicated in single-stage compression.

Example 6.52

Scouring Velocities in Open Channels and Pressure Pipes

1. Derive the expression for limiting velocities in open channels.

If a flowing liquid is to bring about movement of particles, then the average intensity of retractive force along the wetted surface must atleast be equal to the maximum resistance of particles:

$$\tau_w = f\left[\frac{D_p}{\delta}, (\rho_s - \rho), D_p\right] \tag{1}$$

where τ_w is the average shear stress along wetted surface at the velocity where movement of particles is impending lb_f/ft^2, D_p is the particle diameter (ft), δ is the thickness of laminar sublayers (ft), ρ_s is the mass density of particles (lb_{mass}/ft^3), and ρ is the mass density of liquid (lb_{mass}/ft^3)

It has been experimentally found that:

$$\frac{g_c \tau_w}{g(\rho_s - \rho)D_p} = \phi\left(\frac{D_p}{\delta}\right) = K \tag{2}$$

$$\tau_w = \rho\, r_H \frac{h_f}{L} \tag{3}$$

where r_H is the mean hydraulic radius of the wetted surface (ft), and L is the horizontal distance between points 1 and 2 (ft)

Therefore, head loss [h_f]

$$(h_f) = \frac{gK\left[\frac{(\rho_s - \rho)}{\rho}\right]D_p L}{g_c\, r_H} \tag{4}$$

Darcy–Weisbach equation can be modified for open channel:

$$h_f = f\frac{L}{4\, r_H}\frac{V^2}{2g_c} \tag{5}$$

And solving for V:

$$V_L = \left[\frac{8\,gK}{f}\left(\frac{\rho_s - \rho}{\rho}\right)D_p\right]^{0.5} \tag{6}$$

Similarly using Manning equation, [η = Manning Constant]

$$V = \frac{1.486}{n}r_H^{2/3}\left(\frac{h_f}{L}\right)^{0.5} \tag{7}$$

With equation (4) results in

$$V_L = \frac{1.486}{n} r_H^{1/6} \left[K \frac{(\rho_s - \rho)}{\rho} D_p \right]^{0.5}$$ (8)

where K is the constant, g is acceleration due to gravity (ft/ s²), g_c is the conversion factor (lbm–ft/lb$_f$–s²), f is the coefficient of friction, n is the coefficient of roughness, and V_L is the limiting velocity (ft/s).

K = 0.04 for materials of uniform shape and size and for fine non-uniform sand (mean grain size), the value of K for flocculent and sticky materials can be higher.

2. Derive the expression for limiting velocities in pressure pipes.

The velocity at which the solids are kept dispersed evenly across the pipe diameter is called standard velocity, and the minimum velocity is that velocity below which the solids settlement freely.

The gravitational effect is more pronounced in causing the sedimentation solids than it is in separation of an emulsion. Such effect must be considered along with that of turbulence. These are related to:

$F_r = K \dfrac{(\rho_s - \rho)}{\rho} R_e^{0.776}$ [For non-flocculating particles with mean diameter between 0.05 and 05 mm], Fr = Froudie number

$$\frac{V^2}{g\,D_p} = K \left(\frac{\rho_s - \rho}{\rho} \right) \left(\frac{V D \rho_m}{\mu_m} \right)^{0.776}$$ (9)

where $\dfrac{V^2}{g_c D_p}$ = Froude number (gravitational effect), V is the minimum or standard velocity (ft/s), D_p is the particle diameter which is larger than 50% (by weight) of the particles (ft), D is the diameter of the pipe (ft), ρ_m in the mean mass density (lb/ft³), K is the constant (25.3 × 10⁻³ for the minimum velocity and 74.5 × 10⁻³ for standard velocity), and μ_m is the viscosity of the suspension (viscosity of water–lb m/ft.s).

The mean density of the suspension :

$$\rho_m = x_v \rho_v + (1 - x_v)\rho$$ (10)

where x_v is the volumetric fraction of solids in suspension

$$V = \left(K g D_p \frac{\rho_s - \rho}{\rho} \right)^{0.816} \left(\frac{D \rho_m}{\mu_m} \right)^{0.663}$$ (11)

At velocities greater than standard velocity, head loss is given:

$$h_f = \frac{\Delta P}{\rho} = f\left(\frac{L}{D} \right)\left(\frac{V^2}{2\,g_c} \right)$$ (12)

Between the minimum velocity and standard velocity, head loss is given by:

$$\frac{\Delta p^* - \Delta p}{\rho} = 121 x_v \left[\frac{g\, D\,(\rho_s - \rho)}{V^2} \right]^{1.5} \left[\frac{U_t^2}{g\, D_p (\rho_s - \rho)} \right]^{0.75} \tag{13}$$

where ΔP is the pressure drop due to friction estimated for pure water flowing at velocity V, ΔP^* is the pressure drop due to friction in suspension flowing at a velocity V, and U_t is the terminal velocity of solid particle of diameter D_p (ft/s).

3. Determine the width of a channel to ensure no solids deposition alongwith bottom in the channel. Assume the following data:

Specific gravity of sand particles   :   2.65

Mean diameter of sand particles   :   0.15 mm

Wastewater flow rate (Q)   :   6 ft³/s

3.1   Minimum velocity based on Manning equation

$$V = \frac{1.486}{n} r_H^{1/6} \left[K \left(\frac{\rho_s - \rho}{\rho} \right) D_p \right]^{0.5}$$

$$V = \frac{Q}{A} = \frac{Q}{WH} = \frac{1.486}{n} \left(\frac{HW}{W + 2H} \right)^{1/6} \left[K \left(\frac{(\rho_s - \rho)}{\rho} \right) D_p \right]^{0.5}$$

$$r_H = \frac{(\text{Cross-sectional perpendicular to flow})}{(\text{Wetted perimeter})} = \frac{HW}{W + 2H}$$

W = Width and H = Depth of water = 2.5 ft

use n = 0.013, K = 0.04

$$\frac{6}{2.5 \times W} = \frac{1.486}{0.013} \left(\frac{2.5 \times W}{W + 5} \right)^{1/6} [0.04 \times 1.65 (1 \times 3.28 \times 10^{-3})]^{0.5}$$

Solving for width (W) = 2.84 ft.

4. Determine the velocity of particles which must be kept in suspension throughout the pipe [D = 0.5 ft] length. Assume the following data:

Fraction of fines in water   : 15% (by weight)

Specific gravity (ρ_s) : 2.0

Mean diameter of particles (D_p) : 0.2 mm

Water density (ρ) : 62.4 lb m/ft³

Viscosity of water (μ_m) : 1 centi-poise

4.1 Mean density of suspension

$$\rho_m \quad = x_v\,\rho_v + (1 - x_v)\rho$$
$$= 0.15(2.5)\,(62.4) + 0.85 \times 62.4$$
$$= 23.4 + 53.04 = 78.44 \text{ lb m/ft}^3$$

4.2 Using equation (11), Standard velocity, ($K = 74.5 \times 10^{-3}$)

$$V = \left[KgD_p \frac{(\rho_s - \rho)}{\rho} \right]^{0.816} \left(\frac{D\rho_m}{\mu_m} \right)^{0.633}$$

$$= [74.5 \times 10^{-3} \times 32.2 \times 6.56 \times 10^{-4} \times 1]^{0.816} \left[\frac{0.5 \times 78.44}{1 \times 6.72 \times 10^{-4}} \right]^{0.633}$$

$$= 5.16 \times 10^{-3} \times 1040 = 5.4 \text{ ft/s}$$

Note :

1. At R_e greater or equal to 4000, separation of phases (emulsion) is prevented by keeping the velocity high.

Example 6.53

Limiting Velocities to Keep Solids in Suspension

A suspension of one fines in water (10 percent by volume), is to be pumped through a horizontal pipe 6 in. in diameter. The particles have a specific gravity of 2 and a mean diameter of 0.2 mm. If the water has a density of 62.3 lb/ft³ and a viscosity of 1 Cp, how high a velocity must be maintained to keep the particles evenly dispersed throughout the suspension?

Solution

Mean density (ρ_m) of suspension can be calculated using:

$$\rho_m = x_v\,\rho_s + (1 - x_v)\rho$$

where ρ_s is the mass density of solids particles (lb/ft³), x_v is the volumetric fraction of solids in suspension, and ρ is the mass density of water (lb/ft³)

$$\rho_m = 0.10(2 \times 62.3) + 0.90 \times 62.3 = 68.53 \text{ lb/ft}^3$$

Standard velocity ($\overline{V}$) is calculated using:

$$\overline{V}^* = \left(KgD_p \frac{\rho_s - \rho}{\rho} \right)^{0.816} \left(\frac{D\,\rho_m}{\mu_m} \right)^{0.633}$$

$$= (74.5 \times 10^{-3} \times 32.17 \times 6.56 \times 10^{-4} \times 1)^{0.816} \left(\frac{0.5 \times 68.53}{1 \times 6.72 \times 10^{-4}} \right)^{0.633}$$

$$= 4.9 \text{ ft/s}.$$

Note :

The gravitational effect is more pronounced in causing the sedimentation of solids than it is in separation of an emulsion

In dealing with solids-liquid separations, such an effect must be considered alongwith turbulence

For non-flocculating particles with mean diameters between 0.05 and 0.5 mm, this effect have been corrected with limiting velocities to give:

$$Fr = K \frac{(\rho_s - \rho)}{\rho} R_e^{0.775} \tag{1}$$

$$\frac{\bar{V}^2}{g D_p} = K \frac{(\rho_s - \rho)}{\rho} \left(\frac{\bar{V} D \rho_m}{\mu_m} \right)^{0.775} \tag{2}$$

and solving equation (2) for the velocity :

$$\bar{V} = \left(K g D_p \frac{(\rho_s - \rho)}{\rho} \right)^{0.816} \left(\frac{D \rho_m}{\mu_m} \right)^{0.663}$$

where Fr is the Froude number (a measure of the gravitational effect), $\bar{V}$ is the minimum or standard velocity (ft/s), D is the pipe diameter (ft), D_p is the mean particles diameter (ft) [the diameter of that particles which is larger than 50 percent (by weight) of the particles], ρ is the mass density of liquids (lb/ft^3), ρ_m is the mean density of suspension (lb/ft^3), K is the constant (= 25.3 × 10^{-3} for the minimum velocity and 74.5 × 10^{-3} for the standard velocity), ρ_s is the mean density of particles (lb/ft^3), μ_m is the viscosity of the suspension (equal to the viscosity of the liquid (lb mass/ft-sec), and gravity constant.

At velocities greater than the standard velocity, frictional head loss can be computed with Darcy–Weisbach equation $\left(h_f = \frac{\Delta p}{p} = f \left(\frac{L}{D} \right) \frac{\bar{V}^2}{2g} \right)$. In computing R$_e$, the density and viscosity are taken as those of the suspension

Between the minimum velocity and standard velocity, friction losses are considerably higher than those for pure liquids at the same velocity. For closely graded particles suspended in water, the pressure drops due to friction can be estimated with an empirical formula:

$$\frac{\Delta p' - \Delta p}{\rho} = 121 x_v \left[\frac{g D(\rho_s - \rho)}{\bar{V}^2} \right]^{1.5} \left[\frac{U_t^2}{g D_p (\rho_s - \rho)} \right]^{0.75}$$

where Δp, $\Delta p'$ are pressure drop due to friction computed for pure water flowing at a velocity $\bar{V}$ and pressure drop due to fraction in the suspension flowing at a velocity , $\bar{V}$, U_t is the settling velocity of solids particles of diameter D_p (ft/s)

The frictional pressure drop in a suspension of solids having a wide range in particle size is best determined from experiments with particular suspension.

Example 6.54

Determination of Sump Size

A limiting for in the determination of the size of a pump will be the frequency with which the pumps will have to start and stop.

Consider a sump, volume V with an inflow Q_i and pumps designed to remove the flow at a flow rate of Q_o. The cycle of operation consists of a filling and an emptying phase.

Solution

1. No. of pump starts per unit time (N)

$$\text{Time taken to fill the sump} = \frac{V}{Q_i} = t_i$$

$$\text{Time taken to empty the sump} = \frac{V}{(Q_i - Q_o)} = t_o$$

$$\text{Cycle time} = (t_i + t_o) = V\left[\frac{1}{Q_i} + \frac{1}{(Q_i + Q_o)}\right]$$

$$= V\left[\frac{Q_i + (Q_o - Q_i)}{Q_i(Q_i - Q_o)}\right]$$

$$= \frac{V Q_o}{(Q_i)(Q_i - Q_o)}$$

$$\text{No. of pump starts per unit time (N)} = \frac{1}{\text{Cycle time}}$$

$$= \frac{1}{V}\left[Q_i - \left(\frac{Q_i}{Q_o}\right)^2\right]$$

2. Maximum starts and stops (Worst case) and sump size

$$N = \frac{1}{V}\left[Q_i - \left(\frac{Q_i}{Q_o}\right)^2\right]$$

$$\frac{dN}{dQ_i} = 0 = 1 - \frac{2Q_i}{Q_o}$$

$$Q_i = \frac{Q_o}{2}$$

Therefore,

$$N = \frac{1}{V}\left(Q_i - \frac{Q_i}{2}\right)$$

or Sump Volume (V) $= \dfrac{Q_i}{2N}$

[No. of starts and stops should be within the makers recommendations].

Note :

1. The cyclic nature of pumping regime in sewage pumping stations, the flow in the rising main is either at the design flow or at rest resulting in:

    Biological problem : In flat or sparsely populated areas, the flow in rising mains can be at rest for several hours (anaerobic conditions).

    Water hammer (the continual stopping and starting of the pumps). Negative pressure moves up the pipe inducing a suction on the pipe when the pump stops. As the direction of flow changes in the pipe, the returning water hits a non–return valve and a positive pressure wave is transmitted back along the pipe. Therefore, slow starting and slow stopping of the pump motor is recommended. Alternatively, use of fly wheels on the pumps themselves is recommended to ensure continued delivery after the electrical supply to the pump has cut out. In situations, where surge pressures are excessive, it will be necessary to use water towers or pressure vessels.

Example 6.55

Parshall Flume

Design a Parshall flume to measure a rate of flow for a maximum of 1.70 m³/s. Also determine the invert level of the outgoing sewer if the invert level of the incoming sewer is set at an elevation of 100 ft.

Solution

1. Required expression of flow measurement through a Parshall flume.

    In Parshall flume, the flow enters the flume through a converging zone, then passes through the throat, and out into the diverging zone. For the flume to be a measuring device, the depth somewhere at the throat must be critical (and it is possible because the converging and the subsequent diverging as well the downward sloping of the throat make this happen).

    Application of equation for flow measurement (Q) for rectangular weir to the Parshall flume between any point upstream of the flume and its throat is given by:

$$Q = 0.385 \, L \, (2 \, g)^{0.5} \, H^{1.5}$$

H may be replaced by H_a, the water surface elevation above flume floor level in the converging zone, and L may be replaced by W, the throat width yielding:

$$Q = K \, W \, (2 \, g)^{0.5} \, H_a^{1.5}$$

where K = Coefficient; and this equation applies only if the flow is not submerged. There are, two points: One is labeled H_a in the converging zone, and the other is labeled H_b in the throat. These points measure the elevations H_a and H_b, and the submergence criterion is given by the ratio $\left(\dfrac{H_b}{H_a}\right)$ [H_b/H_a ratio is greater than 0.70, flume is said to be submerged and the flow equation is not applicable].

The value of the flow coefficient (K) is given as (Figure 7) :

$$K = 0.488 \text{ for } \frac{H_a}{W} \text{ between 0.20 and above}$$

$$K = 0.40 \text{ to } 0.488 \text{ for } \frac{H_a}{W} \text{ between 0.025 and 0.20, Table 19.}$$

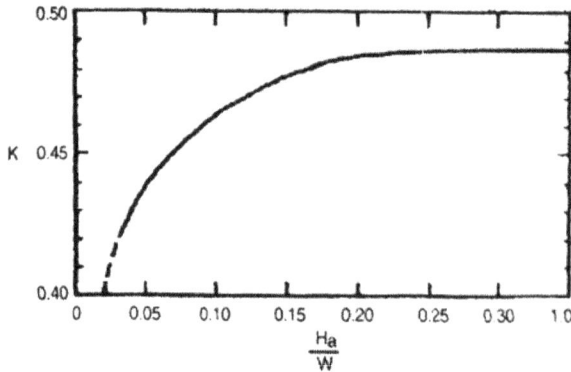

Figure 7 : Discharge coefficient for the Parshall flume.

2. Required design of a Parshall flume

    Refer Figure 8 for plan and sectional view of the Parshall flume

    Refer Table 20 for the standard Parshall flume dimensions

    For a throat (W) of 1 ft to 8 ft, the depth E is equal to 3 ft.

    Ha = (3.0 − 0.5) ft. = 2.5 ft. [for free board of 0.5 ft.]

    For a throat (W) of 9 in (= 0.75 ft), the depth E = 2.5 ft.

    Ha = (2.5 − 0.5) ft. = 2.0 ft. [for free board of 0.5 ft.]

    Various values of K can be obtained for various throat width (W):

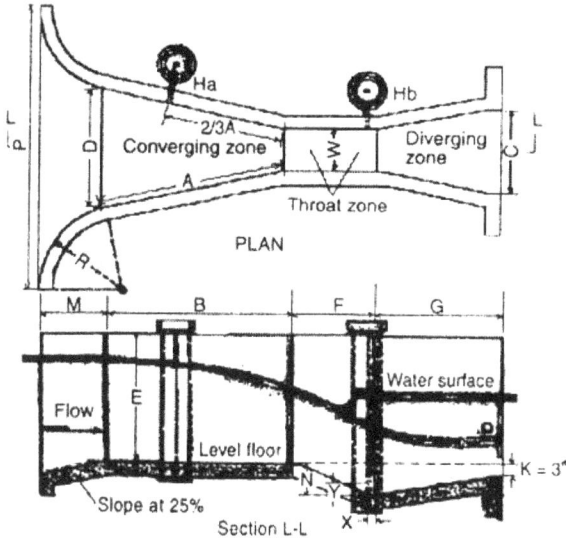

Figure 8 : Plan and sectional view of the Parshall flume

Table 19 : Value π of K for various widths (W)

Throat width (W)		E	Ha		H_d/W	K
ft	m	(ft)	ft	m		
0.75	0.23	2.5	2.0	0.61	2.67	0.488
2.00	0.61	3	2.5	0.76	1.25	0.488
3.00	0.91	3	2.5	0.76	0.83	0.488
4.00	1.22	3	2.5	0.76	0.63	0.488
5.00	1.52	3	2.5	0.76	0.50	0.488
6.00	1.83	3	2.5	0.76	0.42	0.488
7.00	2.13	3	2.5	0.76	0.34	0.488
8.00	2.44	3	2.5	0.76	0.31	0.488

Determine flow for various W and H_a using:

$$Q = K H_a W (2g)^{0.5} H_a^{1.5}, \ [m^3/s]$$

Table 20 : Required Q, With Various H and H_a.

W (m)	Ha (m)	Q (m³/s)
0.23	0.61	0.237
0.61	0.76	0.874
0.91	0.76	1.300
1.22	0.76	1.750

* For a flow of 1.750 m^3/s, choose the throat width of 1.22 m (4.0 ft), Table 1

* For a throat width (W) of 4.00 ft, choose the length of approach floor (M) = 1 ft 6 in (1.5 ft). The entrance to the flume is slopping upwards at 25% [1 Vertical : 4 Horizontal], thereby, the elevation of the floor level is [100 − (1.5) (0.25) ft] = 99.625 ft.

K, the difference in elevation between lower end of the flume and the crest of floor level is = 3 in. Therefore, the invert level of the outgoing sewer should be set at = (99.625 − 3/12) ft = 99.375 ft.

Note :

1. Standard Parshall Flume Dimensions

Table 21 : Standard Dimensions of Parshall Flume

W		A		2/3A		B		C		D		E		F	
ft	in.	ft	in.	ft	in.	ft	in.	ft	in.	ft	in.	ft	in.	ft	in.
0	6	2	$\frac{7}{16}$	1	$4\frac{5}{16}$	2	0	1	$3\frac{1}{2}$	1	$3\frac{3}{8}$	2	0	1	0
0	9	2	$10\frac{5}{8}$	1	$11\frac{1}{8}$	2	10	1	3	1	$10\frac{5}{8}$	2	6	1	0
1	0	4	6	3	0	4	$4\frac{7}{8}$	2	0	2	$9\frac{1}{4}$	3	0	2	0
1	6	4	9	3	2	4	$7\frac{7}{8}$	2	6	3	$4\frac{3}{8}$	3	0	2	0
2	0	5	0	3	4	4	$10\frac{7}{8}$	3	0	3	$11\frac{1}{2}$	3	0	2	0
3	0	5	6	3	8	5	$4\frac{3}{4}$	4	0	5	$1\frac{7}{8}$	3	0	2	0
4	0	6	0	4	0	5	$10\frac{5}{8}$	5	0	6	$4\frac{1}{4}$	3	0	2	0
5	0	6	6	4	4	6	$4\frac{1}{2}$	6	0	7	$6\frac{5}{8}$	3	0	2	0
6	0	7	0	4	8	6	$10\frac{3}{8}$	7	0	8	9	3	0	2	0
7	0	7	6	5	0	7	$4\frac{1}{4}$	8	0	9	$11\frac{3}{8}$	3	0	2	0
8	0	8	0	5	4	7	$10\frac{1}{8}$	9	0	11	$1\frac{3}{4}$	3	0	2	0

Contd...

ft	in.	ft	in.	ft	in.	ft	in.	ft	in.	ft	in.	cfs	cfs
0	6	2	0	1	0	0	$4\frac{1}{2}$	2	$11\frac{1}{2}$	1	4	0.05	3.9
0	9	1	6	1	0	0	$4\frac{1}{2}$	3	$6\frac{1}{2}$	1	4	0.09	8.9
1	0	3	0	1	3	0	9	4	$10\frac{3}{4}$	1	8	0.11	16.1
1	6	3	0	1	3	0	9	5	6	1	8	0.15	24.6
2	0	3	0	1	3	0	9	6	1	1	8	0.42	33.1
3	0	3	0	1	3	0	9	7	$3\frac{1}{2}$	1	8	0.61	50.4
ft	in.	ft	in.	ft	in.	ft	in.	ft	in.	ft	in.	cfs	cfs
4	0	3	0	1	6	0	9	8	$10\frac{3}{4}$	2	0	1.3	67.9
5	0	3	0	1	6	0	9	10	$1\frac{1}{4}$	2	0	1.6	85.6
6	0	3	0	1	6	0	9	11	$3\frac{1}{2}$	2	0	2.6	103.5
7	0	3	0	1	6	0	9	12	6	2	0	3.0	121.4
8	0	3	0	1	6	0	9	13	$8\frac{1}{4}$	2	0	3.5	139.5

As defined by Chow (1959), the letter designation for the dimensions are described as follows:

W = size of flume (in terms of throat width)

A = length of side wall of converging section

2/3A = distance back from end of crest to gage point

B = axial length of converging section

C = width of downstream end of flume

D = width of upstream end of flume

E = depth of flume

F = length of throat

G = length of diverging section

K = difference in elevation between lower end of flume and crest of floor level = 3 in.

M = length of approach floor

N = depth of depression in throat below crest at level floor

P = width between ends of curve wing walls

R = radius of curved wing walls

X = horizontal distance to H_b gage point from low point in throat

Y = vertical distance to H_b gage point from low point in throat.

Example 6.56

Equalization Basin

Determine the concentration–time relationship for an equalization basin of constant volume, having a constant influent rate and an influent concentration variation given as:

$$C_1 = \overline{C}_1 + k\overline{C}_1 \sin wt$$

[C_i = Influent concentration at time, t; $\overline{C}_1$ = Average influent concentration, k = constant (k$\overline{C}_1$ = Maximum influent concentration), W = $2\pi/T$, and T = Period for one cycle of the sin function].

Also calculate the steady–state maximum effluent concentration from a equalization basin with a detention period of 10 hours. Assume the following:

k = 0.5

T = 12 hours

1. Required material balance

Assume completely mixed equalization basin

[Accumulation] = [Inputs] – [Outputs] ± [Reaction]

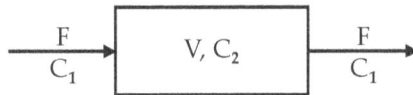

Figure 9 : Schematics of equalization basin.

$$V\frac{dC_2}{dt} = FC_1 - FC_2$$

$$V\frac{dC_2}{dt} = F(\overline{C}_1 + \overline{C}_1 k \sin wt) - FC_2$$

At t = 0, $C_2 = \overline{C}_1$

Therefore,

$$\frac{C_2}{C_1} = 1 + \left[\frac{k\theta w}{(1+\theta^2 w^2)}\right]e^{-t/g} + \frac{k\sin(wt - y)}{(1+\theta^2 w^2)^{0.5}}$$

where θ = V/F

y = $\tan^{-1}$ (w θ)

The solution consists of transients and steady-state terms.

In continuous system, the transient portion of the Solution may be neglected, so that basin performance is expressed as:

$$\frac{C_2}{\overline{C}_1} = 1 + k\frac{Sin(wt-y)}{(1+\theta^2 w^2)^{0.5}}$$

2. Required maximum effluent concentration

$$sin\ (wt - y) = 1$$

$$C_2(max) = \overline{C}_1\left[1 + \frac{k}{(1+\theta^2 w^2)^{0.5}}\right]$$

$$\frac{C_2(max)}{\overline{C}_1} = 1 + \frac{0.5}{\left[1+(10)^2\left(\frac{2\pi}{12}\right)^2\right]^{0.5}}$$

$$= 1.094$$

The equalization basin has damped the peak influent concentration from 50% greater than the average down to 9.4 percent of the average for the effluent concentration.

Example 6.57

Equalization Basin–Flow Equalization

The diurnal flow variation (instantaneous flow) of wastewater to a treatment plant is shown in Figure 10. Estimate the total volume of wastewater to be treated per day, as also the average flow rate and the detention time required.

Solution

1. Required total daily volume of wastewater

$$\text{Total daily value} = \int_0^t Q_i\ dt \cong \sum_{i=0}^t (\text{Area under the curve, Figure 10})$$

$$= [A_1 + A_2 + - - - - + A_8]$$

$$\text{Area of a trapezoid} = \frac{1}{2}\ [\text{Sum of parallel sides}]$$

Area $A_1 = 5h\ (60\ min./h)\ [1/2\ (200 + 260)]\ gal/min. = 69{,}000\ gal$

         $A_2 = 3\ h\ (60\ min./h)\ [1/2\ (260 + 700)]\ gal/min. = 86{,}400\ gal$

         $A_3 = 2\ h\ (60\ min./h)\ [1/2\ (700 + 650)]\ gal/min. = 81{,}000\ gal$

         $A_4 = 2\ h\ (60\ min./h)\ [1/2\ (650 + 640)]\ gal/min. = 77{,}400\ gal$

A_5 = 2 h (60 min./h) [1/2 (640 + 600)] gal/min. = 74,400 gal

A_6 = 4 h (60 min./h) [1/2 (600 + 100)] gal/min. = 84,000 gal

A_7 = 5 h (60 min./h) [1/2 (100 + 300)] gal/min. = 60,000 gal

A_8 = 1 h (60 min./h) [1/2 (300 + 200)] gal/min. = 15,000 gal

Total daily volume = 547,200 gal

2. Required average flow rate (Q_{avg})

$$Q_{avg} = \frac{Total\,daily\,volume\,of\,wastewater}{24}$$

$$= \frac{547,200\,gal/d}{24\,hr\,(60\,min./hr)} = 380\,gal/min$$

[It is a constant rate of equalized flow (slope = 380 gal/min), shown by line (i)]

3. Required cumulative inflow to the basin

At t = 5 h : A = 69,000 gal

t = 8 h : $A_1 + A_2$ = (69,000 + 86,400) = 155,400 gal

t = 10 h : $(A_1 + A_2) + A_3$ = (155,400 + 81,000) = 236,400 gal

t = 12 h : $(A_1 + A_2 + A_3) + A_4$ = (236,400 + 77,400) = 313,800 gal

t = 14 h : $(A_1 + A_2 + A_3 + A_4) + A_5$ = (313,800 + 74,400) = 388,200 gal

t = 18 h : $(A_1 + A_2 + A_3 + A_4 + A_5) + A_6$ = (388,200 + 84,400) = 472,200 gal

t = 23 h : $(A_1 + A_2 + A_3 + A_4 + A_5 + A_6) + A_7$ = (472,200 + 60,000) = 532,200 gal

t = 24 h : $(A_1 + A_2 + A_3 + A_4 + A_5 + A_6) + A_7) + A_8$ = (532,200 + 15,000) = 547,200 gal.

[These values represent line (ii), Figure 11].

4. Required withdrawal based on average flow rate

At t = 5h : 380 gal/min (5 h) (60 min./h) = 113,000 gal

t = 8h : 380 gal/min (8 h) (60 min./h) = 182,400 gal

t = 10h : 380 gal/min (10 h) (60 min./h) = 228,000 gal

t = 12h : 380 gal/min (12 h) (60 min./h) = 273,600 gal

t = 14h : 380 gal/min (14 h) (60 min./h) = 320,000 gal

t = 18h : 380 gal/min (18 h) (60 min./h) = 410,400 gal

t = 23h : 380 gal/min (23 h) (60 min./h) = 524,400 gal

t = 24h : 380 gal/min (24 h) (60 min./h) = 547,200 gal

[These values represent line (i), Figure 12]

5. Required maximum positive and maximum negative change in volume (Figure 11), equalization basin volume, and average detention time

Table 22 : Maximum +ve and maximum –ve change in volume.

Time (h) (1)	Ordinates of Line(i) (× 10³ gal) (2)+	Ordinates of Curve (ii) (× 10³ gal) (3)	[Line (i) – Curve(ii)] Values (× 10³ gal) (4) = (2) – (3)	Basic Contents 44 – (4) (× 10³ gal) Curve–(iii)(5)
0	0	0	0	44
1	23	10	13	31
2	45	20	25	19
3	69	35	34	10
4	90	50	40	4
5*	113	69*	44 [Max. positive change]	0[Min. level]
6	136	93	43	1
7	160	120	40	4
8*	182	155.4*	26.6	17.6
9	205	190	15	29
10*	228	236.4*	– 8.4	52.4
11	250	280	– 30	74
12*	273	313.8*	– 40.8	84.8
13	296	352	– 56	100
14*	320	388.2*	– 68.2	112.2
15	340	410	– 70	114
15.5	352	423	– 71	115
16	363	434	– 71[Max. negative change]	115[Max. level]
16.5	376	445	– 69	113
17	386	455	– 69	113
18*	410	472.2*	– 62.2	106.2
19	430	486	– 56	100
20	455	500	– 45	89
21	480	512	– 32	76
22	500	522	– 22	66
23*	523	532.2*	– 9.2	53.2
24*	547.2	547.2*	0	44

+ Refer section 4

* Data Figure 10

Maximum positive change occurs at 5 hours [(2) – (3)] = 4410³ gal

Maximum negative change occurs at 16 hours [(2) – (3)] = – 7110³ gal

At the start of the operation basin volume must be at least equal to maximum positive volume change

Therefore, basin volume (V) = (44,000 + 71,000) = 115,000 gal

$$\text{Average detention time} (t) = \frac{\text{Basin volume (V)}}{\text{Constant withdrawal rate}}$$

$$= \frac{115,000\,\text{gal}}{380\,\text{gal/min}}$$

$$= 5.04 \text{ hours.}$$

In practice, a somewhat higher storage capacity should be recommended so that the level will never drop entirely to zero.

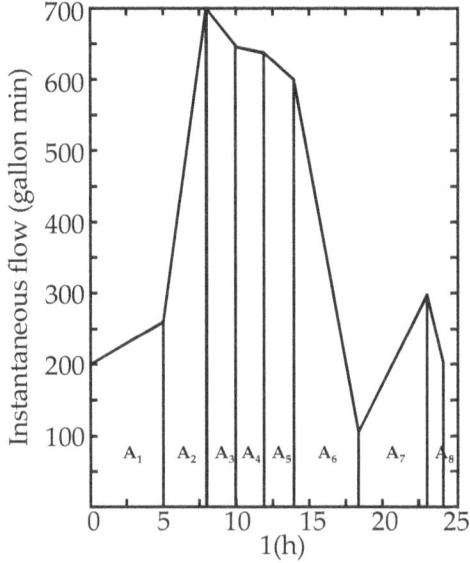

Figure 10 : Graph of instantaneous flow versus time.

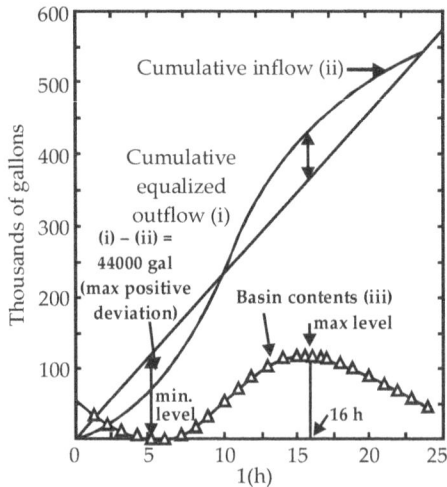

Figure 11 : Graphical solution.

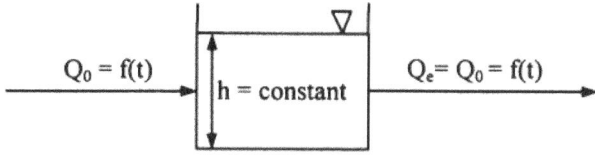

(*a*) Constant Level

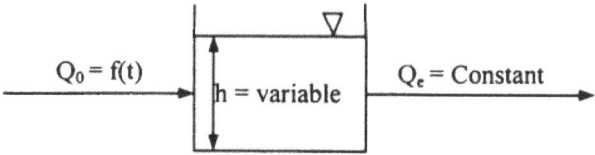

(*b*) Variable level

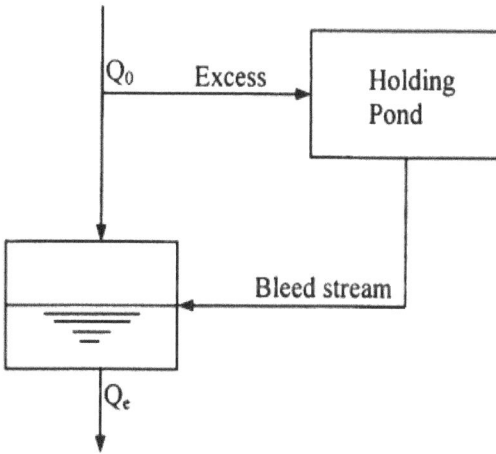

(*c*) Holding pond method

Figure 12 : Schematics of Equalization Basin Systems

Note :

1. Types of flow equalization basins (Figure 12)

    1.1 Constant level equalization basin (Figure 12a)

    – Level of water in the equalization basin is kept constant.

    – Variation in rate of inflow of influent affects the effluent flow to maintain constant water depth.

    – Large fluctuation in influent flow requires the effluent flow from constant level equalization to be diverted to another equalization basin having as objective flow equalization.

- It is more like a method of neutralization than a flow equalization technique.

1.2 Variable level equalization basin (Figure 12b)

- Effluent is withdrawn at a constant rate.

- Water level of the equalization basin is variable.

- Useful for flow equalization and neutralization.

1.3 Holding pond method of equalization (Figure 12c)

- Diverting the excess of the incoming stream to a holding pond from which a bleed stream is fed to the equalization basin.

- Useful for BOD and flow equalization and not suitable for neutralization.

Example 6.58

Equalization Basin

Develop a plot of peaking factor [(PF) the ratio of maximum concentration to the average concentration of the effluent from an EqB] versus volume for a variable–volume equalization basin and determine the volume to yield a peaking factor of 1.2 based on mass discharge. Assume the following effluent discharge from a chemical plant[*]:

Table 24 : Flow data with BOD.

Time Period (hrs)		Average	
	Flow (gal/hr)	(gal/min)	BOD (mg/L)
0800–1000	27000	450	920
1000–1200	37200	620	1130
1200–1400	50400	840	1475
1400–1600	48000	800	1525
1600–1800	20400	340	910
1800–2000	16200	270	512
2000–2200	34200	570	1210
2200–2400	66000	1100	1520
0000–0200	72000	1200	1745
0200–0400	48000	800	820
0400–0600	30600	510	410
0600–0800	34200	570	490

Eckenfelder W.W., Ind. Water Pollu. Control, p 73, McGraw Hill, N.Y., 1999.

Solution

1. Required design procedure for variable volume equalization system

    No evaporation for EqB

    No degradation from EqB

    Completely mixed EqB contents

The governing differential equations for the constant effluent flow system for each time interval are:

Water balance:

$$\frac{dV_i}{dt} = (Q_{0i} - Q_{ei}) = (Q_{0i} - Q_{0avg}) \tag{1}$$

Mass balance:

$$\frac{d(V_i C_i)}{dt} = \left[V_i \frac{dC_i}{dt} + C_i \frac{dV_i}{dt} \right] = V_i \frac{dC_i}{dt} + C_i (Q_{0i} - Q_{0avg})$$

[Mass rate change] = [Mass input] − [Mass output]

$$\frac{d(V_i C_i)}{dt} = V_i \frac{dC_i}{dt} + C_i (Q_{0i} - Q_{0avg}) = (Q_{0i} C_{0i} - Q_{e,i} C_i)$$

where V_i = Volume in the basin at time t for the time interval i (gal)

   Q_{0i} = Influent flow rate at time interval i (gal/min)

   Q_{ei} = Effluent flow rate at time interval i (gal/min.)

   Q_{0avg} = Daily average influent flow rate (gal/min)

   C_{0i} = Influent concentration at time interval i (mg/L)

   C_i = Concentration in the basin and effluent concentration at time t for time interval *i* (mg/L)

Integrating and rearranging equations (1) and (2) yield the following, assuming Q_{0i}, Q_{ei} and C_{0i} are constant during end time interval [Separation of variables]:

$$V_i(f) = V_{(i-1)}(f) + (Q_{0i} - Q_{0avg}) \Delta t_i \tag{3}$$

$$C_i(f) = C_{0i} - \frac{A}{(1 + B\Delta t_i)^D} \tag{4}$$

where $A = C_{0i} - C_{(i-1)}(f)$

   $B = \dfrac{Q_{0i} - Q_{0avg}}{V_{(i-1)}(f)}$

   $D = \dfrac{Q_{0i}}{Q_{0i} - Q_{0avg}}$

$V_{(i-1)}(f)$ = Volume in the basin at the end of time interval (i–1)(gal)

Δt_i = Time interval i (min).

$C_{(i-1)}(f)$ = Concentration in the basin and effluent concentration at the end of time interval(i–1)(mg/L)

BOD mass $M_{ei}(f)$ that leaves the equalization basin during each time interval can be calculated as;

$$M_{ef}(f) = Q_{0avg} \int_0^{\Delta t_i} C_i dt$$

$$= Q_{0avg}\left[C_{0i}\Delta t_i - \frac{A[(1+B\Delta t_i)^{1-D}-1]}{B(1-D)} \right] \qquad (5)$$

The method for calculating the masses leaving the basin at each time interval for a given volume of basin is as follows:

Step 1 :

The magnitude for the first interval (First row in Table 25) is estimated as follow:

Influent volume and mass of BOD are calculated with the mean influent flow and concentration.

Effluent flow rate is calculated by dividing the total influent volume by the total volume, and is the same for all times intervals.

Final volume is calculated by using equation (3), for an initial basin volume (guess).

Final concentration is calculated by using equation (4), for an initial concentration (guess).

Mass of BOD that left the basin is calculated by using equation (5).

Step 2 :

The values for other time intervals are calculated in a similar fashion but the initial volume and concentration for each time interval are the final volume and concentration of the previous time interval.

Step 3 :

The volume guessed for the first interval is modified until the maximum initial or the final volume in the basin is the selected volume.

Step 4 :

The concentration guessed for the first time interval is modified until it matches the final concentration for the last time interval.

Step 5 :

The maximum and average masses leaving the basin are determined and the peaking factor is calculated as the ratio of the maximum to average values.

Refer to the column(s) (Table 25)

First interval (0800 to 1000 hours):

Influent volume = (450 gal/min) (2 hr) = 450 × 120

$$= 54000 \text{ gal}$$

Mass of BOD = (54000 gal) (920 mg/L) (8.3410^{-6})

$$= 414 \text{ lb BOD.}$$

Average effluent flow rate

$$= \frac{450+620+840+800+340+270+570+1100+1200+800+510+570}{12}$$

$$= 672.5 \text{ gal/min.}$$

Final basin volume $[V_i \, (f)]$

Assume initial basin volume $V_{(i-1)}(f)$ = 278,200 gal

$$V_i(f) = V_{(i-1)}(f) + (Q_{oi} - Q_{oavg})\Delta t_i$$

$$= 278200 + (450 - 672.5) \, (120)$$

$$= 251500 \text{ gal}$$

Final basin concentration $[C_i \, (f)]$ and BOD mass left $[M_{ei} \, (f)]$

Assume initial basin concentration $[C_{(i-1)}(f)]$ = 1009 mg/L

$$A = C_{0i} - C_{(i-1)}(f)$$

$$= 920 - 1009 = -89$$

$$D = \frac{Q_{0i}}{Q_{0i} - Q_{0avg}} = \frac{450}{450 - 672.5} = -2.022$$

$$B = \frac{Q_{0i} - Q_{oavg}}{V_{(i-1)}(f)} = \frac{450 - 672.5}{278200} = -0.000799$$

- Mass of BOD that left the basin :

$$M_{ei}(f) = Q_{oavg} \left[C_{0i}\Delta t_i - \frac{A[(1+B\Delta t_i)^{1-D} - 1]}{B(1-D)} \right]$$

$$= 672.5 \left[920(120) - \frac{(-89)[(1-0.000799)(120)]^{3.022} - 1]}{(-0.000799)(3.022)} \right]$$

$$= 672.5[920(120) + 9689] = 673 \text{ lb BOD}$$

– Final basin concentration:

$$C_i(f) = C_{0i} - \frac{A}{(1 + B\Delta t_i)^D}$$

$$= 920 - \frac{(-89)}{[(1 - 0.000799)(120)]^{-2.022}}$$

$$= 920 + \frac{89}{1.226} = 993 \text{ mg/L}$$

Values for other time intervals are calculated in a similar fashion

Average BOD leaving the basin is:

$$= \frac{673 + 681 + 745 + 828 + 842 + 791 + 767 + 843 + 960 + 946 + 829 + 725}{12}$$

= 802 Ib BOD

$$\frac{Max - lb\,BOD}{Avg - lb\,BOD} = \frac{960\,lb}{802\,lb} = 1.2 \quad [= \text{Peaking factor}]$$

The results of calculations from a basin volume of 310,000 gal. are given in Table 25. Repetition of the calculations for different volumes allows the plotting of the peaking factor as a function of basin volume, Figure 13.

In most cases with a variable flow, a variable volume basin will be most effective, Figure 14.

Note :

1. Equalization basin and activated sludge basin

   Where a completely mixed basin is to be used for treatment, such as in an activated sludge basin or an aerated lagoon, the volume can be considered as a part of the equalization basin. For example, if the completely mixed aeration basin retention time is 8 hours, and the total required equalization retention time is 16 hours, then the equalization basin needs to have a retention time of only (16–8) = 8 hours.

2. Patterson and Meneze's equation for equalization requirements when both the flow and strength vary randomly.

   A mass balance can be established for the equalization basin:

   $$C_i QT + C_0 V = C_2 QT + C_2 V$$

   Effluent concentration after each time interval is:

   $$C_2 = \frac{C_i T + C_0 \left(\dfrac{V}{Q}\right)}{T + \left(\dfrac{V}{Q}\right)}$$

[If time intervals are appropriately spaced, the effluent concentration can be assumed to be constant during one time interval].

The range of effluent concentrations can then be calculated for a range of equalization basin volumes (V). A peaking factor (PF) is computed for the influent strength and flow. The effluent PF for design purposes is the ratio of the maximum concentration to the average concentration.

3. The equalization basin may be designed with variable volume to provide a constant flow or with a constant volume and an effluent flow that varies with the influent. The variable–volume basin is applicable to the chemical treatment of wastes having a low daily volume. The variable–volume basin is applicable to the chemical treatment of wastes having a low daily volume. This type of basin may also be used for discharge of wastes to municipal sewers. It may be desirable to programme the effluent pumping rate to discharge the maximum quality of waste during periods of normally low flow to the municipal treatment facility. Ideally, organic loading to the treatment plant is maintained constant over a 24 hour period.

The cummulative flow is plotted verses time over the equalization period (e.g., 24 hrs). The maximum volume with respect to the constant discharge time is the equalization volume required.

Table 25 : Equalization Basin Design (Variable Volume)

Time	Mean flow (gal/min)	Mean BOD (mg/L)	Influent Volume (gal)	Influent BOD (lb)	Basin Volume Initial (gal)	Basin Volume Final (gal)	Basin Concentration Initial (mg/L)	Basin Concentration Final (mg/L)	Effluent Flow (gal/min)	Effluent BOD (lb)
0800–1000	450	920	54000	414	**278200**	251500	**1009**	993	672.5	673
1000–1200	620	1130	74400	701	251500	245200	993	1028	672.5	681
1200–1400	840	1475	100800	1240	245200	265300	1028	1174	672.5	745
1400–1600	800	1525	96000	1221	265300	280600	1174	1278	672.5	828
1600–1800	340	910	40800	310	280600	240700	1278	1225	672.5	842
1800–2000	270	512	32400	138	240700	192400	1225	1125	672.5	791
2000–2200	570	1210	68400	690	192400	180100	1125	1151	672.5	767
2200–2400	1100	1520	132000	1673	180100	231400	1151	1327	672.5	843
0000–0200	1200	1745	144000	2096	231400	294700	1327	1504	672.5	**960**
0200–0400	800	820	96000	657	294700	**310000**	1504	1318	672.5	946
0400–0600	510	410	61200	209	**310000**	290500	1318	1150	672.5	829
0600–0800	570	490	68400	280	290500	**278200**	1150	1009	672.5	725

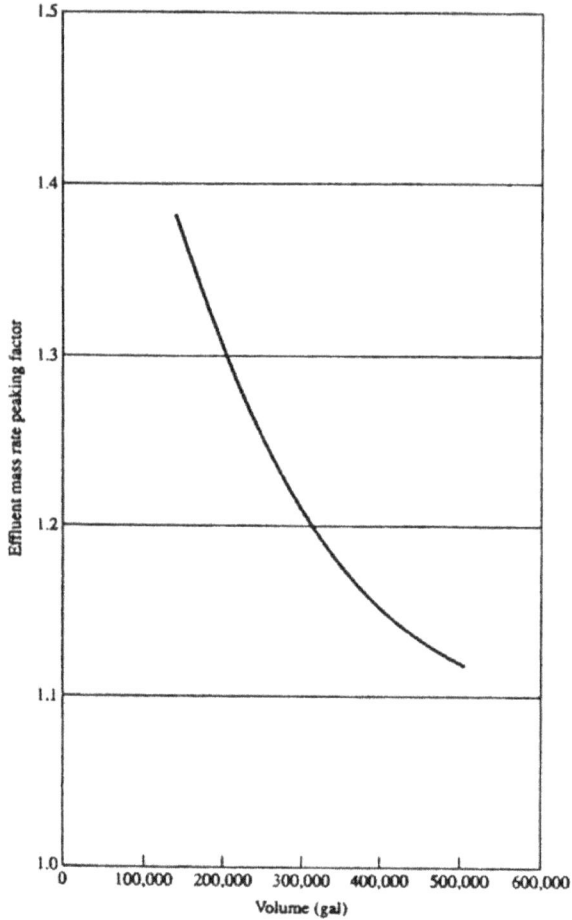

Figure 13 : Variation in effluent mass rate peaking factor with basin volume for a variable-volume equalization basin.

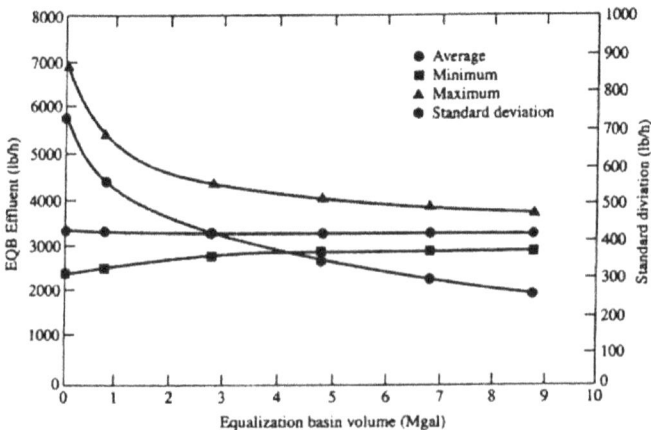

Effect of variable-volume equalization on EQB effluent BOD loading.

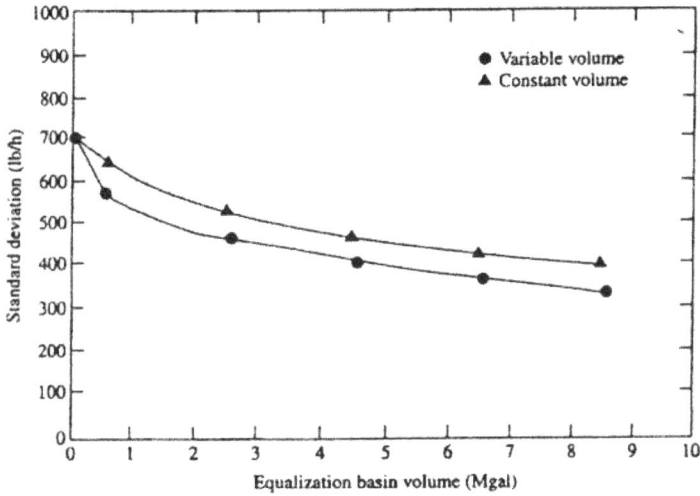

Figure 14 : Comparison of variable-volume and constant-volume.

Example 6.59

Equalization Basin with Constant Outflow

Determine the storage required for an equalization basin.

Solution

1. Required treatment rate

$$= \frac{\text{Total cumulative flow}}{\text{Time}}$$

$$= \frac{19300 \text{ gal/d}}{1440 \text{ min/d}} = 134 \text{ gal/min}$$

2. Required equalization basin volume (V)

$$V = 7580 + 45640 = 53220 \text{ gal}$$

$$= [\text{Extreme minimum} + \text{Extreme maximum}]$$

Plot cumulative inflow with outflow, time of flow measurements. Draw tangents at the maximum and minimum values parallel to the average flow rate (134 gal/min) and the required basin value is:

$$= (41000 + 8000) \text{ gal} = 49000 \text{ gal}.$$

[Read the vertical difference volume between the cumulative inflow and outflow at maximum. and minimum. values].

Table 26 : Equalization basin with constant outflow

Time Period	Inflow	Inflow	Cumulative flow (gal)		
(hr)	(gal/min)	(gal)	Inflow	Outflow	Difference
0800–0900	50	3000	3000	8040	(–) 5040
0900–1000	92	5520	8500	16080	**(–) 7580**
1000–1100	230	13800	22300	24120	(–) 1820
1100–1200	310	18600	40900	32160	(+) 8740
1200–1300	270	16200	57100	40200	(+) 16900
1300–1400	140	8400	65500	48240	(+) 17260
1400–1500	90	5400	70900	56280	(+) 14620
1500–1600	110	6600	77500	64320	(+) 13180
1600–1700	80	4800	82300	72360	(+) 9940
1700–1800	150	9000	91300	80400	(+) 10900
1800–1900	230	13800	105100	88440	(+) 16660
1900–2000	305	18300	123400	96480	(+) 26920
2000–2100	380	22800	146200	104520	(+) 41680
2100–2200	200	12000	158200	112560	**(+) 45640**
2200–2300	80	4800	163000	120600	(+) 42400
2300–2400	60	3600	166600	128640	(+) 37960
0000–0100	70	4200	170800	136680	(+) 34120
0100–0200	55	3300	174100	144720	(+) 29380
0200–0300	40	2400	176500	152760	(+) 23740
0300–0400	70	4200	180700	160800	(+) 19900
0400–0500	75	4500	185200	168840	(+) 16360
0500–0600	45	2700	187900	176880	(+) 11020
0600–0700	55	3300	191200	184920	(+) 6280
0700–0800	35	2100	193300	192960	(+) 340

Example 6.60

Equalization Basin

Determine each equalized BOD value of BOD at every time interval when the tank is filling. Assume the following data (Table 27) :

Table 27 : Flow data with times.

Time	Flow (m^3/hr)	BOD (mg/L)	Time	Flow (m^3/hr)	BOD (mg/L)
0800	27	52	1000	39	100
1200	52	175	1400	22	235
1600	18	175	1800	19	151
2000	27	181	2200	39	135
0000	26	75	0200	22	50
0400	18	42	0600	19	42

Solution

1. Required expressions for concentration and volume

    Concentration

    The concentration of the values of the quality parameters [BOD, N, P, etc.] should be done right before the tank starts filling from when it was originally empty.

$$C_{i,i+1} = \frac{C_{i-1,i}(V_{\text{rem}-1,i}) + \left(\frac{C_i + C_{i+1}}{2}\right) + \left(\frac{Q_i + Q_{i+1}}{2}\right)(t_{i+1} - t_i) - C_{i,i+1}Q_{\text{mean}}(t_{i+1} - t_i)}{(V_{\text{rem}-1,i}) + \left(\frac{Q_i + Q_{i+1}}{2}\right)(t_{i+1} - t_i) - Q_{\text{mean}}(t_{i+1} - t_i)}$$

    which reduces to the following expression:

$$C_{i,i+1} = \frac{C_{i-1}(V_{\text{rem}-1,i}) + \left(\frac{C_i + C_{i+1}}{2}\right) + \left(\frac{Q_i + Q_{i+1}}{2}\right)(t_{i+1} - t_i)}{(V_{\text{rem}-1,i}) + \left(\frac{Q_i + Q_{i+1}}{2}\right)(t_{i+1} - t_i)}$$

    where : $C_{i-1,i}$ = Quality value of the parameter in the equalization basin during previous interval between times t_{i-1} and t_i and $C_{i, i+1}$ during the forward interval times t_i and t_{i+1},

$V_{\text{rem } i-1,i}$ and $V_{\text{rem}, i+1}$ = Corresponding volumes of water remaining in the tank, respectively

C_i = Quality value of the parameter from inflow at time t_i

C_{i+1} = Quality value of the parameter from the inflow at

Q_i = Inflow at t_i

Q_{i+1} = Inflow at

$V_{\text{rem } i-1}$ = Volume of wastewater remaining in the equalization basin at the end of the previous time interval, t_{i-1} to t_i and, therefore, the volume at the beginning of the forward time interval, t_i to t_{i+1}

$C_{i-1,i} (V_{\text{rem}-1, i})$ = Total value of the quality inside the basin at the end of the previous interval [Total value of the quality at the beginning of the forward interval]

$\left(\dfrac{C_i + C_{i+1}}{2}\right)$ = Average value of the parameter in the forward interval

$\left(\dfrac{Q_i + Q_{i+1}}{2}\right)$ = Average value of the inflow in the interval

$$\left[\left(\frac{C_i + C_{i+1}}{2}\right)\left(\frac{Q_i + Q_{i+1}}{2}\right)\right](t_{i+1} - t_i) = \text{Total value of the quality coming}$$

from the inflow during the forward interval

$C_{i, i+1}$ = Equalized quality value during the time interval from t_i to

$C_{i, i+1}(Q_{mean})(t_{i+1} - t_i)$ = Value of quality withdrawn from the basin during the interval from t_i to t_{i+1}.

The sizing of the equalization basin shall be based on an identified cycle (any length of time, but, most likely would be the length of the day).

As pumping continues at Q_{mean}, the level of water in the tank goes down, if the inflow rate is less than the average. The limit of the going down of the water level is the bottom of the tank. If the inflow rate exceeds pumping as this limit is reached, the level will start to rise. The volume of the basin during the levelling down process starting from the highest level until water level hits bottom is the basin volume (V_{basin}).

At any time interval between and t_i when the tank is now filling, the volume is:

$$V_{rem-1, i} = (V_{rem-2, i}) + \left(\frac{Q_{i-1} + Q_i}{2} - Q_{mean}\right)(t_i - t_{i+1})$$

The value of $V_{rem-1, i}$ will always be positive or zero. It is zero at the time interval between t_{ibot} (when the water level hits the bottom and inflow exceeds pumping). At interval between t_{ibot-1} and t_{ibot}, the accumulation of volume in the tank $V_{rem-1, i} = V_{rem\ i\ bot-i,\ ibot}$, is zero.

The calculation for the equalized quality should be started at the precise moment. (Figure 15).

2. Required equalized BOD value

$$C_{i, i+1} = \frac{C_{i-1}(V_{rem-1, i}) + \left(\dfrac{C_i + C_{i+1}}{2}\right) + \left(\dfrac{Q_i + Q_{i+1}}{2}\right)(t_{i+1} - t_i)}{(V_{rem-1, i}) + \left(\dfrac{Q_i + Q_{i+1}}{2}\right)(t_{i+1} - t_i)}$$

$$V_{rem-1, i} = (V_{rem-2, i}) + \left(\frac{Q_{i-1} + Q_i}{2} - Q_{mean}\right)(t_i - t_{i-1})$$

The values are tabulated below for calculation of equalized concentration (C_i), i.e. BOD :

Assume : Q_{mean} = 37.7 m³/h

The tank starts filling at 09:30 hours, therefore, calculation will be started at this time.

Table 28 : Filling up the basin.

t_i	C_i	Q_i	$\left(\dfrac{C_i + C_{i+1}}{2}\right)$	$\left(\dfrac{C_{i-1} + C_i}{2}\right)$	$\left(\dfrac{Q_i + Q_{i+1}}{2}\right)$	$\left(\dfrac{Q_{i-1} + Q_i}{2}\right)$	$(t_{i+1} - t_i)$	$V_{rem\ i-1,i}$	$C_{i,\ i+1}$
0800	52	27	76	26	33	23	2	3.44	101
1000	100	39	137.5	76	45.5	33	2	−5.96 = 0	137.5
1200	175	52	205	137.5	57	45.5	2	15.6⁺	196.88⁺⁺
1400	235	62	205	205	56.7	57	2	54.2	202.37
1600	175	51	163	205	48	56.5	2	91.8	182.24
1800	151	45	166	163	48	48	2	112.24	174.75
2000	181	51	158	166	45.5	48	2	127.84	167.78
2200	135	40	105	158	33	45.5	2	143.44	148.6
0000	75	26	62.5	105	24	33	2	134.04	125.46
0200	50	22	46	62.5	20	24	2	106.64	103.79
0400	42	18	42	46	18.5	20	2	71.24	82.67
0600	42	19	47	42	23	18.5	2	32.84	61.86

$$^+ V_{rem-1,i} = (V_{rem-2,i}) + \left(\frac{Q_{i-1} + Q_i}{2} - Q_{mean}\right)(t_i - t_{i-1}) = 0 + (45.5 - 37.7)(2) = 15.6$$

$$^{++} C_{i,i+1} = \frac{C_{i-1}(V_{rem-1,i}) + \left(\dfrac{C_i + C_{i+1}}{2}\right) + \left(\dfrac{Q_i + Q_{i+1}}{2}\right)(t_{i+1} - t_i)}{(V_{rem-1,i}) + \left(\dfrac{Q_i + Q_{i+1}}{2}\right)(t_{i+1} - t_i)}$$

$$= \frac{137.5(15.6) + 205(57)(2)}{15.6 + 57(2)} = 196.88$$

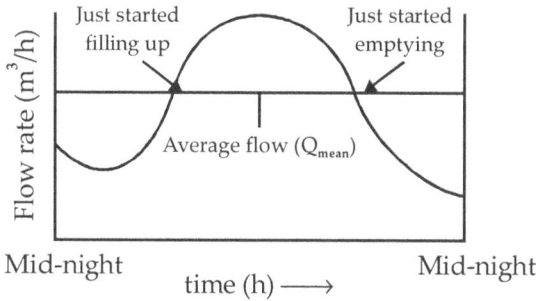

Figure 15 : Diurnal flow rate pattern.

Example 6.61

Equalization Basin

Determine the mean (equalized) flow, as also the size of the equalization basin. Assume the following data:

Table 29 : Basic data for the equalization task.

Time (hr)	Flow (m³/hr)	Time (hr)	Flow (m³/hr)	Time (hr)	Flow (m³/hr)
0000	26	0200	22	0400	18
0600	19	0800	27	1000	39
1200	52	1400	62	1600	51
1800	45	2000	51	2200	40
0000	52	–	–	–	–

Solution

1. Required expressions for mean flow (Q_{mean}) and basin volume (V_{basin})

   Mean flow (Q_{mean})

   $$Q_{mean} = \frac{1}{(t_s - t_1)} \sum_{i=2}^{s} \frac{Q_i + Q_{i-1}}{2}(t_i - t_{i-1})$$

   where   s = Total number of measurements of flow rate

         Q_i = Flow rate at time t_i

         t_s = Time of sampling of the last measurement

         Q_{mean} = Equalized flow rate

   Basin volume (V_{basin})

   $$V_{basin} = \sum_{i=2}^{i=s} positive\, of\left(\frac{Q_i + Q_{i-1}}{2} - Q_{mean} \right)(t_i - t_{i-1})$$

   where $positive\left(\frac{Q_i + Q_{i-1}}{2} - Q_{mean} \right)$ = Only positive values are to

   summed up.

2. Required mean (equalized) flow (Q_{mean})

   $$Q_{mean} = \frac{1}{(t_s - t_1)} \sum_{i=2}^{s} \frac{Q_i + Q_{i-1}}{2}(t_i - t_{i-1})$$

   $$= \frac{1}{24 - 0}\begin{bmatrix} \frac{22+26}{2}(2) + \frac{18+22}{2}(2) + \frac{19+18}{2}(2) + \frac{27+19}{2}(2) \\ + \frac{39+27}{2}(2) + \frac{52+39}{2}(2) + \frac{62+52}{2}(2) + \frac{51+52}{2}(2) \\ + \frac{45+51}{2}(2) + \frac{51+45}{2}(2) + \frac{40+51}{2}(2) + \frac{40+51}{2}(2) \end{bmatrix}$$

   $$= 37.7 \text{ m}^3/\text{hr}$$

Asume water depth (H) of 4 m and a circular tank :

$$D = \left[\frac{V_{basin}}{(\pi/4)\,H}\right]^{0.5} = \left[\frac{149\,m^3}{(3.14/4)\,4}\right]^{0.5} = 6.8\,(\approx 7\,m)$$

[Final volume of the basin to be adopted may be considered to be average of the positive and negative calculations].

3. Required basin volume (V_{basin})

$$V_{basin} = \sum_{i=2}^{i=s} \text{positive of}\left(\frac{Q_i + Q_{i-1}}{2} - Q_{mean}\right)(t_i - t_{i-2})$$

$$= \text{positive of} \begin{bmatrix} \left(\frac{22+26}{2} - 37.7\right)(2) + \left(\frac{18+22}{2} - 37.7\right)(2) + \\ \left(\frac{19+18}{2} - 37.7\right)(2) + \left(\frac{27+19}{2} - 37.7\right)(2) + \\ \left(\frac{39+27}{2} - 37.7\right)(2) + \left(\frac{52+39}{2} - 37.7\right)(2) + \\ \left(\frac{62+52}{2} - 37.7\right)(2) + \left(\frac{51+62}{2} - 37.7\right)(2) + \\ \left(\frac{45+51}{2} - 37.7\right)(2) + \left(\frac{51+45}{2} - 37.7\right)(2) + \\ \left(\frac{40+51}{2} - 37.7\right)(2) + + \left(\frac{26+40}{2} - 37.7\right)(2) + \end{bmatrix}$$

Now consider only the positive members (greater than $Q_{mean} = 37.7\ m^3/$ h) from the above expression (to determine the large capacity) of the equalization basin.

$$V_{basin} = \left(\frac{52+39}{2} - 37.7\right)(2) + \left(\frac{62+52}{2} - 37.7\right)(2) + \left(\frac{51+62}{2} - 37.7\right)(2)$$

$$+ \left(\frac{45+51}{2} - 37.7\right)(2) + \left(\frac{51+45}{2} - 37.7\right)(2)\left(\frac{40+51}{2} - 37.7\right)(2)$$

$$= 15.6 + 38.6 + 37.6 + 20.6 + 20.6 + 15.6$$

$$= 149\ m^3$$

Assume water depth (H) of 4 m and a circular tank:

$$D = \left[\frac{V_{basin}}{(\pi/4)H}\right]^{0.5} = \left[\frac{149\,m^3}{(3.14/4)4}\right]^{0.5} = 6.8\,(\approx 7\,m)$$

[Final volume of the basin to be adopted may be considered to be average of the positive and negative calculations].

Example 6.62

Scour of Bottom Deposits in Sedimentation Basins, Sewers and Grit Chambers

1. Sedimentation Basin

   The sludge (fine, light and flocculent) solids may be lifted from the sludge zone of coagulated waters, biologically treated wastewater, and the like when:

   $$v_s = \left(\frac{\tau}{\rho}\right)^{0.5}$$

   where τ is the shear stress at the liquid–sludge zone, and ñ is the density of the supernatant water. Using Weisbach–Darcy relationship:

   $$\left(\frac{\tau}{\rho}\right)^{0.5} = v_s = v_d \left(\frac{f}{8}\right)^{0.5}$$

   where v_d is the displacement velocity, v_d is expressed as:

   $$v_d = \left[\left(\frac{8}{f}\right)\left(\frac{\tau}{\rho}\right)\right]^{0.5} = \left(\frac{8}{f}\right)^{0.5} v_s$$

   v_d should be less than 18 v_s (f = 2.5 × 10^{-2}), and a useful relationship because:

   $$v_d = 10 v_s$$

   The ratio of length (L_o) to depth (h_o) of rectangular basins or of surface area (A) to cross–sectional area (a) must be kept below:

   $$\frac{A}{a} = \frac{L_o}{h_o} = \frac{v_a\, t_d}{v_o\, t_o} \left(\frac{8}{f}\right)^{0.5}\left(\frac{t_d}{t_o}\right)$$

   or $\dfrac{A}{a} = \dfrac{L_o}{h_o} = 10\left(\dfrac{t_d}{t_o}\right)$ and $\dfrac{h}{h_o} = \dfrac{v_s\, t_o}{v_o t_o} = \dfrac{v_3}{v_o} = v_o\left(\dfrac{Q}{A}\right)$.

2. Grit channel and sewers

   The channel velocity initiating scour of particles deposited in sewers, as also in grit chamber is given by:

   $$v_d = \left[\left(\frac{8k}{f}\right)g\,(S_s - 1)d\right]^{0.5}$$

   $$= \left(\frac{1.49}{n}\right)r^{1/6}\,[k\,(S_s - 1)d]^{0.5}$$

$v_d = v$ must not be exceeded. The values of k are 0.04 (unigranular sand) and 0.06 (non-uniform, interlocking sticking materials. The ratio of length to depth or surface area to cross–sectional area must be kept below the following value to avoid scouring velocities:

$$\frac{L_o}{v_o} = \frac{A}{a} = \frac{v_d t_d}{v_o t_o} = \frac{t_d}{t_o}\left[\left(\frac{6k}{f}\right)C_D\right]^{0.5}$$

where $t_d = t_o$ for an ideal channel.

3. Determine the displacement velocities (t_d) for the alum flocs, and anthracite coal dust at which the flocs and coal dust can be removed without re–suspension. Assume the following data:

$f = 3 \times 10^{-2}$

Temperature = 20°C

S_s for alum flocs = 1.1

Diameter of alum floc = 0.1 cm

S_s for anthracite coal dust = 1.5

Diameter of anthracite coal dust (d) = 0.01cm.

3.1 Alum flocs : displacement velocities [Settling chamber]

$$\frac{v_d}{v_s} = \left(\frac{8}{f}\right)^{0.5} = \left(\frac{8}{3\times10^{-2}}\right)^{0.5} = 16.3$$

Using Stoke's law for determination of settling velocities (laminar, quiescent conditions) in water:

$$v_s = \frac{g}{18}\left[\frac{(S_s - 1)}{v}\right]d^2$$

$$= \frac{981}{18}\left[\frac{(1.1-1)}{1.0105\times10^{-2}}\right](0.1)^2$$

$$= 5.4 \text{ cm/sec}$$

$v_d = 16.35.4 = 87.9$ cm/s = 2.88 ft/s

$$\frac{L_o}{h_o} = \frac{v_d\ t_d}{v_o\ t_o} = (16.3)\left[\frac{t_d}{t_o}\right]$$

For ideal sidimentation basin, $\dfrac{t_d}{t_o} = 1$ or less

$$\frac{L_o}{h_o} = 16.3 \ ; \ \frac{v_d}{v_s} = \frac{87.9}{5.4} = 16.3$$

For poor sedimentation basin $\dfrac{t_d}{t_o} = 2$ or greater

$$\frac{L_o}{h_o} = 2 \times 16.3 = 32.6$$

where v_s is the settling velocity of the particles, v_o is the velocity of particle following through the full depth (h_o) of the settling zone in the detention time (t_o).

3.2 Anthracite coal : displacement velocities (Grit chamber)

$$v_d = \left[\left(\frac{8k}{f} \right) g (S_s - 1) d \right]^{0.5}$$

$$= \left[\frac{8 \times 0.04}{3 \times 10^{-2}} \times 981 \ (1.5 - 1)(0.01) \right]^{0.5} = 7.23 \ \text{cm/sec}$$

Using Stoke's law

$$v_s = \left(\frac{g}{18} \right) \left[\frac{(S_s - 1)}{v} \right] d^2$$

$$= \frac{981}{18} \times \frac{0.5}{1.01 \times 10^{-2}} \times (0.01)^2 = 0.27 \ \text{cm/sec}$$

$$\frac{L_o}{h_o} = \left(\frac{v_d}{v_s} \right) \left(\frac{t_d}{t_o} \right) = \left(\frac{7.23}{0.27} \right) \left(\frac{t_d}{t_o} \right)$$

For ideal basin ($t_d = t_o$)

$$\frac{L_o}{h_o} = (26.78)(1) = 26.70$$

For non-ideal basin ($t_1 = 2t_o$)

$$\frac{L_o}{h_o} = 26.76 \, (2) = 53.52$$

For ideal basin (n = 0)

$$\frac{y}{y_o} = 1 - \exp\left(-\frac{t}{t_o} \right)$$

$$= 1 - \exp(-1)$$

$$= 1 - 0.368 = 0.632 \ (63.2\% \ \text{removal}).$$

4. Efficiency of ideal settling basin

In order to construct a framework for the formulation of sedimentation in continuous-flow basins, certain simplifying assumptions must be introduced. These include the following:

Within the settling zone, sedimentation takes place exactly as in a quiescent container of the same depth.

Flow is steady and, upon entering the settling zone, the concentration of suspended particles of each size is uniform throughout the cross-section at right angles to flow.

Once in the bottom zone, particles are and stay removed.

The paths taken by discrete particles settling in a horizontal–flow, rectangular or circular basin are determined by the vector sums of the settling velocity v_s of the particle and the displacement velocity v_d of the basin. All particles with a settling velocity $v_s \geq v_o$ are removed, v_o being the velocity of the particle falling through the full depth h_o of the settling zone in the detention time t_o. Also, $v_o = h_o/t_o$, $t_o = C/Q$, and $C/h_o = A$, where Q is the rate of flow, C the volumetric capacity of the settling zone, and A its surface area. Therefore $v_o = Q/A$ is the surface loading or overflow velocity of the basin. In vertical–flow basins, particles with velocity $v_s < v_o$ do not settle, out. By contrast, such particles can be removed in horizontal–flow basins if they are within vertical striking distance $h - vt_o$ from the sludge zone. If y_o particles possessing a settling velocity $v_s \leq v_o$ compose each size within the suspension, the proportion y/y_o of particles removed in the horizontal–flow tank becomes

$$\frac{y}{y_o} = \frac{h}{h_o} = \frac{(v_s t_o)}{(v_o/t_o)} = \frac{v_s}{v_o} = \frac{v_s}{(Q/A)} \tag{1}$$

For a rectangular basin of width b, $dh/dl = (v_s \, dt)/(v_d \, dt) = $ constant because both v_s and v_d are constant. Hence $h = (v_s/v_d)l$, and $h/h_o = (v_s/v_d)(l/h_o) = (v_s \, lb)(v_d h_o b) = v_s/(Q/A)$ as before. For a circular basin of radius r, $v_d = Q/(2A \, rh_o)$ s variable, and $dh/hr = v_s/v_d = 2A \, rh_o \, v_s/Q$ or $h/h_o = v_s \, (r_o^2 - r_1^2)/Q = v_s/(Q/A)$ once again.

Derived by Hazen in somewhat different fashion, equation (1) states that, for discrete particles and unhindered settling, basin efficiency is solely a function of the settling velocity of the particles and of the surface area of the basin relative to the rate of flow, which, in combination, constitute the surface loading or overflow velocity. The efficiency is otherwise independent of basin depth and displacement time or detention period. It follows that particles with settling velocity $v_s \geq v_o$ are removed and that particles with velocity $v_s < v_o$ can be fully captured in horizontal–flow basins if false bottoms or trays are inserted at intervals $h = t_s t_o$. The larger the number of trays, the smaller can be the settling velocity of particles. Conceptually, therefore, filters are approached by settling basins with a very large number of trays. Structurally, the possible number of trays is constrained by required clearances and needed cleaning operations.

Example 6.63

Constant Velocity Grit Channel

To maintain constant velocity for all rates of flow. The best Solution is to locate a standing wave (venturi) flume immediately downstream of a grit channel of a parabolic cross-section. This Solution depends on the following two points:

Provided that it is not drowned, a venturi flume produces an upstream depth that is independent of conditions downstream and which is controlled by the magnitude of flow

If the geometry of the grit chamber is such that its cross-sectional area is proportional to the flow then the velocity of flow through the channel will be constant at all flows (if v = velocity, q = flow and a = cross–sectional area, then $v = \dfrac{q}{a}$; but if a is proportional to q, i.e., a = k q, then

$v = \dfrac{q}{kq} =$ a constant).

It is now shown that to comply with second point, the channel should be of parabolic section. The flow, q through a venturi flume is given by:

$$q = k\, b\, h^{3/2} \tag{1}$$

where k is the constant, b is the throat width, and h is the upstream depth differentiating

$$dq = \frac{3}{2}\, k\, b\, h^{1/2} dh \tag{2}$$

The flow dq through a cross–sectional element (X and dh) of the channel is given by

$$dq = v\, X\, dh \tag{3}$$

Equating equations (2) and (3) and re–arranging:

$$X = \left(\frac{3kb}{2v}\right) h^{1/2}$$

where, X is the width at any depth–h. This is an equation of a parabola (or in practice, for ease of construction, a trapezoidal cross–section is used).

If v = 0.3 m/s and X and H are the channel dimensions (m) at maximum flow Q (m³/s), then:

$$Q = k\, b\, H^{3/2}$$

$$X = 5\, k\, b\, H^{1/2}$$

Therefore, $X = \dfrac{5Q}{H}$

Thus the top width of the channel is simply determined from the maximum flow and the corresponding depth.

In practice, however, at least two channels are provided so that one may be closed for manual grit removal.

The channel length is determined by settling velocity of grit particles:

$$\text{Length of Channel} = \frac{\text{Channel depth} \times \text{velocity of flow}}{\text{Settling velocity of grit particles}}$$

Grit particles settle through sewage at about 0.03 m/s, so that when the velocity of flow is controlled to 0.3 m/s:

$$\text{Channel Length} = \frac{\text{Channel depth} \times 0.3 \text{ m/s}}{0.03 \text{ m/s}}$$

$$= 10 \times \text{maximum channel depth} \times f$$

where $f = 2$ (to account for inlet turbulence and variations in settling velocity)

Therefore, channel depth = 20 × maximum depth of flow

Note :

1. Grit collected may be as high as $\dfrac{0.17 \text{ m}^3}{1000 \text{ m}^3}$, $\left[\dfrac{0.05 - 0.10 \text{ m}^3}{1000 \text{ m}^3}\right]$ of sewage and disposal off in landfill.

2. Grit (detritus) has a specific gravity of 2.5 and, therefore, has a much higher settling velocity than organic sewage solids (about 0.03 mm/s as against 3 mm/s). It is this differential in settling velocities that is exploited in grit chambers.

Example 6.64

Grit Chamber

Estimate the grit chamber size for a maximum wastewater flow of 5000 m³/d. Assume the following data:

Particle size (d_p)	: 0.5 mm and 0.2 mm
Specific gravity (r_s)	: 2.65
A flow through velocity	: 0.3 m/s (Horizontal component of flow velocity)
Water temperature	: 20°C

Solution

1. Required settling velocity (v_t) of the particles of $d_p = 0.5$ mm using Stoke's law (assume laminar flow):

$$v_t = \frac{g(\rho_s - \rho_w)}{18\mu} d^2$$

$$= \frac{9.81(2650 - 998.2) \text{ kg/m}^3 (5 \times 10^{-4})^2 \text{ m}^2}{18(1.002 \times 10^{-3} \text{N.s/m}^2)}$$

$$= 0.22 \text{ m/s [Units of N are kg.m/s}^2\text{]}$$

$$R_e = \frac{\rho_w v_t d_p}{\mu}$$

$$= \frac{(998.2 \text{ kg/m}^3)(0.22 \text{ m/s})(5 \times 10^{-4} \text{ m})}{1.002 \times 10^{-3} \text{N.s/m}^2}$$

$$= 112 \text{ [transitional regmine]}$$

$$C_D = \frac{24}{R_e} + \frac{3}{R_e^{0.5}} + 0.34$$

$$= \frac{24}{112} + \frac{3}{(112)^{0.5}} + 0.34$$

$$= 0.84$$

$$v_t = \left[\frac{4}{3} g \frac{(\rho_s - \rho)}{C_D \rho_w} d \right]^{0.5}$$

$$= \left[\frac{4}{3}(9.81) \frac{(2650 - 998.2)(5 \times 10^{-4})}{(0.84)(998.2)} \right]^{0.5}$$

$$= 0.11 \text{ m/s}$$

Repeat this procedure with $v_t = 0.11$ m/s

$$R_e = 55$$

$$C_D = 1.18$$

$$v_t = 0.10 \text{ m/s} \approx 0.11 \text{ m/s.}$$

2. Required grit chamber dimensions

    Assume a rectangular cross-section with depth = 1.5 × width at maximum flow

    Cross–sectional area (A_x) = W(1.5w) = W × D

    $$= 1.5 \, w^2 = \frac{Q_x}{v_h} = \frac{5000 \text{ m}^3/\text{d}}{0.30 \text{ m/s}}$$

    $$= \frac{5000 \text{ m}^3/\text{d}}{0.3 \text{ m/s}} \left(\frac{d}{1440 \text{ min}} \right) \left(\frac{\min}{60 \text{ s}} \right)$$

    $$= 0.193 \text{ m}^2$$

$$\text{Detention time (t)} = \frac{\text{Depth (D)}}{v_t} = \frac{0.5 \text{ m}}{0.10 \text{ m/s}} = 5 \text{ seconds}$$

Length = detention time (t) × Horizontal flow velocity (v_h)

= (0.3 m/s)(5 s)

= 1.5 m [Grit chamber not suitable].

3. Required settling velocity (v_t) of the particles with $d_p = 0.2$ mm using Stoke's law (assume laminar flow):

$$v_t = \frac{g(\rho_s - \rho_w)}{18\mu} d_p^2$$

$$= \frac{9.81(2650 - 998.2)(2 \times 10^{-4})^2}{18(1.002 \times 10^{-3})}$$

$$= 0.036 \text{ m/s}$$

$$R_e = \frac{998.2(0.036)(2 \times 10^{-4})}{1.002 \times 10^{-3}}$$

$$= 7.18 \text{ [OK, less than 10]}.$$

3.1 Required dimensions of grit chamber

Area (A_x) = W × 1.5W = 1.5 W²

$$= \frac{Q}{v_h} = \frac{5000 \text{ m}^3/\text{d}}{0.3 \text{ m/s}} \left(\frac{d}{86400 \text{ s}} \right)$$

$$= 0.193 \text{ m}^2$$

* Width (W) = $\left[\dfrac{0.193 \text{ m}^2}{1.5} \right]^{0.5} = 0.39$ m

* Depth (D) = $\left[\dfrac{0.193 \text{ m}^2}{0.39 \text{ m}} \right] = 0.50$ m

Detention time (t)

$$t = \frac{\text{Depth (D)}}{v_t} = \frac{0.50 \text{ m}}{0.036 \text{ m/s}} = 14 \text{ s}$$

Length (L) = $(v_h)(t)$

= (0.3 m/s) (14 s) = 4.2 m.

Note :

1. Same considerations for the design of grit chamber

    Type–1 settling (discrete particle settling)

Particle size (d$_p$) $\approx$ 0.2 mm

Channel horizontal flow to be maintained is around 0.3 m/s [25% increase may result in washout of the grit, while a 25% decrease may result in retention of non-target organics].

Particles larger than 0.2 mm result in a very small grit chamber length as shown in above examples

Typical terminal (settling) velocities (v$_t$) are in the range of 0.016 to 0.022 m/s

Typical grit value of 15 m^3/10^6 m^3 of wastewater is collected from the grit chamber [4 to 200 m^3/10^6 m^3 of wastewater].

2. Settling (terminal) velocities of various particle sizes (for spheres with specific gravity of 2.65 in water at 20°C)

Table 30 : Settling velocity versus particle diameter.

Diameter (mm)	Type	Settling Velocity (m/s)
10	Pebble	0.73
1	Course sand	0.23
0.1	Fine sand	1.0×10^{-2} (0.6 m/min)
0.01	Silt	1.0×10^{-4} (8.6 m/d)
0.0001	Large colloid	1.0×10^{-8} (0.3 m/yr)
0.000001	Small colloid	1.0×10^{-13} (3 m/million years)

Example 6.65

Grit Chambers

Design the cross–section of a grit chamber consisting of four identical channel to remove grit for a peak of 60000 m^3/d, an average flow of 40,000 m^3/d, and a minimum flow of 15,000 m^3/d. It is desired that there should be a minimum of three channels in operations.

Assume the following data:

Flow through velocity   : 0.3 m/s

Channel are to be control through : (a) Parshall flumes

(b) Proportional flow weirs

Solution

1. Required expression for tank depth and cross section area for control devises (Prashall flume and proportional flow weir orifice)

Proportional flow weir

$$Q = k_0 l h^{1.5}$$

$Q = [k_0 1 h^{1.5}]h$

$Q = $ Constant h

* Cross sectional area (A) of the tank is given by:

$A = k\ W\ H$

* Flow through the tank $= v_h(A) = v_h = (kWH)$

Therefore, $vh(kWH) = [k_0 1 h^{1.5}]h = k_0 1 h^{0.5}\ H = $ constant [H]

$$\text{Sloving for W} = \frac{\text{Constant}}{v_h\ k} = \text{Constant}$$

[Grit chambers controlled by a proportional flow weir, the width of the tank must be constant (rectangular–cross section)].

Grit chamber controlled by Parshall flume [Proportional flow weir is also a flow device]

$$v_h(kWH) = k_0 1 h^{1.5}h = k_0 1 H^{1.5}$$

Solving for H = constant $(W^2) = Z_0$

[Cross–section of flow should be parabolic]

$$\text{Area of parabola (A)} = \frac{2}{3}WH = \frac{2}{3}WZ_0$$

Co-ordinates for the proportional-flow weir orifice

$$1 h^{0.5} = \text{constant} = 1_p\ h_p^{0.5}$$

or $\qquad h = \dfrac{1_p^2\ h_p}{1}$

Assume $\quad lp = \dfrac{1}{3}W$ and $h_p = Z_p$

Therefore, $\quad h = \dfrac{\left(\dfrac{1}{3}W\right)^2 Z_p}{1} = \dfrac{1}{9}\dfrac{W^2 Z_p}{1}$

Co-ordinates for the parabolic cross-section

$$A_p = \frac{2}{3}W_p Z_0 p$$

$$C = \frac{3}{2}\frac{A_p}{W_p^3}$$

$$Z_0 = \frac{3}{2}\frac{A_p}{W_3}(W^2)$$

$$A = \frac{2}{3} W Z_0 = \frac{2}{3} W \left[\frac{3}{2} \frac{A_p}{W_p^3} W^2 \right]$$

$$W^2 = \left(\frac{W_p^3}{A_p} \right)^{2/3} A^{2/3}$$

$$W = \left[\left(\frac{W_p^3}{A_p} \right)(A) \right]^{1/3}$$

and

$$Z_0 = \frac{3}{2} \frac{A_p}{W_p^3} W^2 = \frac{3}{2} \frac{A_p}{W^3} \left(\frac{W_p^3}{A_p} \right)^{2/3}$$

$$A_p^{2/3} = \frac{3}{2} \frac{A^{1/3}}{W_p} A^{2/3}$$

where Q = Flow through orifice

K_0 = Orifice constant

l = Width of flow over the weir

h = Head over the crest

v_h = Flow through velocity

H = Tank depth

L_p = Orifice opening at the maximum peak flow and corresponding h would be h_p

Q_p = Maximum peak flow through the channel

lp = Orifice opening at the maximum peak flow, and corresponding h would be h_p

A_p = It is the area corresponding to Q_p

2. Required cross–section of grit chamber with a Parshall flume control device

$$A_p \text{ [Three channels]} = \frac{60000\,\text{m}^3/\text{d}}{3(0.3\,\text{m/s})\ 24(60)(60)} = 0.77\ \text{m}^2$$

$$A_p \text{ [Four channels]} = \frac{60000\,\text{m}^3/\text{d}}{4(0.3\,\text{m/s})24(60)(60)} = 0.56\ \text{m}^2$$

$$A_{ave} \text{ [Average, four channels]} = \frac{40000\,\text{m}^3/\text{d}}{4(0.3\,\text{m/s})24(60)(60)} = 0.39\ \text{m}^2$$

$$A_m \text{ [Minimum, four channels]} = \frac{15000 \, \text{m}^3/\text{d}}{4(0.3 \, \text{m/s})24(60)(60)} = 0.145 \ \text{m}^2$$

$$Z_0 = \frac{3}{2} \frac{A_p^{1/3}}{W_p} A^{2/3}$$

Assume $W_p = 1.5$ m

Water depth (H_0) flow four channels where at least three channels are in operation:

$$H_0 = Z_0 = \frac{3}{2} A^{*1/3} \frac{[\text{Maximum peak flow}] A_p^{2/3}}{W_P} \quad \text{[Max. peak flow in]}$$
four channels]

$$= \frac{3}{2} \frac{[0.77 \, \text{m}^2]^{1/3} [0.56]^{2/3}}{1.5 \text{m}}$$

$$Z_0 = 1.5 \left(\frac{0.92}{1.5}\right) 0.68 = 0.624 \, \text{m}$$

$$Z_0 = 1.5 \frac{[0.77 \, \text{m}^2]^{1/3}}{1.5 \text{m}} [0.39 \, \text{m}^2]^{2/3} = 0.49 \, \text{m [Average flow]}$$

$$Z_0 = 1.5 \frac{[0.77 \, \text{m}^2]^{1/3}}{1.5 \text{m}} [0.145 \, \text{m}^2]^{2/3} = 0.25 \, \text{m [Minimum flow]}.$$

3. Required cross–section of grit chamber with a proportional flow weir control device

    Cross–section of the grit chamber should be rectangular

$$W = \frac{\text{Constant}}{v_h \, k}; \, W = 1.5 \, \text{m [Assumed]}$$

$$H = \frac{A}{W}$$

$$H_0 = \frac{A}{W} = \frac{0.77 \, \text{m}^2}{1.5 \text{m}} = 0.52 \ \text{m [Three channels, maximum flow]}$$

$$H_0 = \frac{A}{W} = \frac{0.56 \, \text{m}^2}{1.5 \text{m}} = 0.37 \ \text{m [Four channels, maximum flow]}$$

$$H_0 = \frac{A}{W} = \frac{0.39 \, \text{m}^2}{1.5 \text{m}} = 0.26 \, \text{m [Four channels, average flow]}$$

$$H_0 = \frac{A}{W} = \frac{0.145 \, \text{m}^2}{1.5 \text{m}} = 0.09 \ \text{m [Four channels, minimum flow]}.$$

Note :

1. Parshall flume

    In practice, the co–ordinates of parabolic cross–section should be checked against the flow conditions of chosen dimensions of Parshall flumes. If the flumes are shown to the submerged forcing them not to be critical flows, other co–ordinates of the parabolic cross–sections must be tried until the flumes show critical flow conditions or un-submerged.

Example 6.66

Neutralization of Acidic Wastewater

A highly acidic wastewater (pH = 1.8) is to the neutralized prior to treatment through PST, and secondary units. The neutralization (titration curve–pH versus mg of lime per litre of wastewater) curve reveals that at a pH of 3.8, lime requirement is 2250 mg/L and at a pH of 7.0, lime requirement is 3500 mg/L. Use the following data:

Acidic wastewater flow : 0.025 MGD

Determine the lime dosage for first and second stage neutralization, as also the neutralization basin volume.

Solution

1. Required lime dosage

    For first–stage neutralization:

    $$= (0.025 \text{ mil–gal/d})(8.34 \text{ lb/mil–gal/mg/L})(2250 \text{ mg/L})$$

    $$= 469.1 \text{ lb/d}$$

    For second–stage neutralization:

    $$= (0.025)(8.34)(3500 - 2250)$$

    $$= 260.6 \text{ lb/d}$$

    Total lime dosage = (469.1 + 260.6) = 729.7 lb/d

    or 0.025 × 8.343500 = 729.7 lb/d.

2. Required neutralization basin volume (V)

    Assume a retention time of 5 min

    $$V = \left(0.025 \text{ mil–gal/d} \times \frac{1 \text{ d}}{24 \text{ hr}} \times \frac{\text{hr}}{60 \text{ min}} \right) (5 \text{ min})$$

    $$= 87 \text{ gal}$$

3. Required power level

    Assume power level of = 0.2 hp/1000 gal

    Therefore, use 0.1 hp mixer.

Note :

Table 31 : Neutralization system design parameters

Item	Details
Chemical storage tank	Liquid–use store supply vessel dry–dilute in a mix or day tank
Reaction tank:	
Size	Cubic or cylindrical with liquid depth equal to diameter
Retention time	5 to 30 min (30 min–time)
Influent	Locate at tank top
Effluent	Locate at tank effluent
Agitator:	
Propeller type	Under 1000 gal–tanks
Axial–flow type	Over 1000 gal–tanks
Peripheral speeds	12 ft/s for large tanks
	25 ft/s for tanks less than 1000 gal
pH sensor	Submersible preferred to flow–through type
Metering pump or control value	Pump delivery range limited to 10 to 1; valves have greater ranger

Table 32 : Neutralization factors for common alkaline and acid regents.

Chemical	Formula	Equivalent Weight	To Neutralize 1 mg/L Acidity or Alkalinity (Expressed as $CaCO_3$) Requires nmg/L)	Neutralization Factor, Assuming 100%Purity of all Compounds
				Basicity
Calcium carbonate	$CaCO_3$	50	1.000	1.000/0.56 = 1.786
Calcium oxide	CaO	28	0.560	0.560/0.56 = 1.000
Calcium hydroxide	$Ca(OH)_2$	37	0.740	0.740/0.56 = 1.321
Magnesium oxide	MgO	20	0.403	0.403/0.56 = 0.720
Magnesium hydroxide	$Mg(OH)_2$	29	0.583	0.583/0.56 = 1.041
Dolomitic quick lime	$(CaO)_{0.6}(MgO)_{0.4}$	24.8	0.497	0.497/0.56 = 0.888
Dolomitic hydrated lime	$[Ca(OH)_2]_{0.6}$ $[Mg(OH)_2]_{0.4}$	33.8	0.677	0.677/0.56 = 1.209
Sodium hydroxide	NaOH	40	0.799	0.799/0.56 = 1.427
Sodium carbonate	Na_2CO_3	53	1.059	1.059/0.56 = 1.891
Sodium bicarbonate	$NaHCO_3$	84	1.680	1.680/0.56 = 3.000
				Acidity
Sulphuric acid	H_2SO_4	49	0.980	0.980/0.56 = 1.750
Hydrochloric acid	HCl	36	0.720	0.720/0.56 = 1.285
Nitric acid	HNO_3	62	1.260	1.260/0.56 = 2.250
Carbonic acid	H_2CO_3	31	0.620	0.620/0.56 = 1.107

Table 33 : Chemical coagulant applications

Chemical process	Dosage Range (mg/L)	pH	Comments
Lime	150–1500	9.0–11.0	For colloid coagulation and P removal Wastewater with low alkalinity, and high and variable P Basic reactions: $Ca(OH)_2 + Ca(HCO_3)_2 \rightarrow 2CaCO_3 + 2H_2O\ MgCO_3 + Ca(OH)_2 \rightarrow Mg(OH)_2 + CaCO_3$
Alum	75–250	4.5–7.0	For colloid coagulation and P removal Wastewater with high alkalinity and low and stable P Basic reactions: $Al_2(SO_4)_3 + 6H_2O \rightarrow 2Al(OH)_3 + 3H_2SO_4$
$FeCl_3$, $FeCl_2$	35–150	4.0–7.0	For colloid coagulation and P removal
$FeSO_4.7H_2O$	70–200	4.0–7.0	Wastewater with high alkalinity and low and stable P where leaching of iron in the effluent is allowable or can be controlled Where economical source of waste iron is available (steel mills, etc.) Basic reactions: $$FeCl_3 + 3H_2O \rightarrow Fe(OH)_3 + 3HCl$$
Cationic polymers	2–5	No change	For colloid coagulation or to aid coagulation with a metal, where the buildup of an inert chemical is to be avoided
Anionic and some non-ionic polymers	0.25–1.0	No change	Use as a flocculation aid to speed flocculation and settling and to toughen floc for filtration
Weighting aids and clays	3–20	No change	Used for very dilute colloidal suspensions for weighting

Example 6.67

Hydrocyclone

Describe the mathematical expressions for sizing of a hydrocyclone and the number of stages needed based on feed stream flow rate.

Solution

1. Required expressions for sizing of the hydrocyclone and number of stages

   Hydrocyclone should be selected by considering the recovery of a particle of specific size. The limit is decided as where 50% of a particle of a particular feed size reports in the overflow while the other 50% reports to the underflow [defined as cut size d_{50C}]

2. Table 34, Summary of properties for typical neutralization chemicals.

Table 34

Property	Calcium carbonate ($CaCO_3$)	Calcium hydroxide $Ca(OH)_2$	Calcium oxide (CaO)	Hydrochloric acid (HCl)	Sodium carbonate (Na_2CO_3)	Sodium hydroxide ($NaOH$)	Sulphuric Acid (H_2SO_4)
Available form	Powder, crushed (various sizes)	Powder, granules	Lump, pebble, ground	Liquid	Powder	Solid flake, ground flake, liquid	Liquid
Shipping container	Bags, barrel, bulk	Bags (50 lb),+ bulk	Bags (80 lb), barrels, bulk	Barrels, drums, bulk	Bags (100 lb), bulk	Drum (735), 100, 450 lb)	Carboys, drums (825 lb), bulk
Bulk weight (lb/ft³)	Powder 48 to 71; crushed 70 to 100	25 to 50	40 to 70	27.9%; 0.53 lb/gal++; 31.45%, 9.65 lb/gal	34 to 62	Varies	106, 114
Commercial strength	—	Normally 13% $Ca(OH)_2$	75 to 99% normally 90% CaO	27.9, 31.45, 35.2%	99.2%	98%	60° Be, 77.7%; 66° Be, 93.2%
Water solubility (lb/gal)	Nearly insoluble	Nearly insoluble	Nearly insoluble	Complete	0.58 @ 32°F, 1.04 @ 50°F, 1.79 @ 68°F, 3.33 @ 86°F,	3.5 @ 32°F, 4.3 @ 50°F, 9.1 @ 68°F, 9.2 @ 86°F,	Complete
Feeding form	Dry slurry used in fixed beds	Dry or slurry	Dry or slurry [must be slaked to $Ca(OH)_2$]	Liquid	Dry, Liquid	Solution	Liquid
Feeding type	Volumetric pump	Volumetric metering pump	Dry-volumetric, wet slurry (centrifugal pump)	Metering pump	Volumetric feeder, metering pump	Metering pump	Metering pump
Accessory equipment	Slurry tank	Slurry tank	Slurry tank, slaker	Dilution tank	Dissolving tank	Solution tank	—
Suitable handling materials	Iron, steel	Iron, steel, plastic, rubber hose	Iron, steel, plastic, rubber hose	Hastelloy A, selected plastic and rubber types	Iron, steel	Iron, steel	—
Comments	—	—	Provide means for cleaning slurry transfer pipes	—	Can cake	Dissolving solid forms generate much heat	Provide for spill cleanup and neutralization

+ lb × 0.4536 = kg;   ++ lb/gal × 0.1198 = kg/L; § 0.555(°F − 32) = °C.

Various correlations are available for the determination of d_{50C}:

$$d_{50C} = \frac{14.2\,D^{0.46}\,d_i^{0.6}\,d_o^{0.21}\,\exp(0.063\ V)}{d_u^{0.71}\,h^{0.38}\,Q^{0.45}\,(\rho_p - \rho_f)^{0.5}}$$

where
Q = Slurry feed rate (m³/h)

d_i = Internal diameter of inlet (cm)

d_o = Internal diameter of the overflow (cm)

d_u = Internal diameter of the underflow (cm)

D = Diameter of cyclone (cm)

V = Volume fraction of solids in feed

h = Distance between bottom of the vortex to the top of the underflow orifice (cm)

ρ_p, ρ_f = Density of solids and liquid, respectively (kg/m³)

Inlet orifice area should be 6 to 8% of the cross-sectional area of the feed chamber.

Vortex finder diameter = 35 to 40% of cyclone diameter

Cylindrical section length = Diameter of feed chamber

Angle of conical section = 12° for cyclone diameter less than 300 mm

= 20° for cyclone diameter greater than 300 mm

Underflow orifice diameter = 25% of vortex finder diameter

If the hydrocyclone have above geometry, then d_{50} can also be estimated as:

$$d_{50} = \frac{13.2\,D^{0.675}\,\exp[-0.3 + 0.095\ V - 0.0036\ V^2 + 0.000065\ V^3]}{\Delta P^{0.3}\,(\rho_p - \rho_f)}$$

where
ΔP = Pressure drop across cyclone (k P_a)

Relationship between Q and ΔP is:

$$Q = 9.4 \times 10^{-3}\ \Delta P^{0.5}\ D^2$$

D_{50} base point, can also be determined as:

$$d_{50-b} = 2.86\ D^{0.66}$$

where
D = Cyclone diameter (cm)

d_{50-b} = Base diameter (mm)

Standard configuration of hydrocyclone is:

* Cyclone diameter (D) – Reference

* Inlet slurry diameter   = 0.05 D
* Vortex finder diameter (d_o) = 0.35 D
* Cylindrical section length = 0.35 D
* Cone angle   = 10 – 20°
* Apex orifice   = (0.1 to 0.35) D
* Standard operating conditions are stated as water as fluid, particles (quartz) of specific gravity 2.65, feed solid concentration less than 1% (V/V), and pressure drop (ΔP) of 69 k P_a (maximum)
* If the above conditions cannot be satisfied, correction factors have to be applied:

– Correction factor for feed concentration (C_1)

$$C_1 = \left[\frac{53 - V^{-0.43}}{53} \right]$$

where V = % volume of solids in feed concentration

– Correction factor for pressure drop [C_2]:

$$C_2 = 3.2\ \Delta P^{-0.28}\ [\Delta P \text{ in } kP_a]$$

– Correction factor for specific gravity of solids [C_3]:

$$C_3 = \left[\frac{1.65}{\rho_p - \rho_f} \right]^{0.5}$$

– Overall cut size $D_{50} = (D_{50 -b})(C_1 \times C_2 \times C_3)$

In optimum design of a hydrocyclone, the minimum pressure drop for the desired degrees of separation has to be evaluated

Permissible pressure drop in the hydrocyclone is between 0.2 to 0.7 kg/cm². To design hydrocyclone, initially the dimensions of inlet, overflow, and discharge openings can be assumed and the pressure drop for this configuration is evaluated and the procedure can be continued till the minimum possible ΔP is achieved.

Capacity of a 38° hydrocyclone can be estimated as:

$$Q = K D d_u \left[\frac{\Delta P}{\rho_s} \right]^{0.5}$$

where   Q = Slurry feed rate (L/min)

           D = Cyclone diameter (cm)

d_u = Internal diameter of underflow (cm)

ΔP = Pressure drop between inlet and overflow outlet (kg/cm^2)

ρ_s = Slurry density (kg/cm^3)

K = 0.51 for D = 80 to 100 mm

  = 0.524 for D = 125 to 600 mm

Ratio of $\dfrac{d_u}{d_o} = 0.4$; d_i = Inlet diameter

d_o = Outlet diameter

$(2d_i + d_o) = \dfrac{D}{2}$

$\dfrac{d_o}{d_i} = 1.0$ to 2.0

Optimum diameter of the hydrocyclone is 5 to 8 times the internal diameter of inlet

Length of conical section can be estimated as:

$$= \dfrac{D - d_u}{2\tan\left(\dfrac{\alpha}{2}\right)}$$

Length of the inlet feed nozzle (l) = 4 × diameter

Geometric proportions of the conventional cyclone and devoting hydrocyclone are shown as:

Figure 16 : Typical design considerations.

Table 35 : Typical dimensions

Diameter	Hydrocyclone	
	Conventional (XD)	Deoiling (XD)
D_i	0.14–0.17	0.35 (Twin)
D_o	0.14	0.04–0.15
D_u	0.1–0.36	0.5
D_s	–	2
l	0.67–2.0	2
m	0.33–0.4	Zero
L	–	~ 45

Contd...

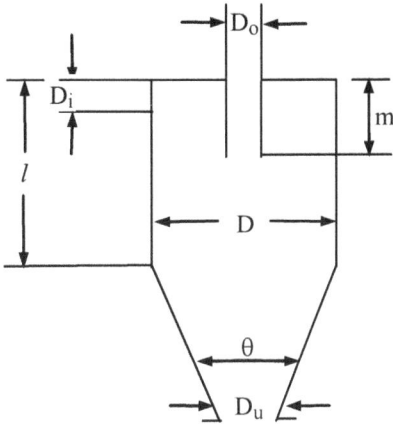

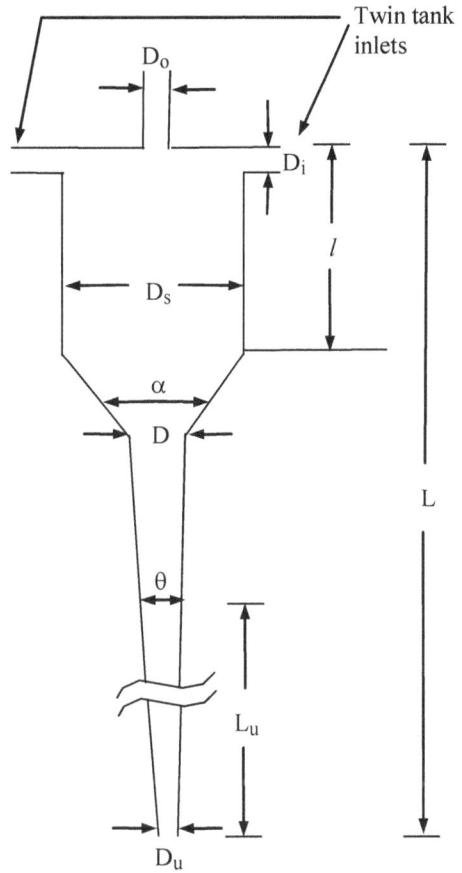

L_u	–	less than 20
θ	10–20°	1.5°
α	–	20°

Hydrocyclone performance totally depends on:

– Solid content of feed

– Feed flow rate

– Pressure at which it enters the hydrocyclone

– Any deviation in these parameters can lead to erratic performance.

Solid contents of feed can cause wear and abrasion of the metal wall of the hydrocyclone. Wear can be minimized by applying soft rubber lining, if the temperature is below 60°C. If the particle size is above 100 mm, wear resistant metals can be used. For small size hydrocyclone polyurethane can be used as material of fabrication.

Example 6.68

Chemical Precipitation

Determine the amount of SO_2 and lime required for precipitation of chromium as $Cr(OH)_3$. Also estimate the sludge generation and reactor size for removal of chromium from plating wastewater containing 150 mg/L CrO_3 as H_2CrO_4. The related stoichiometric relations are:

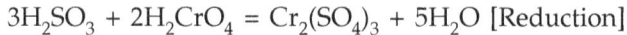

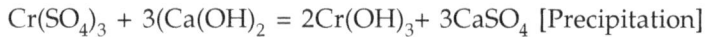

$$3SO_2 + 3H_2O = 3H_2SO_3$$

$$3H_2SO_3 + 2H_2CrO_4 = Cr_2(SO_4)_3 + 5H_2O \text{ [Reduction]}$$

$$Cr(SO_4)_3 + 3(Ca(OH)_2 = 2Cr(OH)_3 + 3CaSO_4 \text{ [Precipitation]}$$

Overall reaction :

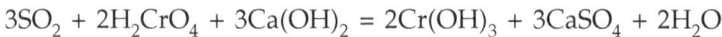

$$3SO_2 + 2H_2CrO_4 + 3Ca(OH)_2 = 2Cr(OH)_3 + 3CaSO_4 + 2H_2O$$

Use the following data :

Waste water flow   : 7500 gal/d

Dry sludge produced   : 1.25 lb/lb of CrO_3

$Ca(OH)_2$ required   : 1.5 lb/lb of CrO_3

SO_2 required   : 1.3 lb/lb of CrO_3

Solution

1. Required mass balance

$$\text{Theoretical : lb } SO_2 \text{ required/lb } CrO_3 = \left(\frac{3 \text{ lb mole of } SO_2}{2 \text{ lb mole of } H_2CrO_4} \right)$$

$$[H_2CrO_4 = 2H_2O + CrO_3]$$

$$\text{Therefore, } \frac{3 \text{ lb mole } SO_2}{2 \text{ lb mole } C_rO_3} = \frac{3 \times 64}{2 \times 100} = 0.96$$

$$\text{Theoretical : lb } Ca(OH)_2/\text{lb } CrO_3 = \frac{3 \text{ lb mole of } Ca(OH)_2}{2 \text{ lb mole of } CrO_3}$$

$$= \frac{3 \times 74}{2 \times 100} 1.11$$

$$\text{Theoretical : lb } Cr(OH)_3/\text{lb } CrO_3 = \frac{2 \text{ lb mole of } Cr(OH)_3}{2 \text{ lb mole of } CrO_3}$$

$$= \frac{2 \times 103}{200} = 1.03$$

[SO_2 actually required 1.3 lb/lb CrO_3 as against 0.96. This may be due to presence of DO.

Ca(OH)$_2$ actual requirement is 1.5 lb Ca(OH)$_2$/lb CrO$_3$, and may be due to demand from other constituents].

Influent mass of pollutant

$$= \frac{(7500 \text{ gal/d})(8.34 \text{ lb/gal})(150 \text{ mg/L})}{10^6} = 9.4 \text{ lb CrO}_3/\text{d}.$$

2. Required Chemicals

SO$_2$: (9.4 lb CrO$_3$/d) (1.3 lb SO$_2$/lb CrO$_3$)

   = 12.2 lb SO$_2$/d

Lime : (9.4 lb CrO$_3$/d) (1.5 lb Ca(OH)$_2$/lb CrO$_3$)

   = 14.1 lb Ca(OH)$_2$/d

$$= \frac{14.1 \text{ lb Ca(OH)}_2/\text{d}}{0.05} = 282 \text{ lb of Ca(OH)}_2/\text{d at 5\%}$$

or 282 lb of Ca(OH)/d 5% slurry = (282) (0.1198 gal/lb)

$$= 33.78 \text{ gal of Ca(OH)}_2/\text{d}$$

3. Required Sludge Generation

   = (9.4 lb CrO$_3$/d) (1.25 lb sludge produced/lb CrO$_3$)

   = 11.75 Ib of sludge produced/d and is equivalent to be

   concentration of $\dfrac{11.75 \text{ lb Cr(OH)}_3/\text{d}}{(7500 \text{ gal/d})(8.34)} \times 10^6$

   = 187 mg/L as solids.

4. Required Waste Storage

   Chromium rinse waters are not continuously discharged, but discharged intermittently. A storage tank of 15,000 to 20,000 gallon may be provided, with downstream continuous treatment using a standard treatment package unit.

5. Required reactor size

   Reaction is complete in 60 min

   Reactor volume = (7500 gal/d)(60 min) $\left(\dfrac{d}{1440 \text{ min}} \right)$

   = 313 gal

   Use two 450 gal reactors in series; one for reduction and the other for precipitation.

6. Required mixing levels

   = [313 gal][5 to 10 hp/1000 gal]

   = 1.5 to 3.13 hp.

Note :

1. Solubility product

    Product of molar concentrations of reactants exceeding the solubility product, precipitation or super saturation occurs.

    Reactions that result in relatively insoluble products are easy to precipitate. However, consideration must be given to the residual concentration, which under the best of conditions may still be greater than the statutory requirements

    Solubility product is temperature sensitive

    [Values generally increase with increasing temperature. Therefore, formation of precipitate is more difficult at elevated temperatures]

    Stability of the precipitate will be affected by pH if hydrolysis is a part of the reaction

    Precipitation reaction is dependent on available reacting components which must be present in adequate amounts and in ionic form.

2. Guide lines governing solubility (feasibility of precipitation as a treatment system)

    Nitrates (NO_3^-)   : Soluble

    Chlorides (Cl^-)   : Soluble except Hg_2Cl_2 and $AgCl$

    Sulphates (SO_4^{-2})   : Soluble except $CaSO_4$, SrO_4, Hg_2SO_4 $HgSO_4$ $PbSO_4$, and Ag_2SO_4

    Carbonates (CO_3^{-2}) : Insoluble except for those of the IA elements– L_I, Na, K, Cs, Fr, and NH_4^+

    Hydroxides (OH^-)   : Insoluble except for three of the I–A elements, $Sr(OH)_2$, $Ba(OH)_2$, and $Ca(OH)_2$ [Slightly soluble]

    Sulphides (S^{-2})   : Insoluble except for IA elements and 2A elements, Be, Mg, Ca, Sr, Ba, Ra, and NH_4^+

    Effect of Hydrolysis : $[H^+] [OH^-] = K_w = 10^{-14}$ at 25°C

    Precipitate stability is effected by the pH system.

    For example

    $$H_2O = H^+ + OH^- ; K_w = 10^{-14}$$
    $$Zn^{+2} = Zn(OH)_2$$
    $$K = 5 \times 10^{-17}$$
    $$[Zn] [OH^-] = 5 \times 10^{-17}$$
    $$[Zn] [10^{-14}/[H^+] = 5 \times 10^{-17}$$
    $$[Zn] [10^{(-14 + pH)}] = 5 \times 10^{-17}$$

As alkalinity increase, the Zn concentration is further reduced. Zn is one of the many metals whose hydroxide can exhibit both acidic and basic characteristics (amphoteric hydroxides):

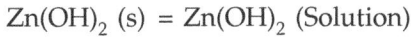

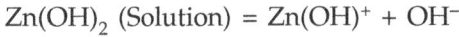

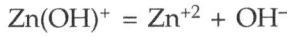

$$Zn(OH)_2 \text{ (s)} = Zn(OH)_2 \text{ (Solution)}$$

$$Zn(OH)_2 \text{ (Solution)} = Zn(OH)^+ + OH^-$$

$$Zn(OH)^+ = Zn^{+2} + OH^-$$

Precipitation reaction is expressed as:

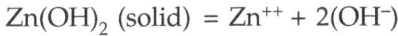

$$Zn(OH)_2 \text{ (solid)} = Zn^{++} + 2(OH^-)$$

[Data defining the reaction path are not always available]

Table 36 : pH effect on zinc precipitation.

pH	Zn (moles/L)	Zn(ppm)
1	5×10^{-4}	33
2	5×10^{-5}	3
3	5×10^{-6}	0.3
4	5×10^{-7}	3×10^{-2}
5	5×10^{-8}	3×10^{-3}
6	5×10^{-9}	4×10^{-4}
7	5×10^{-10}	5×10^{-5}
8	5×10^{-11}	3×10^{-6}

3. Common precipitation treatment system

    Heavy metals as the hydroxide

    Phosphate precipitation with aluminium sulphate or chlorohydrate, ferric chloride (sulphate), or lime

    Fluoride removal using calcium salts, aluminium sulphate or activated alumina

    Sulphate removal using aluminium salts

    Sulphide removal using iron salts

    Chromium removal as trivalent hydroxide.

4. Heavy metal removal as hydroxide

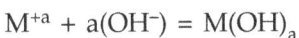

    $$M^{+a} + a(OH^-) = M(OH)_a$$

    Intermediate formation makes it difficult to predict effluent quality

    Amphoteric qualities of some metals result in their dissolving at elevated pH, making controlling reactor alkalinity critical

    Actual hydroxide molecule may contain water, so that the basic formula may not describe the precipitated compounds

Flocs formed, excess reagent, or complex compounds formed with other constituents of waste may enhance adsorption of metals.

5. Metal removal by sulphide precipitation

   S^{-2} solubility low

   H_2S is toxic, excess sulphide may violate effluent standards.

6. Phosphate precipitation

   Alum and ferric phosphate complexes formed are pH sensitive (amphoteric characteristics)

   Calcium phosphate is less sensitive to residue dissolving (greater production of sludges).

Example 6.69

Coagulation and Flocculation

1. Determine the daily amount of alum [$Al_2(SO_4)_3.14H_2O$], and polyelectrolyte requirements for the treatment of raw water. Assume the following data :

   Raw water to be treated : 50 MGD

   Alum requirement determined through Jar testing (2 liter solution) 10 mL alum containing 6 mg Al^{+3}/mL

   Polyelectrolyte requirement determined through Jar testing (2 liter solution) 1 mL containing 1 mg/mL of polyelectrolyte

   1.1 Required coagulant dosages

   $$\text{Alum} = \frac{(6 \text{ mg/mL})(10 \text{ mL})\left(\dfrac{594.4}{54.0}\right)}{2\,L}, \quad [MW = 594.4]$$

   $$= 330 \text{ mg/L}$$

   $$\text{Polyelectrolyte} = \frac{(1 \text{ mg/mL})(1 \text{ mL})}{2\,L} = 0.5 \text{ mg/L}$$

   1.2 Required total daily dosages of coagulants for 50 MGD plant

   $$\text{Alum} = (50 \text{ MGD})(330 \text{ mg/L})(8.34 \text{ lb/MGD.mL})$$

   $$= 137,610 \text{ lb/d}$$

   $$\text{Polyelectrolyte} = (50 \text{ MGD})(0.5 \text{ mg/L})(8.34)$$

   $$= 208.5 \text{ lb/d}$$

2. Determine the amount of alkalinity that will be destroyed by using a dose of 150 mg/L of bulk ferric sulphate containing 20 percent weight of Fe. Also determine the amount of lime requirement if the raw has an alkalinity of 50 mg/L for a plant of 50 MGD capacity.

2.1 Required amount of alkalinity destruction

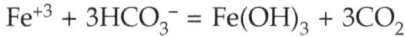

$$Fe^{+3} + 3HCO_3^- = Fe(OH)_3 + 3CO_2$$

Three equivalents of (bicarbonate) alkalinity reacts with 1 mol weight of Fe^{+3}.

Equivalent weight of $CaCO_3 = 50$.

Therefore, 150 mg of alkalinity reacts with 55.8 mg of Fe^{+3}.

$$Fe^{+3} \text{ applied} = (150 \text{ mg/L})(0.2)$$
$$= 30 \text{ mg/L}$$

$$\text{Amount of alkalinity destroyed} = (30 \text{ mg/L})\left(\frac{150}{55.8}\right)$$
$$= 80.64 \text{ mg/L}$$

2.2 Required amount of lime

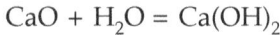

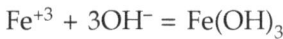

$$CaO + H_2O = Ca(OH)_2$$
$$Fe^{+3} + 3OH^- = Fe(OH)_3$$

Minimum alkalinity required = 80.64 mg/L

Alkalinity imparted by addition of lime = $(80.64 - 50.0)$ mg/L
$$= 30.64 \text{ mg/L}$$

$$\text{Equivalent of } OH^- \text{ required} = \frac{30.64}{50} = 0.613 \text{ m} - \text{eq/L}$$

$$CaO \text{ concentration} = \frac{(0.613)}{0.70}\left(\frac{56.1}{2}\right) \text{ [Bulk lime contains 70\% CaO]}$$
$$= 24.56 \text{ mg/L}$$

Amount of bulk lime required = (50 MGD)(24.56 mg/L)(8.34 lb/MG.mg/L)
$$= 10240 \text{ lb/d}.$$

3. Determine the maximum amount of dry sludge production. Assume the following data:

Plat capacity	: 30 MGD
Required alum dose	: 150 mg/L

$[Al_2(SO_4)_3.14H_2O]$

Influent suspended solids : 20 mg/L

3.1 Required amount of dry solids production

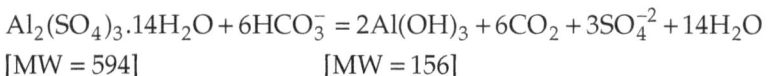

$$Al_2(SO_4)_3.14H_2O + 6HCO_3^- = 2Al(OH)_3 + 6CO_2 + 3SO_4^{-2} + 14H_2O$$
$$[MW = 594] \qquad\qquad [MW = 156]$$

Assume that the all Al^{+3} precipitates as $Al(OH)_3$

Amount of alum required = (30 MGD)(150 mg/L)(8.34)

$$= 37530 \text{ lb/d}$$

Alum precipitated is $Al(OH)_3 = \dfrac{(37530)(156)}{(594)} = 9856 \text{ lb/d}$

Influent suspended solids = (30 MGD)(20 mg/L)(8.34)

$$= 5004 \text{ lb/d}$$

$$\text{Total dry solids} = 9856 + 5004$$

$$= 14860 \text{ lb/d.}$$

3.2   Required sludge volume

Wet sludge water volume = (14860)(98/2)

$$= 728,140 \text{ lb/d}$$

[Wet sludge concentrated to 2% by weight]

Wet sludge volume:

$$= \left(\frac{14860}{8.34}\right)(2.0) + \left(\frac{728140}{8.34}\right)(1) \quad \text{[Specific gravity of solids = 2]}$$

$$= 3563.5 + 87307$$

$$= 90870.5 \text{ gal/d.}$$

4.   Compare the coagulant doses of alum, ferric chloride, or lime for precipitation of phosphate and removal of organic phosphorus for an STP of 5 MGD capacity. Assume the following data:

Experimental data:

Lowering phosphorus below 0.5 mg/L, dosage of alum and iron

$$= 1.5 \text{ to } 2.0 \text{ mol Al/mol P}$$

$$= 1.0 \text{ to } 7.5 \text{ mol Fe/mol P}$$

Line requirement is pH and alkalinity dependant. At pH–11, $Ca(OH)_2$ required = 400 mg/L

Wastewater alkalinity   = 200 mg/L

Total influent phosphorus concentration = 15 mg/L as P.

4.1   Required alum, iron salt and lime dosages

$$Al^{+3} \text{ required} = \frac{(1.5 \text{ mol Al/mol P})(15 \text{ mg P/L})}{(31 \text{ mg P/m} - \text{mol P})}$$

$$= 0.726 \text{ m mol Al/L or } 0.726 \times 27 = 19.6 \text{ mg/L}$$

$$\text{Alum dose} = (19.6 \text{ mg/L})(594/54) = 215.6 \text{ mg/L}$$

$$[Al_2(SO_4)_3.14H_2O]$$

Alum dose for 5 MGD plant:

$$= (5 \text{ MGD})(215.6 \text{ mg/L})(8.34)$$

$$= 8989 \text{ lb/d}$$

$$\frac{Fe^{+3} \text{ required}}{\left[FeCl_3.6H_2O\right]} = \frac{(1.0 \text{ mol} - \text{Fe/mol P})(15 \text{ mg/L} - P)}{(31 \text{ mg P/m} - \text{mol P})}$$

$$\text{Ferric dose required} = (27 \text{ mg/L})\left(\frac{270}{55.8}\right) = 130.6 \text{ mg/L}$$

Ferric dose for 5 MGD plant :

$$= (5 \text{ MGD})(130.6)(8.34)$$

$$= 5446 \text{ lb/d}$$

Lime required for 5 MGD plant:

$$= (400 \text{ mg/L})(5 \text{ MGD})(56.1/74.1)(8.34) \text{ as CaO}$$

$$= 12628 \text{ lb/d.}$$

5. Determine the amount of dry sludge removed from the clarifier using alum or ferric chloride. Assume the following data:

Wastewater flow : 5 MGD

Influent suspended solid : 200 mg/L

P–content associated with suspended solids : 5 mg/L

Total P – content in the influent : 15 mg/L

5.1 Required dry solids removal using alum

Assume that all suspended solids and all Al–precipitates are removed

$$Al^{+3} + PO_4^{-3} = AlPO_4$$

Amount of $AlPO_4$ precipitated $= (15 - 5) \text{ mg/L as P}$

$$= \frac{10}{31} = 0.32 \text{ m} - \text{mol P/L}$$

$$= (0.32)(122) = 39 \text{ mg/L as } AlPO_4$$

$$= (5 \text{ MGD})(39 \text{ mg/L})(8.34)$$

$$= 1641 \text{ lb/d}$$

Al^{+3} available for precipitation as $Al(OH)_3$

$$= \text{Total P} - \text{Removal as P}(AlPO_4)$$

$$= \frac{(1.5 \text{ mol Al/mol P})(15 \text{ mg} - \text{P/L})}{31} - 0.32 \text{ m mol P/L}$$

$$= 0.726 - 0.32 \text{ m mol P/L} = 0.406 \text{ m mol/L}$$

Precipitate as $Al(OH)_3 = (0.406)(78) = 31.65 \text{ mg/L}$

Amount of $Al(OH)_3 = (5 \text{ MGD})(31.65 \text{ mg/L})(8.34)$

$$= 1320 \text{ lb/d}$$

Removal of SS $= (5 \text{ MGD})(200 \text{ mg/L})(8.34)$

$$= 8340 \text{ lb/d}$$

Total dry solids removed $= 1641(AlPO_4) + 1320[\text{as } Al(OH)_3]$
$$+ 8340 \text{ (as SS)}$$

$$= 11301 \text{ lb/d}$$

5.2 Required dry solids removed using $FeCl_3$

$$Fe^{+3} + PO_4^{-3} = FePO_4$$

$$FePO_4 \text{ precipitate} = \frac{(10 \text{ mg/L})}{31} = 0.32 \text{ m mol P/L}$$

$$= (0.32)(150.8) = 48.4 \text{ mg FePO}_4/\text{L}$$

$$= (5 \text{ MGD})(48.4 \text{ mg/L})(8.34)$$

$$= 2018 \text{ lb/d}$$

Fe^{+3} available for precipitation as $Fe(OH)_3$

$$= (1.0) \, (15/31) - 0.32$$

$$= 0.164 \text{ m mol/L}$$

$Fe(OH)_3$ precipitate $= (0.164)(106.9) = 17.53 \text{ mg/L}$

$$= (5 \text{ MGD})(17.53 \text{ mg/L})(8.34)$$

$$= 731 \text{ lb/d}$$

Total dry solids removed

$$= 2018 \text{ lb/d (as FePO}_4) + 731 \text{ lb/d (as Fe(OH)}_3$$
$$+ 8430 \text{ lb/d (as SS)}$$

$$= 11179 \text{ lb/d}$$

6. Determine the value of shear rate (G) and the detention time for a flash mixing operation. Assume the following data:

Wastewater flow   : 10 MGD

Power input through a turbine : 4 hp

Wastewater temperature   : 50°F

Mixer volume   : 400 ft^3

6.1 Required shear rate and detention period for flash mixing operation

$$G = \left(\frac{P}{V\mu}\right)^{0.5}$$

$$G = \left[\frac{(4 \text{ hp})(550 \text{ lb} - \text{ft/hp.s})}{(400 \text{ ft}^3)(2.74 \times 10^{-5} \text{ lb} - \text{s/ft}^2)}\right]^{0.5}$$

$$= 448 \text{ s}^{-1} \text{ [OK]}$$

$$\theta = \frac{V}{Q} = \frac{400 \text{ ft}^3}{(10 \text{ MGD})(1.547 \text{ ft}^3/\text{s.MGD})}$$

$$= 25.86 \text{ s}^{-1} \text{ [OK]}$$

7. Determine the point for introducing chemical coagulant in a circular pipe (acting as a flash mixing operation with a detention time of 30 s prior to a flocculation. Assume the following data :

Wastewater flow   : 5 MGD

Pipe diameter   : 2 ft

Water temperature   : 50°F $\left(\mu = 2.74 \times 10^{-5} \dfrac{\text{lb} - \text{s}}{\text{ft}^2}\right)$

Hazen–William coefficient for pipe (C) : 120

7.1 Required slope in the pipe

$$Q = 194CD^{2.63} S^{0.54} \text{ [Hazen–William formula]}$$

Velocity of water in circular pipe $(V) = \dfrac{Q}{A}$

$$A = \frac{\pi}{4}\left(\frac{10}{12}\right)^2 = 0.545 \text{ ft}^2$$

$$V = \frac{(5 \text{ MGD})(1.547 \text{ ft}^3/\text{s.MGD})}{(0.545 \text{ ft}^2)}$$

$$= 14.2 \text{ ft/s}$$

Injection point for chemical introduction = (V)(t)

$$= (14.2 \text{ ft/s})(10 \text{ s})$$

$$= 142 \text{ ft upstream of the flocculator}$$

$$\text{Slope (S)} = \left[\frac{Q}{194 \, CD^{2.63}}\right]^{1.85}$$

$$= \left[\frac{(5 \times 10^6 \ gal/d)\left(\dfrac{d}{24 \times 60 \ min} \right)}{194(120)\left(\dfrac{10}{12} \right)^{2.63}} \right]^{1.85}$$

$$= 0.072 \ ft/ft$$

$$hp = 0.072 \times 142 = 10.2 \ ft$$

7.2 Required power consumed (head loss), and shear rate (G)

$$P = rQ \ h_f$$

$$G = \left[\frac{\rho Q h_f}{V \mu} \right]^{0.5} = \left[\frac{\rho h_f}{\theta \mu} \right]^{0.5}$$

$$= \left[\frac{(62.4)(10.2) \ ft}{(10 \ s)(2.74 \times 10^{-5} \ lb - s/ft^2)} \right]^{0.5}$$

$$= 1524 \ s^{-1}$$

8. Determine the quantity of ferrous sulphate, chlorine, lime, and sludge disposal. Ferrous sulphate can be oxidized to ferric sulphate at normal pH with the use of chlorine, or at about pH of 9.5 where Fe^{+2} is oxidized to Fe^{+3} by dissolved oxygen. Assume the following data:

Plant capacity   : 10 MGD

Fe^{+2} dose at normal pH   : 15mg/L

Fe^{+2} dose at pH 9.5   : 10 mg/L

Lime requirement to raise the pH to 9.5 : 150 mg/L as $CaCO_3$ with the subsequent precipitation of 180 mg/L of $CaCO_3$

Low pH operation produces sludge of 1.8% by weight

High pH operation produces sludge of 4.5% by weight

8.1 Required assumptions

Neglect the amount of suspended solids removed during flocculation

Assume iron precipitates as $Fe(OH)_3$ with dry specific gravity of 1.95

Specific gravity of dry precipitate at high = 2.0

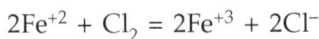

$$2Fe^{+2} + Cl_2 = 2Fe^{+3} + 2Cl^-$$

[111.7] [70.9]   [111.7]

8.2 Required amounts of $FeSO_4$, Cl_2, and sludge generated at normal pH

$$FeSO_4 = (10 \ MGD)(15 \ mg/L)\left(\frac{151.8}{55.8} \right)(8.34)$$

$$= 3403 \text{ lb/d}$$

$$Cl_2 = (10 \text{ MGD})(15 \text{ mg/L})\left(\frac{70.9}{111.7}\right)(8.34)$$

$$= 794 \text{ lb/d}$$

$$\text{Dry Fe(OH)}_3 \text{ produced} = (10 \text{ MGD})(15 \text{ mg/L})\left(\frac{106.9}{55.8}\right)(8.34)$$

$$= 2397 \text{ lb/d}$$

$$\text{Water in sludge} = 2397\left(\frac{98.2}{1.8}\right) = 130770 \text{ lb/d}$$

$$\text{Wet sludge volume} = \frac{2398}{(1.95)(8.33)} + \frac{130770}{8.33(1)}$$

$$= 148 + 15699$$

$$= 15847 \text{ gal/d}$$

8.3 Required amount of $FeSO_4$, lime and sludge generated at high pH

$$FeSO_4 = (10 \text{ MGD})(10 \text{ mg/L})\left(\frac{151.8}{55.8}\right)(8.34)$$

$$= 2269 \text{ lb/d}$$

$$\text{Lime} = (10 \text{ MGD})(150 \text{ mg/L})\left(\frac{56.9}{100}\right)(8.34)$$

$$= 7118 \text{ lb/d}$$

$$\text{Dry Fe(OH)}_3 \text{ produced} = (10 \text{ MGD})(10 \text{ mg/L})\left(\frac{106.9}{55.8}\right)(8.34)$$

$$= 1598 \text{ lb/d}$$

$$\text{Dry CaCO}_3 \text{ produced} = (10 \text{ MGD})(180 \text{ mg/L})(8.34)$$

$$= 15012 \text{ lb/d}$$

$$\text{Total solids (dry) produced} = 15012 + 1598$$

$$= 16610 \text{ lb/d}$$

$$\text{Water in the sludge} = \frac{16610(95.5)}{4.5} = 352501 \text{ lb/d}$$

$$\text{Wet sludge volume} = \frac{16610}{(8.33)(2.0)} + \frac{352501}{(8.33)(1)}$$

$$= 997 + 423170$$

$$= 424167 \text{ gal/d.}$$

Example 6.70

Design of Rapid Mix and Flocculation Basins

Design a rapid mix and flocculation basin. Use the following data

Wastewater flow	: 2.5 m^3/s
Rapid mix (t_o)	: 20 s
Rapid mix (G)	: 1000 s^{-1}
Floc (t_o)	= 1 hour
Floc (G)	= 30 s^{-1}
Wastewater temperature	= 18°C

Solution

1. Rapid mix system

   1.1 Required rapid mix reactor volume

   $$V = Q \, (t_o)$$

   $$= (2.5)(20 \text{ s}) = 50 \text{ m}^3 \text{ [Recommended volume} = 8 \text{ m}^3\text{]}$$

   In order to achieve the guideline of 8 m^3, it is necessary to provide tanks in parallel. Since we are also constrained by the availability of mixers, it is necessary to assess the limitation imposed by the power available from standard mixers.

Table 37 : Available models for mixers.

Model	Rotational Speed (rpm)	Power (New)
JTQ–50	30, 45	0.37
JTQ–75	45, 70	0.56
JTQ–100	45, 110	0.75
JTQ–150	45, 110	1.12
JTQ–200	70, 110	1.50
JTQ–300	110, 175	2.24
JTQ–500	175	3.74

Place two impellers on a shaft to achieve a multiplier of 1.9, the largest available mixer can achieve a water power of :

(3.74 kW) (0.8 efficiency) (1.9 factor) = 5.68 kW

$$G^2 = \frac{P}{\mu V}$$

$$\text{or } V = \frac{P}{\mu G^2} = \frac{5.68 \times 10^3 \text{ W}}{(1.053 \times 10^{-3} \text{ P}_a.s)(1000 \text{ s}^{-1})^2}$$

$$= 5.39 \text{ m}^3$$

No of mixing tanks required are:

$$\frac{50 \text{ m}^3}{5.39 \text{ m}^3} = 9.27 \text{ or } 9 \text{ rapid mixing tanks}$$

With 9 tanks, the volume per tank = 5.5 m³, therefore

$$G = \left[\frac{5.68 \times 10^3}{(1.053 \times 10^{-3} \times 5.5)} \right]^{0.5}$$

$$= 990 \text{ s}^{-1} \text{ (OK)} = 1000 \text{ s}^{-1} \text{ (Optimum value)}$$

1.2 Required diameter of the impeller

$$P = \frac{K n^3 D^5 \rho}{g}$$

where P is the agitator required power (W), K is the impeller constant, n is the rotational speed (rps), D is the impeller diameter (m), r is the liquid density (kg/m³), and g is the gravity constant (9.8 m/s²)

$$D = \left[\frac{(P)(g)}{K(n^3)\rho} \right]^{0.2}$$

$$= \left[\frac{(5.68 \times 10^3 \text{ W})(9.8 \text{ m/s}^2)}{(6.30)(3)^3 (1000 \text{ kg/m}^3)} \right]^{0.2} \; ; n = 3 \text{ rps}$$

$$= 0.79 \text{ m}$$

1.3 Required tank dimension

$$\text{Ratio of } \frac{\text{Im peller diameter}}{\text{Tank diameter}} = 0.5 \text{ (Recommended)}$$

$$\text{Tank diameter} = \frac{\text{Im peller diameter}}{0.5} = \frac{0.79 \text{ m}}{0.5} = 1.58 \text{ m}$$

$$\text{Tank surface area} = \frac{\pi}{4}(1.58)^2 = 1.96 \text{ m}^2$$

With 9–5.5 m³ tanks, the tank depth is;

$$\text{Tank depth} = \frac{\text{Volume of tank}}{\text{Surface area of tank}}$$

$$= \frac{5.5 \text{ m}^3}{1.96 \text{ m}^2} = 2.8 \text{ m}$$

$$\text{Ratio of } \frac{\text{Liquid depth}}{\text{Tank depth}} = 1.6 \text{ [Maximum recommended value]}$$

Actual ratio is $\dfrac{2.8 \text{ m}}{1.58 \text{ m}} = 1.77$ and is quite close to 1.6

[It is within the guide line for two impellers on a shaft].

2. Flocculator system

   2.1 Flocculator volume (V)

$$V = (Q)(t_o)$$

$$= (2.5 \text{ m}^3/\text{s})(60 \text{ s}/\text{min})(60 \text{ min})$$

$$= 9000 \text{ m}^3$$

It is logical to divide this flocculator volume into 9 flocculator chambers (9 rapid mix–chambers). Each tank will have a volume of each 1000 m^3.

Each of these tanks will be sub–divided into there compartments to achieve a tapered flocculation. The design G for compartments will be 40 s^{-1}, 30 s^{-1}, and 20 s^{-1} to yield an average [(40 + 30 + 20)/3] = 30 s^{-1}.

$$V = \frac{P}{G^2 \mu} = \frac{5.68 \times 10^3 \text{ W}}{(30 \text{ s}^{-1})^2 (1.053 \times 10^{-3} \text{ P}_a\text{s})}$$

$$= 5993 \text{ m}^3 \text{ [At this 'G' value, mixer power will not be limiting]}.$$

   2.2 Required power

Using 1000 m^3 tank and dividing it into 3 compartments of 333.4 m^3, the required power is:

$$\text{Net} - P = G^2 \mu V$$

$$= (30)^2 (1.053 \times 10^{-3})(333.4)$$

$$= 316 \text{ W } (0.316 \text{ kW})$$

$$\text{Gross Power} = \frac{0.316 \text{ kW}}{0.8} = 0.395 \text{ kW}$$

   2.3 Required flocculator dimensions

Assume a liquid depth in the flocculator chamber of 4 m, and a square configuration:

$$\text{Cross} - \text{sectional area} = \frac{333.4 \text{ m}^3}{4 \text{ m}} = 83.35 \text{ m}^2$$

Each side = $(83.35)^{0.5}$ = 9.13 m an each side

   2.4 Required impeller diameter

$$\text{Ratio of } \frac{\text{Diameter of inpeller}}{\text{Width}} = 0.3 \quad \text{[Recommended]}$$

Diameter of impeller = 0.3×9.13 m = 2.74 m

2.5 Required rotational speed (n)

Use propeller impeller with 3 blades, K = 1.00 (with a pitch of 2)

$$n^3 = \frac{(P)(g)}{(K)(D)^5(\rho)} = \frac{(395 \text{ W})(9.8 \text{ m/s}^2)}{(1.0)(2.74)^5(1000 \text{ kg/m}^3)}$$

n = 0.29 rps [17.6 rpm]

[The same motor may be used and the power input may be altered by adjusting the rotational speed].

3. Determine volume required for a rapid mix basin that converts ferric chloride dose of 50 mg/L to 40 mg/L of ferric hydroxide. Assume that a CSTR will be used and the conversion of ferric chloride follows first order kinetics with a reaction rate constant of 0.006 s^{-1}. Flow to the coagulation unit is 0.20 m^3/s.

3.1 Required detention period

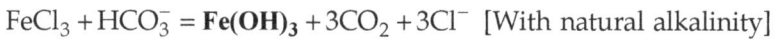

$FeCl_3 + HCO_3^- = \mathbf{Fe(OH)_3} + 3CO_2 + 3Cl^-$ [With natural alkalinity]

$FeCl_3 + 3H_2O = \mathbf{Fe(OH)_3} + 3HCl$ [Without natural alkalinity]

One mole of $FeCl_3$ produce 1 mole of $Fe(OH)_3$ Therefore, 40 mg/L of $Fe(OH)_3$ will require 40 mg/L of $FeCl_3$ be converted.

Required detention period for first order reaction in CSTR system is:

$$kt = \left[\frac{C_o}{C}\right] - 1$$

$$= \left[\frac{50}{40}\right] - 1$$

$$t = \frac{0.25}{k} = \frac{0.25}{0.006} = 42 \text{ s}$$

3.2 Required volume for rapid mix basin (V)

$$V = (Q)(t)$$

$$= (0.2 \text{ m}^3/\text{s})(42 \text{ s})$$

$$= 8.4 \text{ m}^3 \text{ (OK)}.$$

4. Coagulation efficiency is determined by conducting jar test six beakers are filled with water, and than each is mixed and flocculated uniformly by a stirrer. A test is conducted by first dosing each jar with the same alum dose and varying the pH in the each jar. The test can then be repeated by holding the pH constant and varying the coagulant dose. Determine the optimum pH, the theoretical amount of alkalinity that would be consumed at the optimum dose.

4.1 Optimum pH and alum dose

The experimental data states that optimum value of pH is 6.5 (by keeping alum dose constant), and the alum dose in 15 mg/L (by keeping pH constant 6.0). It is necessary to repeat the test using a pH of 6.5 and varying the alum dose between 12 to 20 mg/L to determine exact optimal conditions.

4.2 Required amount of alkalinity consumed

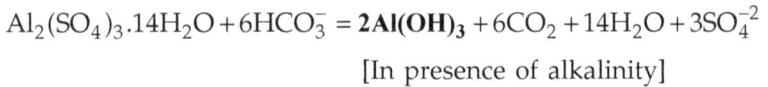

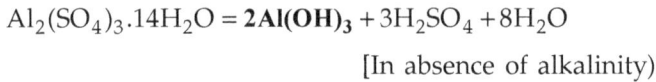

$$Al_2(SO_4)_3.14H_2O + 6HCO_3^- = 2Al(OH)_3 + 6CO_2 + 14H_2O + 3SO_4^{-2}$$

[In presence of alkalinity]

$$Al_2(SO_4)_3.14H_2O = 2Al(OH)_3 + 3H_2SO_4 + 8H_2O$$

[In absence of alkalinity)

1mole of alum consumes 6 moles of alkalinity (HCO_3^-)

$$\text{Moles of alum added} = \frac{15.0 \times 10^{-3} \text{ g/L}}{594 \text{ g/mole}}$$

$$= 2.5 \times 10^{-5} \text{ moles/L}$$

Which will consume $6(2.5 \times 10^{-5}) = 1.50 \times 10^{-5}$ moles of HCO_3^-/L

$$= (1.5 \times 10^{-4} \text{ moles } HCO_3^-/L)(61 \text{ g/moles})(10^3 \text{ mg/g})$$

$$(MW : HCO_3 = 61]$$

$$= 9 \text{ mg/L } HCO_3^- \text{ are consumed}$$

$$= (9.0)\left(\frac{50}{61}\right) = 7.38 \text{ mg/L as CaCO}_3 .$$

Note :

1.

Table 38 : G values for rapid mixing

For a given Detention Time (t_o) (s)	Shear Rate G (s^{-1})
0.5 (in line blending)	3500
10–20	1000
20–30	900
30–40	800
Longer	700

2.

Table 39 : G to values for flocculation

Type	G (s^{-1})	(G to)
Low turbidity, colour removal coagulation	20–70	60,000–200,000
High turbidity, solids removal coagulation	50–150	90,000–180,000
Softening, 10% solids	130–200	200,000–250,000
Softening, 39% solids	150–300	390,000–400,000

3. For rapid tanks have detention times of 10 to 30 s with velocity gradients of 600 to 1000 s^{-1}. The volume of the rapid mix tank seldom exceed 8 m^3 because of mixing equipment and geometry constraints. The mixing equipment consists of an electrical motor, gear–type speed reducer, and either a turbine or axial flow impeller. The turbine impeller provides more turbulence and is preferred for rapid mixing. The tanks should horizontally buffed into at least two and preferably three compartments in order to provide sufficient residence time. They are also vertically baffled to minimise vortexing. Chemicals should be added below the impeller, the point of most mixing.

Some rules of thumbs for the rapid mix system

Design liquid depth is 0.5 to 1.1 times the diameter or width

Impeller diameter be between 0.30 and 0.50 times the tank diameter or width

Vertical baffles extend into the tank about 10% of the tank, width a diameter. Although they may be larger, impeller diameter normally donot exceed 1m in diameter

Liquid depth may be increased to between 1.1 and 1.6 times the tank diameter if dual impellers on the shaft are employed

When dual impellers are employed on gear driver mixers, they are spaced approximately two impeller diameters apart.

Two straight turbines impart about 1.9 times as much power to the water as one turbine alone for the same molar power.

4. Flocculation is normally accomplished with an axial flow impeller, a paddle flocculator or a baffled chamber. Axial flow impellers impart a nearly constant G throughout the tank. The flocculator basin should be divided into at least three compartments. The velocity gradients is tapered so that the G decreases from the first compartment to the last and that the average of the compartment is the design value selected. Some rules of thumb for axial flow impellers are:

Diameter of the impeller is between 0.2 and 0.5 times the width of the chamber

Maximum impeller diameter is about 3 m.

Table 40 : Values of the impeller constant (k)

Impeller Type	K
Propeller, pitch of 1, 3 blades	0.32
Propeller, pitch of 2, 3 blades	1.00
Turbine, 6 flat blades, Vaned disc	6.30
Turbine, 6 curved blades	4.80
Far turbine, 6 blades at 45°	1.65
Shrouded turbine, 6 curved blades	1.08
Shrouded turbine, with stator, no baffles	1.12

6. In an un–baffled tank, the power imparted may be as low as one–sixth of that predicted by the equation.

Example 6.71

Cyclone and Paddle Mixers [Figure 17]

1. Design a cyclone mixer for the wastewater flow of 2000 m^3/h.

   1.1 Required design consideration for mixers

   Open channels with baffles perpendicular to the direction of flow

   The baffles have rectangular flooded free space openings [holes; ϕ = 20 to 100 mm], which cause turbulence and eddies, and promote homogenization of the wastewater flux. These kinds of mixers are efficient for mixing of the chemicals in the dissolved form, and should not be used with when calcium milk is added [sedimentation of the suspension cannot be prevented].

   Cyclone types can be used for both the suspended and dissolved chemicals. The inlet of the wastewater is located at the bottom part of the chamber, and outlet is through the collecting channel located at the upper part of the cyclone. The chemicals are introduced at the bottom part of the chamber, perpendicular to the vertical axis of the cyclone mixer.

   A part of the pipe (of approximate length) is eventually connected to the venturi. The chemicals are introduced into the sucking tube of the pump. To prevent entering of air into the pipe, introduction of the chemicals should be accomplished through the funnel, and with the float valve connected to the sucking pipe of the pump.

   A condition must be satisfied regarding the placement of funnel [it should be located above the line of piezometric pressure of the main pipe, and there should not be any valve on the segment between the place of introduction of the chemical and the end of the pipe]. The length of the segment at which mixing occurs, should be calculated as not shorter than 50 times the diameter of the pipe. To increase the effectivity of mixing and to shorten the length of the mixing segment, a venturi can be placed. The headloss due to venturi should not exceed 0.3 to 0.4 m.

The pipe conducting the chemicals into the pipe should reach the axis of the pipe, and its end should have an inclination of 45°.

If the location of the main pipe and of the basin containing the chemicals cannot ensure gravitational flows, introduction of the chemical can be accomplished through an injector.

Design principles

* Mixing chambers are designed on the basis of complete mixing (preventing floc formation in case of coagulation – clarification).

* Time of mixing in various mixing devices should not exceed:

    – 60 to 120 s for hydraulic mixers.

    – 10 to 60 s for mechanical mixers [propellers (10 to 13 s); paddle mixers (30 to 60 s)].

    – Nominal flow into the mixing chambers should not exceed:

        ■ 600 m^3/h (hydraulic mixers with baffles with rectangular openings).

        ■ 1000 m^3/h (hydraulic mixers with open holes).

        ■ 1500 m^3/h (vertical cyclones).

* Following rules should be observed for hydraulic mixers:

    – Velocity of wastewater in the open channel should be lower than 0.6 m/s and through the holes and openings in barriers should around 1 m/s.

    – Number of barriers should be equal to 3, and the distance between them, twice the width of the channel.

    – In the first and the third, barrier, the free opening should be placed at the center them, and in the second barrier two spacings placed near the wells of the mixer.

    – If the barriers are with the holes, their diameter should be within the range of 20 to 100 mm.

    – Depth of the water flow in the channel after the last barrier should not be smaller than 0.4 to 0.5 m.

- Highest located openings and higher edge of the free opening should be situated 0.1 to 0.15 m beneath the water level.

- Pressure loss during the wastewater flow through the channel opening or holes in one barrier can be calculated using:

$$h_l = K \frac{V^2}{2g}, \ [m]$$

where : V = Velocity of water through the openings (m/s)

g = Gravity constant (m/s²)

K = Coefficient of resistance (1.78 to 2.3)

* Following rules should be observed for cyclone mixers

- Mixing chambers should have a form of a part cone (bottom part) and of a cylindrical shape [upper part]

- Conic walls should have an inclination of 30 to 40°

- Velocity of water in the pipe before the inlet to the mixer should be 1.0 to 1.5 m/s; velocity at the proximity of the collecting channel should be around 0.025 m/s, and inside the collecting channel, the velocity should not exceed 0.5 m/s.

* Mechanical mixers :

- Rotational speed of about 0.5 to 0.1 m/s should be ensured

- Reynolds number $\left[= \dfrac{nd^2}{v} \right]$ should not be lower than 10⁴.

  Where v = Rotational speed (s⁻¹), d = Outer diameter of the mixing device (m), and v = Kinematic viscosity (m²/s).

- Standard dimensions and parameters for mechanical mixers. (Table 41)

- Power (P) dissipated [at given standard dimensions] is expressed as:

P = C d⁵ n³ ρ g, (Watt)

where: C = Coefficient of resistance as a function of R_e (Table 41)

ρ = Specific mass of wastewater (kg/m³)

g = Gravity constant

- while designing mechanical mixers having dimensions different from those of standard ones (Table 41), geometric similarity is not maintained. The coefficient C (Table 42) should be corrected according to:

$$C_1 = C\, f_D\, f_H\, f_l\, f_b\, f_B\, f_N$$

where $\quad f_D = \left(\dfrac{D}{\alpha d}\right)^d ; f_H = \left(\dfrac{H}{D}\right)^h ; f_l = \left(\dfrac{l}{0.25D}\right)^m$

$$f_b = \left(\dfrac{b}{\beta D}\right)^k ; f_s = \left(\dfrac{s}{d}\right)^p ; f_B = \left(\dfrac{B}{0.1d}\right)^r , \text{ and } f_N = \left(\dfrac{N}{4}\right)^n$$

D = Diameter of mixing chamber

α = D/d ratio for standard dimensions

H = Wastewater depth in side mixing chamber

l = Length of the paddle mixer

b = Width of the paddle mixer

β = b/D ratio for standard dimensions

s = A jump of the propeller (m)

B = Width of the baffle (m)

N = Number of baffles in the chamber

Values of exponents are presented in Table 42.

Table 42 : Values of exponents for different mixing equipment.

Mixing Equipment	a	h	k	m	p	r	n
Paddles in unbaffled chamber	1.1	0.6	0.3	–	–	–	–
Propeller and turbine in unbaffled chamber	0.93	0.6	–	1.5	–	–	–
Propeller in baffled chamber	0	0	–	–	1.7	0.3	0.4
Turbine in baffled chamber	0	0	–	1.5	–	0.3	0.4

- Motor power (P_e) for mixing device (mechanical) can be determined as:

$$P_e = \dfrac{kP}{\eta} \times 10^{-3} ; \text{ [kW]}$$

where P = Power dissipated (W)

η = Gear efficiency (0.9 to 0.95)

k = Safety factor for design [1.1 to 2.5, higher values should be used for small values power dissipated (P)]

Table 41 : Main design parameters of standard mixers and the values of coefficient of the resistance of the mixers.

Mixing Device	Mixing Chamber	Main Design Parameters							Coefficient of Resistance C at the Reynolds Number R_r						
		H/D	D/d	H/d	l/d	B/d	s/d	B/D	10^4	2×10^4	5×10^4	10^5	$2 \cdot 10^5$	$5 \cdot 10^5$	10^5
Two paddle	Unbaffled	1	3	3	–	0.250	–	–	1.050	0.903	0.974	0.600	0.600	0.600	0.600
Two paddle	4 baffles	1	3	3	–	0.167	–	0.1	1.730	1.730	1.730	1.730	1.730	1.730	1.730
Two paddle	Unbaffled	1	2	2	–	0.885	–	–	0.800	0.595	0.550	0.530	0.530	0.530	0.530
Six paddle	Unbaffled	1	1.11	1.11	–	0.066	–	–	1.305	1.160	0.900	0.770	0.650	–	–
Propeller	Unbaffled	1	3	3	–	–	1	–	0.295	0.287	0.266	0.260	0.255	0.245	0.238
Propeller	3 baffles	1	3	3	–	–	1	0.1	0.368	0.354	0.354	0.354	0.354	0.354	0.354
Propeller	Unbaffled	1	3	3	–	–	2	–	0.605	0.570	0.550	0.515	0.510	0.508	0.503
Propeller	4 baffles	1	3	3	–	–	2	0.1	1.000	1.000	0.950	0.900	0.900	0.900	0.900
Turbine with 6 vertical blades	Unbaffled	1	3	3	0.25	0.200	–	–	1.430	1.165	1.148	1.105	1.065	1.000	0.990
Turbine with 6 vertical blades	4 baffles	1	3	3	0.25	0.200	–	0.1	6.120	6.120	6.120	6.120	6.120	6.120	6.120

- Pipes connecting mixing chamber with flocculation chamber, clarifiers or contact filter should be calculated for the velocity flow of 0.8 to 1.0 m/s, and should not exceed 2 min.

The schematics of various mixers.

1.2 Required useful volume of the mixer

As the flow exceeds 1500 m³/h, two mixers will be used for the flow (Q) of 1000 m³/h

Assume detention time of about 100 s

$$V = \frac{Qt}{3600} = \frac{(1000)(100)}{3600}$$

$$= 27.7 \text{ m}^3$$

1.3 Required diameter of the upper (cylindrical) part

$$D = \left[\frac{4Q}{3600\pi V_u}\right]^{0.5}$$

Assume velocity of wastewater near the entrance into the collecting channel V_u of about 0.025 m/s.

$$D = \left[\frac{4(1000)}{3600(\pi)(0.025)}\right]^{0.5}$$

$$= 3.8 \text{ m}$$

1.4 Required height of the conical part of the cyclone

$$h_b = \frac{1}{2}(D-d)\text{Cot}\left(\frac{\alpha}{2}\right)$$

Assume : $\alpha = 40°$

d = Diameter of the inlet pipe to the chamber

$$= \left[\frac{4Q}{3600(\pi)(V_b)}\right]^{0.5}$$

V_b = Flow velocity in the inlet pipe (1 to 1.5 m/s)

$$d = \left[\frac{4(1000)}{3600(\pi)(1.2)}\right]^{0.5} = 0.55 \text{ m}$$

$$h_b = \frac{1}{2}(3.8-0.55)\cot\left(\frac{40}{2}\right)$$

$$= 4.4 \text{ m.}$$

1.5 Required height of the upper part of the chamber

$$h_u = \frac{4[V - V_b]}{\pi D^2}, \text{ (m)}$$

V_b = Volume of the bottom part of the chamber

$$V_b = \frac{1}{3}\pi h_b \left[\frac{D^2}{4} + \frac{dD}{4} + \frac{d^2}{4} \right]$$

$$= \frac{1}{3}(\pi)(4.4)\left[\frac{(3.8)^2}{4} + \frac{0.55(3.8)}{4} + \frac{(0.55)^2}{4} \right]$$

$$= 19.38 \text{ m}^3$$

$$h_u = \frac{4[27.8 - 19.38]}{\pi(3.8)^2} = 0.74 \text{ m}.$$

1.6 Total height of the mixing chamber

$$H = h_b + h_u + h_k$$

where h_k = Height of the edge of the mixing chamber above [0.3 to 0.5 m]

$$H = 0.74 + 4.4 + 0.4 = 5.54 \text{ m}$$

2. Design a mixing chamber with a paddle mixer and determine the dissipated power for a wastewater flow (Q) of 600 m³/h. (Figure 17 and Table 42).

2.1 Required working volume of the chamber

$$V = \frac{Qt}{3600} \qquad [t = 30 \text{ to } 60 \text{ s; assume } t = 30 \text{ s}]$$

$$= \frac{600(30)}{3600} = 5 \text{ m}^3$$

2.2 Required dimensions of the chamber

$$D = \left(\frac{4V}{\pi H} \right)^{0.5}$$

Assume ratio of $\dfrac{H}{D} = \phi$ (0.8 to 1.3)

Therefore, $$D = \left[\frac{4V}{\pi\phi} \right]^{0.33}$$

$$D = \left[\frac{4(5)}{\pi(1.2)} \right]^{0.33}; \phi = 1.2$$

$$= 1.74 \text{ m}$$

Assume D = 1.8 m

$$H = \phi D = 1.2(1.8) = 2.16 \text{ m}.$$

2.3 Required mixer dimensions

Using Table 41 : $d = \dfrac{1}{3}D = \dfrac{1.8}{3} = 0.6$ m; and height of paddle (assuming b/d = 0.25), b = 0.25d = 0.25 (0.6) = 1.5 m.

2.4 Required hydraulic conditions

$$R_e = \dfrac{nd^2}{v}$$

$$= \dfrac{(1)(0.6)^2}{1.31 \times 10^{-6}} \quad \text{[Assume n = 1 s}^{-1}\text{, and v = 1.31} \times 10^{-6} \text{ m}^2\text{/s at 10°C]}$$

$= 0.27 \times 01^6$ greater than 12500.

2.5 Required power dissipated (P)

Geometric ratios : $\dfrac{H}{D} = \dfrac{2.2}{1.8} = 1.22$ (Standard value = 1)

$\dfrac{D}{d} = \dfrac{1.8}{0.6} = 3$ (Equal to standard value = 1)

As one of the ratios is not equal to the standard value, the power dissipated is given as :

$$P = C \, f_H \, d^5 \, n^3 \, r$$

$$= C \left(\dfrac{H}{D} \right)^h d^5 n^3 \rho$$

Assuming the value of C = 0.6; h = 0.6 and r = 1000 kg/m³, the power required is:

$P = 0.6(1.22)^{0.6} \, (0.6)^5 \, (1)^3 (1000)$

$= 52.57$ W

Required actual power is:

$$P_e = \dfrac{kP}{\eta} \times 10^{-3}$$

$$= \dfrac{(2)(52.57)}{0.9} \times 10^{-3} = 0.117 \text{ kW} \quad \text{[k = 2, } \eta = 0.9\text{]}$$

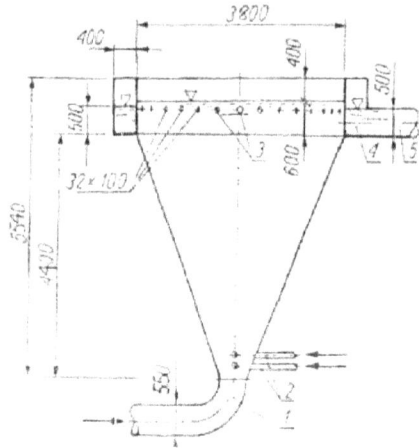

Figure 16 : Vertical mixing cyclone

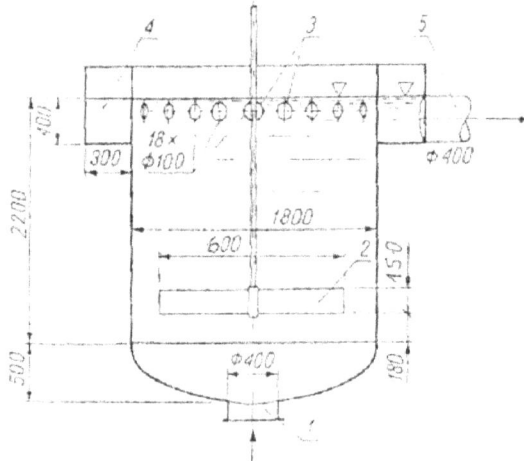

1 Wastewater inlet; 2 Paddle mixer; 3 Openings;
4 Collecting funnel; 5 Wastewater outlet

Figure 17 : Mechanical mixing chamber with a paddle mixer.

Example 6.72

Hydraulic Jump Mixers

Determine the following for the hydraulic jump mixer:

　　Mixing power developed

　　Volume, and detention period

Assume the following data:

Height of the bottom of the sluice gate to the bottom of the channel (y_1) : 4.8 cm

Wastewater flow (Q) : 0.05 m³/s

Channel width (W) : 10 cm

Solution

1. Required expressions for power developed (P), Volume (V) and detention period (t) for the hydraulic jump mixes.

   Power (P) developed for mixing purposes by hydraulic jump is:

   $$P = \frac{Q\gamma(y_2 - y_1)[q^2(y_2 + y_1) - 2g(y_1 y_2)^2]}{2g(y_1 y_2)^2}$$

   $$y_2 = \frac{y_1}{2}[(1 + 8\,Fr_1^2)^{0.5} - 1]$$

   $$y_2 = \frac{y_1}{2}\left[\left(1 + \frac{8\,v_1^2}{gy_1}\right)^{0.5} - 1\right]$$

   $$F_{r_1} = \frac{v_1}{(gy_1)^{0.5}} = \text{Froude number}$$

   $$P = \frac{8\gamma(y_2 - y_1)[(Q/W)^2(y_2 + y_1) - 2g(y_1 y_2)^2]}{2g(y_1 y_2)^2}$$

   Volume (V) of the hydraulic jump is simply the volume of the trapezoidal prism volume and is expressed as:

   $$V = \frac{1}{2}(y_1 + y_2)W = 3y_2 W(y_1 + y_2)$$

   Detention period (t$_o$) is:

   $$t_o = \frac{V}{Q} = \frac{1}{2Q}(y_1 + y_2)LW = \frac{3y_2 W}{Q}(y_1 + y_2)$$

   where L = Length of hydraulic jump is measured from the front face of the jump to a point on the surface of the flow immediately after the roller

   L = 6 y_2 for 4 < F_{r_1} < 20, and for Froude numbers outside this range, L is some what less than 6 y_2

   W = Width of channel

   Q = Flow rate

   γ = Specific weight

   g = Gravity constant

   y_1 and y_2 = Distance at points 1 and 2

   P = Power for mixing = $Q\,\gamma\,h_f$

   h_f = Head loss, and for hydraulic jump it is:

$$= \frac{(y_2 - y_1)[q^2(y_2 + y_1) - zg(y_2 y_1)^2]}{2g(y_2 y_1)^2}$$

F_{r_1} = Froude No. (= $v/(g\,h)^{0.5}$

The schematics of hydraulic jump mixer (Figure 18) is given as:

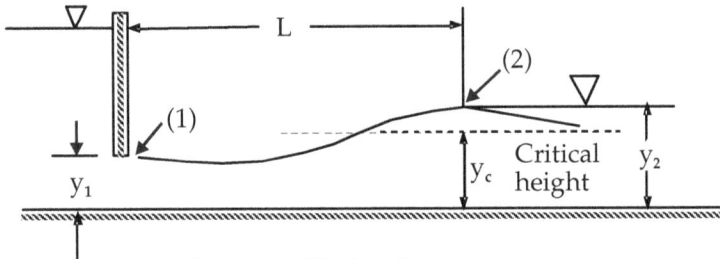

Figure 18 : Hydraulic-jump mixer

2. Required power developed

    • Velocity at point 1 $(v_1) = \dfrac{Q}{y_1(W)}$

    $$= \frac{0.05\ m^3/s}{0.048\,m\,(0.1\ m)}$$

    $$= 10.41\ m/s$$

    Distance downstream (y_2) for the hydraulic jump is:

    $$y_2 = \frac{y_1}{2}\left\{\left(1 + 8\frac{v_1^2}{g\,y_1}\right)^{0.5} - 1\right\}$$

    $$= \frac{0.048}{2}\left[\left(1 + 8\frac{(10.41)^2}{9.81(0.048)}\right)^{0.5} - 1\right]$$

    $$= 1.006\ m.$$

    Power (P) developed is:

    $$P = Q\gamma h_f = Q(\rho\,g)h_f$$

    $$P = \frac{Q\,\gamma(y_2 - y_1)[(Q/W)^2(y_2 + y_1) - 2g(y_2 y_1)^2]}{2g(y_2 y_1)^2}$$

    $$P = \frac{0.05(1000)(9.81)(1.006 - 0.048)\left[\left(\dfrac{0.05}{0.10}\right)^2(1.006 + 0.048) - 2g\,(1.006 \times 0.048)^2\right]}{2(9.81)(1.006 \times 0.048)^2}$$

    $$= 2239.5 = 3.0\ hp.$$

3. Required volume (V) and detention time (t) for the hydraulic jump mixer

$$\text{Shear rate}(G) = \left[\frac{P}{\mu V}\right]^{0.5}$$

$$\text{Volume (V)} = 3\, y_2\, (y_1 + y_2)\, W$$
$$= 3(1.006)(0.048 + 1.006)(0.1)$$
$$= 0.32 \text{ m}^3$$

$$G = \left[\frac{2239.5}{(8.8 \times 10^{-4})(0.32)}\right]^{0.5}$$
$$= 2800 \text{ s}^{-1}$$

$$\text{Detention time (t)} = \frac{V}{Q} = \frac{3y_2(y_1 + y_2)W}{Q}$$

$$= \frac{0.32 \text{ m}^3}{0.05 \text{ m}^3/\text{s}}$$

$$= 6.4 \text{ seconds}$$

[G and t are within the range for effective mixing].

Note :

1.

Table 43 : Shear rate (G) and detention time (t) criteria for effective mixing

Detention Time (t, seconds)	Shear Rate (G, s^{-1})
less than 10	4000–1500
10–20	1500–950
20–30	950–850
30–40	850–750
40–130	750–700

As detention time (*t*) increases, shear rate (*G*) tends to decrease for effective mixing.

Example 6.73

Pneumatic Mixers

Determine the pressure at which air is forced into the diffuser. Assume the following data:

Volume of basin (V)	: 10 m³
Power dissipated	: 6 hp
Air flow rate	: 2 m³ air/m³–water treated
Detention time (t)	: 2.5 min

Barometric pressure (P_a)   : 101300 N/m²

Water depth from the diffuser (h)   : 2 m

Water temperature   : 25°C

Also the shear rate for effective mixing.

Solution

1. Required power for pneumatic mixers

Number of bubbles (n):

$$n = \frac{\left(\dfrac{P_i}{P_a} \times Q_i\right)t}{V} = \frac{\left(\dfrac{P_i}{P_a} \times Q_i\right)\left(\dfrac{h}{v_b + v_l}\right)}{V}$$

As v_l is small in comparison to v_b, t can be neglected:

$$n = \frac{\left(\dfrac{P_i}{P_a} \times Q_i\right)h}{V\, v_b} = \frac{P_i\, Q_i\, h}{P_a V\, v_b}$$

Equation for v_b are:

$$v_b = \frac{2(r)^2(\rho_l - \rho_g)}{9\mu} = \frac{2(r^2)\rho_l}{9\mu}; \quad R_e < 2$$

$$v_b = 0.33 g^{0.76}\left(\frac{\rho_l}{\mu}\right)^{0.52}(r)^{1.28}; \quad 2 < R_e < 4.02 G_1^{-2.214}$$

$$v_b = 1.35\left(\frac{\sigma}{\rho_l}\right)^{0.50}; \quad 4.02 G_1^{-2.214} < R_e < 3.10 G_1^{-0.25}$$

$$v_b = 1.53\left(\frac{g\sigma}{\rho_l}\right)^{0.25}; \quad 3.10 G_1^{-0.25} < R_e < G_2$$

where   n = Number of bubbles

P_i = Influent absolute pressure

Q_i = Influent air flow

P_a = Atmospheric pressure

t = Average total rise time [$= h/(v_b+v_l)$]

v_b = Average rise velocity of the bubbles

v_l = Net average upward velocity of water

h = Water depth

$$G_1 = \text{Peebles number } [= g\, \mu^4/\rho_l\, \sigma^3]$$

$$G_2 = \text{Garber number } [= g\, (r)^4 (v_b)^4 /\rho_l^3 /\sigma^3]$$

$$R_e = \text{Reynolds number } [= 2\rho_l\, v_b\, (r)/\mu]$$

$$\rho_l = \text{Fluid mass density}$$

$$\mu = \text{Absolute viscosity}$$

$$r = \text{Average radius of the air bubbles}$$

$$\rho_g = \text{Air mass density}$$

Power dissipation in pneumatic mixers

Assume no. acceleration of the air bubbles, and the force balance becomes:

$$F_B - F_g - F_D = 0;\ F_D = (F_B - F_g)$$

[Buoyancy] [Gravity] [Drag force]

Therefore, Power dissipated $(P) = F_D\, v_b\, n$

$$= \frac{V^*(\gamma_l - \gamma_g) v_b\, P_i\, Q_i\, h}{P_a\, \overline{V}\, v_b}$$

where $\quad V^* = \text{Average volume of single bubble}$

$$\overline{V} = \text{Average volumes of bubbles at the surface}$$

Pressure at water depth $(h) = P_a + \gamma_l\, h$

$$\text{Volume (V) of single bubble} = \left[\frac{P_a}{(P_a + h\gamma_l)}\right] \overline{V}$$

[Corrected for pressure]

Average volume of single bubble $(V) = \dfrac{1}{h}\displaystyle\int V^* dh$

$$V = \frac{P_a \overline{V}}{h} \int_0^h \frac{dh}{P_a + h\gamma_l}$$

$$V = \frac{P_a \overline{V}}{h\gamma_l} \ln\left[\frac{h_a + h\gamma_l}{P_a}\right]$$

Therefore, $\quad P = \dfrac{V^*(\gamma_l - \gamma_g) v_b\, P_i\, Q_i\, h}{P_a\, \overline{V}\, v_b}$

or $\qquad P = \dfrac{P_a \overline{V}}{h\gamma_l} \ln\left(\dfrac{P_a h\gamma_l}{P_a}\right)(\gamma_l - \gamma_g) v_b \dfrac{P_i\, Q_i\, h}{P_a\, \overline{V}\, v_b}$

$$= P_i Q_i \frac{(\gamma_l - \gamma_g)}{\gamma_l} \ln\left[\frac{P_a + \gamma_l h}{P_a}\right]$$

$$= P_i Q_i \ln\left[\frac{P_a + \gamma_l h}{P_a}\right]$$

where γ_l = Specific weight of fluid (= ρ_l g)

γ_g = Specific weight of air (= ρ_g g)

2. Required pressure at which air is forced into the diffuser

$$P = P_i Q_i \ln\left[\frac{P_a + \gamma_l h}{P_a}\right]$$

$$\text{Water flow } (Q_o) = \frac{10 \text{ m}^3}{2.5 \text{ min}} = 4 \text{ m}^3/\text{min}$$

$$= 0.06 \text{ m}^3/\text{s}$$

Air flow rate (Q_i) = (2 m^3 – air/m^3 – water treated)(0.06 m^3/s)

$$= 0.12 \text{ m}^3/\text{s}$$

$$(6 \text{ hp})(746 \text{ W/hp}) = P_i(0.12 \text{ m}^3/\text{s}) \ln\left[\frac{101300 + 2(9.81)(1000)}{101300}\right]$$

$$P_i = \frac{6(746)}{0.02124} = 210684 \text{ N/m}^3 \text{ absolute}$$

$$\text{Shear rate}(G) = \left[\frac{P}{\mu V}\right]^{0.5}$$

$$G = \left[\frac{6 \times 746}{8.8 \times 10^{-4}(10)}\right]^{0.5}$$

$$= 700 \text{ s}^{-1} \text{ [OK]}.$$

Example 6.74

Mixing Power for Weir Mixers

Determine the following for weir mixer:

– Power dissipation (P)

– Volume (V)

– Shear rate (G)

– Detention time (t)

Assume the following data:

Length of the weir (L)	: 1.5 m
Drop provided from the weir crest to the surface of water below (H_D)	: 2.0 m
Wastewater flow	: 0.4 m³/s
Height of weir (p)	: 1 m.

Solution

1. Required expressions for power dissipation (P)

$$P = Q \gamma h_f \; ; \; [\gamma = \rho g]$$

$$h_f = H + H_D$$

where   H = Head over the weir crest

H_D = Drop provided from the crest to the surface of water below

γ = Specific weight

h_f = Head loss

The schematics of power dissipation in weir mixers (Figure 19) :

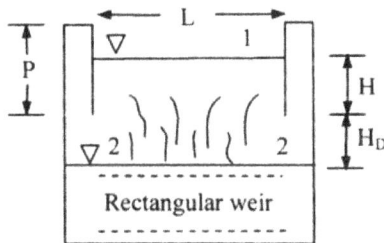

Figure 19 : The schematics of power dissipation in weir mixers.

At the points directly below the falling water there is turbulence. As the particles of water reach point 2, however, turbulence ceases and the velocity becomes zero (energy at the point of turbulence has been dissipated before reaching point 2).

2. Required power dissipated (P), shear rate (G), volume (V), and detention time (t) for a rectangular weir

Power dissipated (P)

$$P = Q(sg)(H + H_D)$$

$$Q = K(2g \; L)^{0.5}H^{1.5} \; ; \; [\text{Rectangular flow in the weir}]$$

$$K = \left[0.40 + 0.05\frac{H}{P*} \right]; \; P* = \text{Height of weir}$$

Therefore,

$$Q = \left[0.40 + 0.05 \frac{H}{(1)} \right] (2 \times 9.81 \times 1.5)^{0.5} H^{1.5}$$

$0.4 \text{ m}^3/\text{s} = (2.17 + 0.27H)H^{1.5}$

$$= 2.17H^{1.5} + 0.27H^{2.5}$$

Applying trial and error method, H is equal to 0.22 m

$$P = 0.4(1000 \times 9.81)(0.22 + 2.0)$$

$$= 8711.28 \text{ N/m}^2$$

Volume of the weir (V)

V = LWD   Length (L) = 2 m

$\quad$ = 2[0.3(2)][0.4(2)];   Width(W) = 0.3 of length of weir

$\quad$ = 0.96 m^3 Depth (D) = 0.4 of the H$_D$

Shear rate (G)

$$G = \left(\frac{P}{\mu V} \right)^{0.5}$$

$$= \left[\frac{8711}{8.8 \times 10^{-4} (0.96)} \right]^{0.5} = 3200 \text{ s}^{-1} \quad [\text{OK}]$$

Detention time (t)

$$t = \frac{V}{Q} = \frac{0.96 \text{ m}^3}{0.4 \text{ m}^3/\text{s}} = 2.4 \text{ seconds.}$$

Note :

- Shear rate (G) for various systems

  * Baffled flocculator (over and under flow)

    $P = \rho \, g \, Q \, h$

    $$G = \left(\frac{P}{\mu V} \right)^{0.5} = \left[\frac{\rho g h Q}{\mu V} \right]^{0.5} = \left[\frac{g h}{v \theta} \right]^{0.5}$$

    where   P = Power input for over and under baffle system (Watt)

    $\qquad$ V = Volume of the system (m^3)

    $\qquad$ ρ = Mass density of fluid (kg/m^3)

g = Gravity constant (9.81 m/s^2)

Q = Fluid flow rate (m^3/s)

h = Head loss

n = Kinetic viscosity (μ/ρ, m^2/s)

θ = Hydraulic detention time [V/Q; s]

* Mechanical paddle system

$$\text{Drag force}(F_D) = C_D A\rho_w \frac{v^2}{2}$$

$$\text{Power input}(P) = F_D v = C_D A\rho_w \frac{v^3}{2}$$

$$G = \left[\frac{C_D A\rho_w v^3}{2V\mu}\right]^{0.5} = \left[\frac{E_F P_r}{\mu V}\right]^{0.5}$$

where F_D = Drag force (N)

C_D = Drag coefficient [= 1.8 for flat blades]

A = Total cross–sectional area of flocculator paddles [m^2]

ρ_w = Fluid mass density [kg/m^3]

v = Fluid velocity [m/s]

E_f = Motor–flocculator efficiency and accounts for losses in gear box and U–joints

P = Rated motor power output.

Example 6.75

Weir Mixer

Size the weir mixer at a temperature of 20°C. Assume the following data:

Shear rate (G)   : 1000 s^{-1}

Minimum downstream velocity (v) : 0.5 m/s

Flow rate (Q)   : 0.324 m^3/s

Downstream channel is rectangular

Solution

1. Required mathematical expressions

    Hydraulic mixers

* Coagulants applied at any point where turbulence is high (locations are below weirs, at the suction side of pumps).

* Head losses for hydraulic devices should be at least 0.6 m to ensure good mixing.

* Minimum detention time of 2 s is recommended when mixing takes place in a pipe or open channel with a velocity greater than 0.5 m/s

* When coagulant is to be added in a channel such as below a weir or at a hydraulic jump, the dispensing pipe for the coagulant should extend across the channel with uniformly spaced parts to achieve effective distribution of coagulant in the flow. Height of the dispensing pipe over the weir should be at least 0.3 m to give the coagulant stream a sufficient velocity to penetrate the nappe thickness (and applies to other hydraulic devices).

Venturi sections and hydraulic jump

* Reduced pressure in the throat of the section aspirates the chemical feed Solution into the flow. Turbulence generated in the throat and as the flow get expands upon exiting the throat causes mixing.

* Head loss $(h_e) = C_d V^2/2g$; C_d = Discharge coefficient.

  C_d depends on the entrances pipe or channel area and the flow area in the throat as well as frictional losses in the section. Parshall flumes are venturi flumes designed for a critical conditions.

* Hydraulic jumps may be designed for dispersion of coagulants. For a flat channel (ignoring friction effects along the wall) the momentum principle is used for analysis:

$$\frac{\Delta(m \overline{V})}{\Delta t} = \Sigma \overline{F}$$

ln X – deviation

$$(\rho Q \overline{v}.\overline{n})\big|_1 + (\rho\, Q \overline{w}.n)\big|_2 = p_1\, A_1 - p_2\, A_2$$

$$= \frac{1}{2}\rho g\, b(y_1^2 - y_2^2)$$

where   n = Unit vector normal to surface

             p = Average pressure on a face

             A = Cross–sectional area

             y = Water depth

             b = Width of the jump

1, and 2 = Locations upstream and downstream of the jump, respectively

Or, $\rho Q v_2 - \rho Q v_1 = \dfrac{1}{2}\rho g\, b(y_1^2 - y_2^2)$, and

$$v = \frac{Q}{A} = \frac{Q}{by}$$

Substituting this into the above expressions, and simplifying it results in:

$$\frac{y_2}{y_1} = \frac{1}{2}[(8\,Fr_1^2 + 1)^{0.5} - 1]; \quad Fr_1 = \frac{v_1}{(g\,y_1)^{0.5}}$$

* To assure the location of the jump, the floor should be dropped and the drop is generally placed at the end of an expression of **supercritical flow. For good jump,** $Fr_1 > 2$ and the depth ratio (y_2/y_1) to be greater than 2.38 (with appreciable energy loss, although not all of the energy is dissipated usefully, there is turbulence to provide good mixing)

* Head loss (h_l) is calculated using Bernoulli's equation applied at the upstream and downstream depths [typical head loss are 0.3 m (1 ft) or greater, and the recommended value is 0.6 m (2 ft)]

$$h_l = \left(y_1 + \frac{v_1^2}{2g}\right) - \left(y_2 + \frac{v_2^2}{2g}\right)$$

2. Required head loss in a weir

    Datum will be the elevation of the downstream channel

    * Best hydraulic section for an open channel satisfies the condition:

    Dimensions of channel :

    W = 2d

    where   W = Channel width

    d = Channel depth

    $$v = \frac{Q}{Wd} = \frac{Q}{2d^2} \quad \text{or} \quad d = \left(\frac{Q}{2v}\right)^{0.5}$$

    $$d = \left[\frac{0.324 \text{ m}^3/\text{s}}{2(0.50 \text{ m/s})}\right]^{0.5} = 0.57 \text{ m}$$

    W = 2(d) = 2(0.57 m) = 1.14 m

Head loss in a channel

$$G = \left[\frac{P}{\mu V}\right]^{0.5}$$

$$P = rg\, h_l\, Q$$

Therefore, $G = \left[\dfrac{\rho g\, h_l Q}{\mu V}\right]^{0.5}$

$$h_l = \left[\frac{\mu G^2\, V/Q}{\rho g}\right]$$

$$= \frac{(1.002 \times 10^{-3}\ \text{kg/m.s})(1000\ \text{s})^2(2\ \text{s})}{(998.2\ \text{kg/m}^3)(9.81\ \text{m/s}^2)}$$

$$= 0.20$$

[At 20°C, ρ = 998.2 kg/m³, μ = 1.002 × 10⁻³ kg/m.s, and

V/Q = θ = 2s where v ≥ 0.5 m/s, g = 9.81 m/s²]

Depth of the upstream water at 20°C = 1.14 m + 0.20 m

$$= 1.34\ \text{m}$$

Height of water over the weir (h)

Flow over the weir (Q) = $\dfrac{2}{3}C_d\, W\,(2g\,h)^{0.5}$

where C_d = Discharge coefficient

$$h = \frac{1}{2g}\left[\frac{3Q}{2C_d W}\right]^{0.5}$$

$$= \frac{1}{2(9.81)}\left[\frac{3(0.324)}{2C_d(1.14)}\right]^{0.5} = \frac{0.0333}{C_d^{0.5}}$$

It is clear that the depth of water over the weir will be less than the height of weir. The discharge coefficient is typically around 0.62 and, therefore,

$$h = \frac{0.0333}{(0.62)^{0.5}} = 0.04\ \text{m}$$

Height of the weir (h_w) above the datum will be:

$$h_w = 1.34\ \text{m} - 0.04\ \text{m} = 1.30\ \text{m}$$

[Design must be compatible with the hydraulic grades in the upstream and down stream section. G cannot be maintained at the optimum value for all times (because of flow variation) and, therefore, similar calculation must be made for other flow conditions, and overall evaluation must be made to assess the final design].

Example 6.76

Mechanical Flocculation Chamber

Design a mechanical flocculation chamber comprising of paddle mixers with horizontal axis for a wastewater flow of 840 m³/s.

Solution

1. Required design principles

    Flocculation chamber should be designed to obtain uniform mixing [not too vigorous to destroy flocs]

    Residence time requirements

    * 20 to 30 min. [hydraulic mixer with baffles]

    * 6 to 10 min. [cyclone type chamber]

    * 15 to 20 min. [screw–cyclone type chamber]

    * 20 to 40 min. [mechanically mixed chambers]

    Design consideration with hydraulic type system

    * No. of changes of wastewater directions should be 8 to 10

    * Wastewater velocity near the inlet should V_{in} = 0.2 to 0.3 m/s, and near the outlet V_{out} = 0.05 to 0.1 m/s

    * Width of the channel should be greater than b = 0.7 m

    * Mean height of wastewater in the middle of each channel should be H = 2.5 m

    * Channel should be constructed in a way allowing to shorten the way covered by the wastewater in the case of a rapid floc forming. This can be done foreseeing a possibility of installing the gates at 50 and 75% of the length of the chamber.

    * Inclination of the bottom of the channel should be i = 1 to 5% towards the outer wall of the chamber.

    * A channel should be provided with a collecting channel perpendicular to the channels forming the labirynth, into which the gates can be opened.

    * Head loss (h_p) in the channel can be calculated using:

        $$hp = (0.15) \ V^2 n, \ (m)$$

where : V = Wastewater flow at an average velocity in the channels

n = No. of times the wastewater changes directions.

Design considerations in cyclone–type chamber.

* Chamber should have a conic shape for the whole working volume.

* Velocity of the wastewater at the bottom should be 0.7 to 1.2 mm/s.

* Velocity of the wastewater at the entrance to the collecting channel should be in the range of 4 to 8 mm/s.

* Angle of the cone should be a = 50 to 70°.

* For small chambers (diameter 1.5 to 2.0 m), the flocculates wastewater should be collected by means of a flooded funnel. For bigger channels, the collection should be designed as a perforated flooded tube or the outschert channel.

* Head loss (hp) is calculated as:

hp = (0.02 to 0.05) H; (m)

where : H = Working height of the chamber [calculated between the inlet and the level of the wastewaters, m].

Design considerations of the screw-cyclone chamber

* Wastewater inlet should be accomplished through the segnor mill (two inlets) or through a nozzle (one inlet) concentrical to the axis of the chamber. In both the cases, the wastewater should be directioned horizontally, perpendicular to the walls of the chamber.

* Velocity of the wastewater at the entrance to the chamber should be V = 2.0 to 3.0 m/s.

* The end of the inlet should be placed 0.5 m below the water level, and at the distance from the side wall of the chamber equal to 20% of the diameter of the chamber, taken in the widest part of the chamber.

* At the bottom part of the chamber, at the outlet, a wheel directing the flux of wastewater should be placed. Schematics of this system. (in Figure 20).

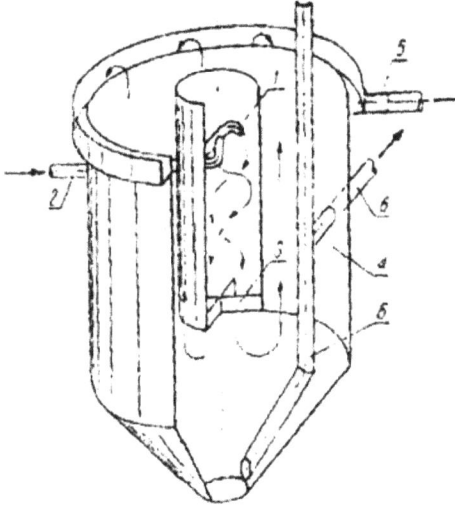

Figure 20 : A screw-cyclone flocculation chamber, coupled with a vertical settlers.

Design considerations for the mechanical flocculating chambers

* Height of water should be in the range H = 3 to 4 m

* Total length of the chamber is calculated as:

$$L = \beta \, H.z$$

where β = Coefficient (1 to 1.5)

H = Wastewater depth (m)

z = No. of axes of mixers

* No. of axis of mixers in one chamber should be

* Rotational speed of mixers should be in the range of 0.015 to 0.08 s^{-1}.

* Dimensions of a paddle mixer should be :

$$\frac{\text{Width}}{\text{Length}} = 1:10 \text{ to } 1:15$$

* Total surface of the paddles should be equal to a minimum 65% of the horizontal cross–section of the chamber to a maximum of 15 to 20% of the vertical cross–section; this is to avoid the movement of the total mass of the wastewater without generating or necessary velocity gradient (G).

* Placement of mixers:

* Minimum distance between the paddle and the wall should be 0.25 m.

* Maximum distance between the paddle and the bottom should be 0.15 m.

* Maximum distance between the paddles of the two mixers should be 0.6 to 0.9 m.

* Power dissipated by the paddle mixer is:

$$P = z\, m\, \pi^3\, \xi\, k^3\, n^3\, \rho l\, (r_2^4 - r_1^4), (W)$$

where z = No. of axes of mixers

m = No. of paddles on one axis

ξ = Coefficient of hydraulic resistance [and is a function of the ratio between the length of the paddle (l) to its width (b)]:

l/b	ξ
5	1.2
20	1.5
∞	1.9

ξ

K = Relative velocity coefficient [wastewater to paddle (0.75 to 1); typical value = 0.75]

n = Rotational speed of the mixer (0.015 to 0.08 s^{-1})

ρ = Specific mass of wastewater (kg/m^3)

l = Paddle length (m)

r_2 = External radius of the paddle (m)

r_1 = Internal radius of the paddle (m)

* Checking for mixing conditions:

– Mean velocity gradient (G) = $\left[\dfrac{P}{v\rho V}\right]^{0.5}$;(s^{-1})

where P = Power dissipated (W)

n = Kinematic viscosity (m^2/s)

ρ = Specific mass of wastewater (kg/m^3)

V = Volume of mixing chamber (m^3)

– Critical number (M) = tG [40000 to 210000]

where t = Residence time of wastewater (s)

* Gross power of the mixer (P_e)

$$P_e = \dfrac{kP}{\eta} \times 10^{-3} ; \text{(kW)}$$

where   P = Net dissipated power (W)

            k = Safety factor (1.1 to 2.5)

            η = Efficiency (0.9 to 0.95)

Outlet of wastewater from the flocculation should be designed avoiding the destruction of flocs; velocity of wastewater in the pipe between flocculation chamber and the settler should not exceed 0.1 m/s.

2. Required dimensions of the flocculation chamber

Assume residence time (t) = 25 min

$$\text{Volume of the chamber} = \frac{(840 \text{ m}^3/\text{h})(25 \text{ min})}{(60 \text{ min}/\text{h})}$$

$$= 350 \text{ m}^3$$

Assume β = 1.2; height of water in the chamber (H) = 3.5 m; and the number of axes of the mixer (z) = 2; the required length of each chamber (L) is given as:

$$L = \beta H z = 1.2(3.5 \text{ m})(2)$$

$$= 8.4$$

$$\approx 9.0 \text{ m}$$

Width of each part of the chamber is:

$$B = \frac{V}{LH} = \frac{350 \text{ m}^3}{(9 \text{ m})(3.5 \text{ m})} = 11.1 \text{ m}.$$

3. Required dimensions of the mixer

Let there be 4 perpendicular paddles on each rotating frame.

External diameter of the paddle is calculated assuming the distance between the paddle and the bottom equal to 0.10 cm; it will be equal to the immersion depth of the paddle:

$$D = H - 0.2 = 3.5 - 0.2 = 3.3 \text{ m}$$

Assuming that there will be two paddles on each axis, the number of mixers in one chamber will be equal to n = 4. Assuming, finally, a distance between the end of a paddle and the wall of the chamber and the distance between the paddles placed on different axes, equal to 0.3 m, the length (l) of the paddle will be:

$$l = \frac{B - (n+1)(0.3)}{n}$$

$$= \frac{11.1 - (4+1)(0.3)}{4} = 2.4 \text{ m}$$

Assuming the ratio between the width (b) of the paddle to its length (*l*) equal to 1:12, the width (b) of the paddle is:

$$b = \frac{l}{12} = \frac{2.4}{12} = 0.2 \text{ m}$$

4. Required power for the mixer

$$z = 2$$

$$m = z(n) = 2(4) = 8$$

$$\frac{l}{b} = \frac{2.4}{0.2} = 12, \ \xi = 1.26;$$

$$k = 0.75; \ n = 0.05 \text{ s}^{-1}; \ \rho = 1000 \text{ kg/m}^3;$$

$$l = 2.4 \text{ m}$$

$$r_2 = \frac{D}{2} = \frac{3.3}{2} = 1.65 \text{ m}$$

$$r_1 = \frac{D}{2} - b = 1.65 - 0.2 = 1.45 \text{ m}$$

$$P = z m \pi^3 \xi k^3 n^3 \rho l (r_2^4 - r_1^4)$$

$$= 2(8)(3.14)^3 (1.26)(0.75)^3 (0.05)^3 (1000)(2.4)(1.64^4) - (1.45)^4]$$

$$= 236.26 \text{ W} = 0.236 \text{ kW (For one mixer)}$$

$$v = 1.31 \times 10^{-6} \text{ m}^2/\text{s at } 10°C$$

$$\text{Shear rate (G)} = \left(\frac{P}{v\rho V}\right)^{0.5} = \left[\frac{236.26}{1.31 \times 10^{-6}(1000)(350)}\right]^{0.5}$$

$$= 32.1 \text{ s}^{-1} \text{ (OK)}$$

Critical dimensionless number (m) = Gt

$$= (32.1)(25 \text{ min } 60 \text{ s/min})$$

$$= 48150 \text{ (OK)}$$

$$\text{Gross power (P}_e) = \frac{kP}{\eta} \times 10^{-3}$$

$$= \frac{2(2 \times 236.26 \text{ W})}{0.9} \times 10^{-3}$$

$$= 1.05 \text{ kW ; } [k = 2, \ \eta = 0.9].$$

Example 6.77

Mean Velocity Gradient in Baffled Tank

1. Required mean velocity gradient (G) and its relation to power (P) G is the ratio of power input to the fluid:

$$G = \left(\frac{P}{\mu V}\right)^{0.5}$$

or $$Gt_d = \frac{V}{Q}\left(\frac{P}{\mu V}\right)^{0.5} = \frac{1}{Q}\left[\frac{PV}{\mu}\right]^{0.5}$$

where V = Flocculent volume of reactor (m³)

  μ = Viscosity (N.s/m²)

  P = Average power requirements (W)

  t_d = Detention time

In baffled flocculation chambers, interparticle unit can be based as:

$$P = Q \rho g h_f, \ [h_f = \text{Head loss of tank}]$$

or $$G = \left[\frac{Q\rho g h_f}{V\mu}\right]^{0.5} = \left[\frac{g h_f}{\upsilon t_d}\right]^{0.5}$$

[υ = Kinematic viscosity]

Interparticle contacts are accomplished by mechanical stirring with rotary pedals:

$$F_D = 0.5[C_D A \rho u^2], \text{ and}$$

$$P = F_D u$$

where F_D = Drag force (Netown)

  A = Cross–sectional area of the pedals

  C_D = Drag coefficient

  u = Relative velocity of the pedal with respect to fluid [0.5 to 0.75 of u_p (pedal tip velocity)

2. Determine the power requirements and pedal area for coagulation if the G value in a tank of 3000 m³ 50 s⁻¹. Assume the following data:

  Dry coefficient (C_D) = 1.8

  Pedal velocity (u_p) = 0.6 m/s.

  2.1 Required power (P) cross-sectional area of the pedals

  $$P = \mu G^2 V$$

$$= (1.139 \times 10^{-3} \text{ N.s/m}^2)(50/\text{s})^2(3000 \text{ m}^3)$$

$$= 8.543 \text{ kW}$$

$$A = \frac{2P}{C_D \rho u^3} = \frac{(2 \times 8543 \text{ kg/m}^2.\text{s}^2)}{(1.8)(999 \text{ kg/m}^3)(0.75 \times 0.6 \text{ m/s})^3}$$

$$= 104.3 \text{ m}^2$$

Example 6.78

Horizontal–Flow Baffled–Channel Flocculator

Design a horizontal–flow baffled channel flocculator for a treatment plant of 10,000 m^3/d capacity. The flocculation basin is to be divided into three segments of equal volume, each section having constant velocity gradient of 50, 35 and 25 s^{-1} respectively. The total flocculation time is to be 21 min and the water temperature is 15°C. The timber baffles have a roughness coefficient of 0.3. A common wall is shared between the flocculation and sedimentation basins; hence the length of the flocculator is fixed at 10.0 m. A depth of 1.0 m is considered reasonable for the horizontal flow flocculators.

Solution

Design the first flocculator section with a velocity gradient of 50 s^{-1} and detention time of 7 min.

Total volume of flocculation is :

$$V = \left(\frac{3 \times 7}{24 \times 60}\right)(10,000) = 146 \text{ m}^3$$

Total width of flocculator is:

$$W = \frac{V}{L \times H} = \frac{146}{(10.0)(1.0)} = 14.6 \text{ m (say 15 m)}$$

Width of each section is:

$$W = \frac{15}{3} = 5.0 \text{ m}$$

Viscosity and density of water at 15°C are:

$$\mu = 1.14 \times 10^{-3} \text{ kg/m.s}$$

$$\rho = 1000 \text{ kg/m}^3, \text{ respectively}$$

The number of baffles in the first flocculator can be estimated using:

$$n = \left\{\left[\frac{(2(\mu t))}{\rho[1.44 + f]}\right]\left[\frac{\text{HLG}}{Q}\right]^2\right\}^{1/3} \quad \text{for horizontal units}$$

and $\quad n = \left\{ \left[\dfrac{(2(\mu t))}{\rho[1.44+f]} \right] \left[\dfrac{WLG}{Q} \right]^2 \right\}^{1/3} \quad$ for vertical units

where n is the number of baffles in the basin, H is the depth of water in the basin (m), L is the length of the basin (m), G is the velocity gradients $(G = (P/\mu V)^{1/2})$ (s^{-1}), P is the power $(P = Q\rho g_c vh)$, h is the head loss, m is the dynamic viscosity (kg m/s), r is the density of water (kg/m^3), V is the volume of the unit (m^3), g_c is the gravitational constant $(9.81 \ m/s^2)$, H is the depth of water in the basin (m), L is the length of the basin (m), Q is the flow rate (m^3/s), f is the coefficient of friction of the baffles, and W is the width of the basin (m)

$$n = \dfrac{[2(1.44 \times 10^{-3})(7)(60)]}{1000(1.44+0.3)} \left\{ \left[\dfrac{1.0 \times 1.0 \times 50}{10,000/86,400} \right]^2 \right\}^{1/3} = 22$$

Spacing between baffles :

$\dfrac{10.0}{22} = 0.45$ m, (Satisfies minimum spacing requirements of 0.45 m)

Head loss (h) is the flocculator section is calculated using:

$$h = \dfrac{\mu T}{\rho g_c} \times G^2$$

$$= \dfrac{[(1.14 \times 10^{-3})(7)(60)]}{1000 \times 9.8} \times (50)^2 = 0.12 \ m$$

where T is the detention time (Q/V), s

The same series of calculation is repeated for remaining two flocculator sections. The results are as follows:

$\quad$ –G = 35 s^{-1}

$\quad$ T = 7 min

$\quad$ n = 17 (Use 16 for simplification)

Spacing between baffles = 0.62 m

Head loss = 0.06 m

$\quad$ –G = 25 s^{-1}

$\quad$ T = 7 min

$\quad$ n = 14 (Use 13 for simplification)

Spacing between baffles = 0.77 m

Head loss = 0.03 m

Total head loss (h_T) in the flocculator is:

$$h_T = 0.12 + 0.06 + 0.03 = 0.21 \text{ m}$$

Note :

The water velocity in both horizontal flow and vertical flow units generally varies from 0.3 to 0.1 m/s.

Detention time varies from 15 to 30 min.

In general, velocity gradients for both types of baffled channel flocculators should vary between 100 to 10 s^{-1}.

Guidelines for design and construction of baffled flocculators:

A : Horizontal Type

1. Distance between baffles should not be less than 45 cm to permit cleaning.
2. Clear distance between the end of the baffle and the wall is about 1.5 times the distance between baffles; should not be less than 60 cm.
3. Depth of water should not be less than 1.0 m.
4. Decay–resistant timber should be used for baffles; wood construction is preferred over metal parts.
5. Avoid using asbestos–cement because they corrode at the pH of alum coagulation.

B : Over and Under–Vertical Flow

1. Distance between baffles should not be less than 45 cm
2. Depth should be two or three times than the baffles
3. Clear space between upper edge of a baffle and water surface or the lower edge of a baffle and the basin bottom, should be 1.5 times the distance between baffles
4. Weep holes shall be provided for drainage.

    Tapered energy flocculation in baffled channel generally is achieved by varying the spacing of the baffles, that is, close spacing of baffles for high velocity gradients, and wider spacing for low velocity gradient. It can be done either by respacing or changing the number of baffles in the flocculation basin to attain desired head loss. It is suggested to have a tapered velocity gradient from about 75 s^{-1} at the inlet to about 10 to 15 s^{-1} at the outlet of the flocculators.

    Baffled flocculators are limited to a relatively large treatment plants, (greater than 10,000 m^3/d) capacity) where the flow rates can maintain sufficient head losses for slow mixing without requiring that baffles be spaced too close together (which make cleaning difficult).

Baffled flocculators operate under plug flow conditions that force them for short circuiting problems.

Horizontal–flow flocculators with around the end baffles are sometimes preferred over vertical–flow ones with over and under baffles because they are easier to drain and clean; also the head loss with which govern the degree of mixing, can be changed more easily by installing additional baffles or removing portions of existing ones.

Recommended G (velocity gradient) and GT values for flocculators:

Table 44 : Turbidity V$_s$ G^{-1} and GT

Type	Velocity Gradient (s^{-1})	GT
Turbidity or colour removal (without solids recirculation)	20–100	20,000–150,000
Turbidity or colour removal (with solids recirculation)	75–175	125,000–200,000
Softeners (solids contact reactors)	130–200	200,000–250,000

$$G = \left(\frac{Q\rho g h}{\mu V}\right)^{1/2} = \left(\frac{\rho g h}{\mu T}\right)^{1/2} \quad \text{for hydraulic flocculators,}$$

$$G = \left(\frac{P}{\mu V}\right)^{1/2} \quad \text{for mechanical flocculators.}$$

where P is the power (watts, kg m^2/s^3), T is the detention time (s), and h is the head loss (h = kV^2/2 g; k (constant) varies from 2.5 to 4; v is the fluid velocity (m/s), and g is the gravitational constant (9.81 cm^2/s).

The major short comings of hydraulic flocculators are:

1. No flexibility to respond to change in raw water quality.
2. Hydraulic and consequent flocculation parameters are a function of the flow and cannot be adjusted independently.
3. Head loss is often appreciable.
4. Cleaning may be difficult.

Example 6.79

Zig–Zag Channel for Flocculation

A Zig–Zag channel has 13 round the end cross walls. Water is passed along with a velocity of 0.2 m/s between the cross walls and 0.5 m/s round the ends. The flow is 0.25 m³/s and the nominal detention period is 20 min. Determine the additional head loss, the power dissipated, and the G and Gθ values. Assume the water temperature of 10°C.

Solution

1. Required head loss

    Head loss is given by $v^2/2$ g and is calculated as:

    $$\frac{14 \times (0.2)^2 + 13(0.5)^2}{2 \times 9.81} = 0.1656 \text{ m}$$

2. Required power dissipated

    $$P = \rho \text{ g h Q}$$
    $$= 0.25(9.8)(10^3)(0.1656) \text{ W} = 406 \text{ W}$$

3. Required G and Gθ

    $$G = \left(\frac{P}{\mu V}\right)^{0.5}$$

    $$= \left[\frac{406}{1.31 \times 10^{-3}(20 \times 60 \times 0.25)}\right]^{0.5}$$

    $$= 32.14 \text{ s}^{-1}$$
    $$G\theta = 32.14(20 \times 60)$$
    $$= 3.86 \times 10^4.$$

Example 6.80

Design of Flocculator

Determine the following for a flocculator:

Dimensions of the flocculator and flocculator compartment [Two flocculators in parallel].

Power requirements.

Dimensions of paddle slots [Four paddle blades per paddle wheel].

Rotational speed of the paddle wheels.

Assume the following data:

Flocculator compartments   : Three

Number of paddles in each compartment   : One paddle wheel

Ratio of the length of the paddle blades of the: 2.6:2 longest compartment to that of the length of the paddle blades of the middle of compartment. Ratio of the length of the paddle blades of the 2:1 middle compartment to that of the length of the paddle blades of the shortest compartment

Water flow rate   : 0.5787 m³/s

Water temperature : 20°C

Motor efficiency   : 90%

Brake efficiency   : 75%

Solution

1. Required expression for power input

   Peripheral velocity $(v_p) = w \, r_p$

   Drag force $(F_o) = \dfrac{C_D \, A_p \, \rho_e \, v_p^2}{2}$

   Power dissipation per blade $(P) = F_D . v_p$

   $$= 0.5 C_D A_p \rho_e \, v_p^3$$

   Total power (P^*) in the flocculator is the sum of the powers in each blade :

   $$P^* = \sum P = \sum \frac{C_D \, A_p \, \rho_e \, v_p^3}{2}$$

   $$= \frac{C_D \, A_p \, \rho_e \, (a \, v_{pt})^3}{2}$$

   or $A_{pt} = \dfrac{2 P^*}{C_D \, \rho_e (a \, v_{pt})^3}$

   Drag coefficient (C_D)

   $$C_D = 0.008 \left(\frac{b}{D} \right) + 1.3; \quad \text{[For a single blade]}$$

   where v_p = Tangential velocity

   r_p = Radius of paddle wheel

   v_{pt} = Paddle tip velocity

   a = A constant (0.75) to convert paddle tip velocity to a conglomerate velocity representing all blades in paddle

   A_p = Projected area of paddle blade

   A_{pt} = Total (sum) of projected areas of paddle blades

   b = Length of paddle blade

   D = Width of paddle blade

2. Required dimensions of flocculator and flocculator compartments

   Assume the value of $G = 30 \text{ s}^{-1}$

Therefore,   Gt = 80,000

$$t = 44.44 \text{ min}$$

Flocculator volume is :

$$= \frac{(50000 \text{ m}^3/d)(44.44 \text{ min})}{2(24)60} = 771.53 \text{ m}^3$$

when depth is equal to width, uniform velocity gradient is ensured. Assume a water depth of 5 m, the each flocculator volume becomes:

D(W)L = V

5(5)L = 771.53; L = 30.86 m

Length of compartment no. 1 = $\dfrac{30.86}{1+2+2.6}$ = 5.51 m [shortest]

Width and depth of compartment Nos. 1, 2 and 3 are 5 m each (assumed)

Length of compartment no. 2 = $\dfrac{2}{5.6}(30.86) = 11.02$ m [middle]

Length of compartment no.3 = 30.86 – 5.51 – 11.02 = 14.33 m [longest].

3. Required power for three flocculators

$$G = \left(\frac{P}{\mu V}\right)^{0.5}$$

Assume the following values of G for three compartments:

G = 40 s^{-1} [Compartment No. 1]

G = 30 s^{-1} [Compartment No. 2]

G = 20 s^{-1} [Compartment No. 3]

Power for three compartments

V_1 = 5(5)(5.51) = 137.75 m^3

P_1 = μ V_1 G_1^2 = (10 × 10^{-4})(137.75)(40)^2

= 326.52 N.m/s = 0.44 hp

Power consumed = $\dfrac{326.52}{0.75(0.9)}$ = 484 N.m/s

V_2 = 5(5)(11.02) = 275.50 m^3

P_2 = (10 × 10^{-4})(275.50)(30)^2 = 0.44 hp

Power consumed = $\dfrac{248}{0.75(0.9)}$ = 367 N.m/s

$$V_3 = 5(5)(14.33) = 358.25 \text{ m}^3$$
$$P_3 = (10 \times 10^{-4})(358.25)(20)^2 = 0.28 \text{ hp}$$

$$\text{Power consumed} = \frac{143.3}{0.75(0.90)} = 212.3 \text{ N.m/s.}$$

4. Required paddle dimensions

$$P = \frac{C_D A_{pt} \rho_e (a\, v_{pt})^3}{2}; \quad C_D = 0.008\frac{b}{D} + 1.3$$

$$= \left[0.008\frac{b}{D} + 1.3\right]\frac{(n\, b\, D)(1000)(0.75)^3 v_{pt}^3}{2}$$

$$= \left[0.008\frac{b}{D} + 1.3\right](n\, b\, D)(210.3)\, v_{pt}^3$$

where   n = Number of blades

$$\text{Assume Reynold's number } (R_e) = 10^5 = \frac{D v_p}{\upsilon} = \frac{D(a\, v_{pt})}{\upsilon} \text{ or } v_{pt} = \frac{10^5 \upsilon}{0.75 D}$$

$$\text{Therefore, } P = 326.52 = \left[0.008\frac{b}{D} + 1.3\right](n\, b\, D)(210.3)\left[\frac{10^5 (\upsilon)}{0.75 D}\right]^3$$

$$\upsilon = \frac{\mu}{\rho} = \frac{10 \times 10^{-4}}{1000} = 10^{-6}$$

Compartment no. 1

$$b_1 = 5.51 - 2(0.3) = 4.91 \text{ m}$$

Solving for D_1 assuming n = 4 yields;

$$D_1 = 0.18 \text{ m}$$

Similarly for compartment nos. 2 and 3:

$$b_2 = 11.02 - 2(0.3) = 10.42 \text{ m}$$
$$D_2 = 0.25 \text{ m [for n = 4]}$$
$$b_3 = 14.33 - 2(0.3) = 13.73 \text{ m}$$
$$D_3 = 0.36 \text{ m [for n = 4].}$$

5. Required rotational speeds (W) for the three flocculators

$$v_{pt} = w\, r;$$

$$R_e = 10^5 = \frac{D v_p}{\upsilon} = \frac{D(a\, v_{pt})}{\upsilon}$$

Compartment no. 1:

$$V_{pt} = \frac{10^5(v)}{D} = \frac{0.10}{D} = \frac{0.10}{0.18} = 0.56 \text{ m/s}$$

$$r = \frac{5 - 0.66}{2} = 2.2 \text{ m}$$

$$w = \frac{0.56 \text{ m/s}}{2.2} = 0.25 \text{ rad/s} = 15.17 \text{ rad/min} = 2.43 \text{ rpm}$$

Compartment no. 2 :

$$V_{pt} = \frac{0.1}{0.25} = 0.40 \text{ m/s}$$

$$w = 1.74 \text{ rpm}$$

Compartment no. 3 :

$$V_{pt} = \frac{0.10}{0.36} = 0.28 \text{ m/s}$$

$$w = 1.22 \text{ rpm.}$$

Note :

1.

Table 44 : Criteria values for effective flocculation

Type of raw-water	G (s^{-1})	G (t)
Low turbidity and coloured	20–70	50,000–250,000
High turbidity	70–150	80,000–190,000

2. Power for mixing using propellers and turbine in the transition range (R_e between 10 and 10,000)

$$P = \frac{1}{2}[N^2 D_i^3 (K_L \mu + K_T N D^2 D_i^2 \rho)]$$

where   P = Power dissipated

           μ = Dynamic viscosity

           K_L = Power coefficient [Laminar region]

           K_T = Power coefficient [Turbulent region]

           D_i = Impeller diameter

           N = Rotational speed

           ρ = Mass density.

Example 6.81

Flocculation in Treated Wastewater

Determine the following:

Weekly alum requirements

Flocculation dimensions

Power requirements

Paddle configuration. Assume the following data:

Treated wastewater flow (Q)	: 50,000 m³/d
Alum dosage (Jar test)	: 40 mg/L
Flocculation factor (Gt)	: 4×10^4
Wastewater temperature (T°C)	: 30°C
Density of water (ρ)	: 1000 kg/m³
Viscosity of water (μ)	: 0.8×10^{-3} N.s/m²
Average shear rate (G)	: 30 s⁻¹

Solution

1. Required amount of alum

    Alum required = (50000 m³/d)(0.04 kg/m³)(7d)

    = 14000 kg/week

2. Required flocculator size

    Detention (t):

    $Gt = 4 \times 10^4$

    $$t = \frac{4 \times 10^4}{G} = \left(\frac{4 \times 10^4}{30 \, s^{-1}}\right)\left(\frac{1 \, min}{60 \, sec}\right) = 22.22 \, min$$

    Flocculator volume (V) = Q (t)

    $$= (50000 \, m^3/d)(22.22 \, min)\left(\frac{1d}{1440 \, min}\right)$$

    $$= 772 \, m^3$$

    * Assume three compartments of the flocculator

    * Equal distribution of velocity gradients can be achieved only if the end area of each compartment is a square [depth = 1/3 length]

    * Assume a flocculator depth of 5 m

      Depth (D) = 5 m

      Length (L) = 3 × 5 m = 15 m

      $$\text{Width (W)} = \frac{V}{A} = \frac{772 \, m^3}{5 \, m \times 15 \, m} = 10.3 \, m$$

Consider diameter of flocculator = 4.2 m, and the schematics of the flocculator is shown in Figure 21.

L = 15 m = 0.4 m + [4.2 m] + 0.4 m + 0.4 m + [4.2 m] + 0.4 m + 0.4 m + [4.2 m] + 0.4 m

W = 10.3 m = 0.4 m + [2.5 m] + 1.0 m + [2.5 m] + 1.0 m + [2.5 m] + 0.4 m

H = 5 m = 0.4 m + [4.2 m] + 0.4 m

A square cross–sectional area for paddle,
LH direction = 4.2 m × 4.2 m

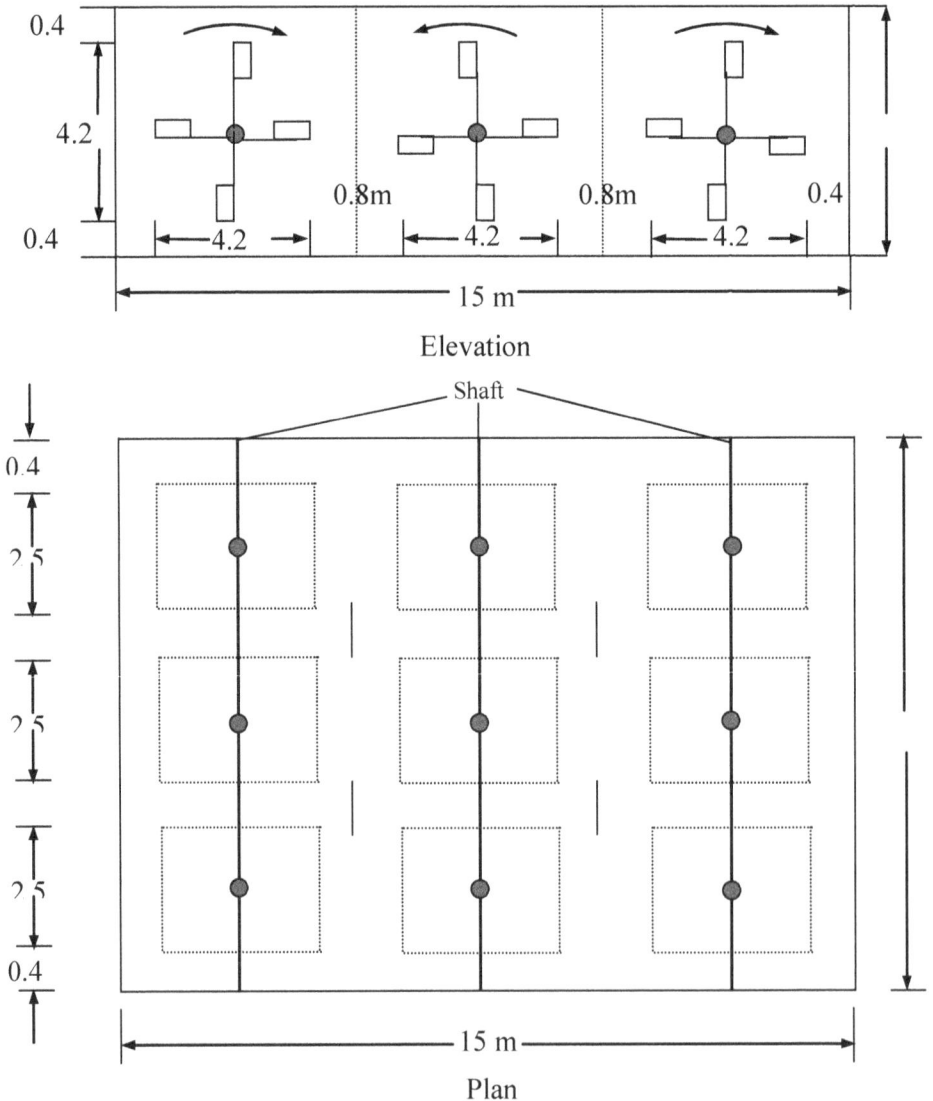

Elevation

Plan

Figure 21 : Schematics of the flocculator.

3. Required power consumption

   Shear rate (G):

   First compartment $\quad = 40\ s^{-1}$

   Second compartment $= 30\ s^{-1}$

   Third compartment $\quad = 20\ s^{-1}$

   Power (P) = $G^2 V \mu$

   Volume (V) of each compartment $= \dfrac{772}{3} = 257\ m^3$

   First compartment $(P_1) = \left(\dfrac{40}{s}\right)^2 (257\ m^3)(0.8 \times 10^{-3}\ N.s/m^2)$

   $$= (329\ N.m/s)\ (10^{-3}\ kW/N.m/s)$$

   $$= 0.33\ kW$$

   Gross power $P_1 = \dfrac{0.33\,kw}{0.8} = 0.41\ kw$

   Second compartment $(P_2) = \left(\dfrac{30}{s}\right)^2 (257\ m^3)(0.8 \times 10^{-3}\ N.s/m^2)$

   $$= (185\ N.m/s)\ (10^{-3}\ kW/N.m/s$$

   $$= 0.19\ kW$$

   Gross power $\qquad P_2 = \dfrac{0.19\,kw}{0.8} = 0.24\,kw$

   Third compartment $(P_3) = \left(\dfrac{20}{s}\right)^2 (257\ m^3)(0.8 \times 10^{-3}\ N.s/m^2)$

   $$= (82\ N.m/s)\ (10^{-3}\ kW/N.m/s$$

   $$= 0.082\ kW$$

   Gross power $\qquad P_3 = \dfrac{0.083\,kw}{0.8} = 0.10\,kw.$

4. Required rotational speed of the paddles [Each paddle has four boards 2.5 m long and W–wide – three paddle wheels per compartment]

   First compartment

   $$P = \dfrac{C_D A_P \rho v_P^3}{2} = 0.10\,kw$$

   Assume $v_p = (0.67\ m/s)\ (0.75) = 0.5\ m/s$

   $\qquad C_D = 1.8$

   $\qquad A_p$ = (Length of board) (W) (No. of boards)

* Board width of the paddles:

  3 paddles at 4 board/paddle = 12 boards

  A_p = (12 Nos.) (2.5 m) (W) = 30 W [Width of the board (W)]

  $$P_1 = 329 \text{N.s/m} = \frac{(1.8)^2 (30\,\text{W})(1000\ \text{kg/m}^3)(\text{N.s}^2/\text{kg.m})(0.5\,\text{m/s})^3}{2}$$

  = 3375 W

  or $W = \dfrac{329}{3375} = 0.1\,\text{m}$

  [Same width of the board is used in all the paddles]

* Rotational speed of the paddles (v_p)

  $$(v_p)_1 = \pi DN$$

  $$0.67 \text{ m/s} = 3.14(4.2\text{m})N$$

  or $N_1 = \dfrac{0.67\,(60\,\text{s/min})}{(3.14)(4.2)} = 3.05\,\text{rpm}$

* Second compartment

  $$P_2 = 185\,\text{N.m/s} = \frac{(1.8)(30\times0.1)(1000)(v_p^3)}{2}$$

  or $(v_p)_2 = \left[\dfrac{2\times185}{(1.8)(30\times0.1)(1000)}\right]^{0.333}$

  = 0.40 m/s

  Velocity of the paddle $= \dfrac{0.40\,\text{m/s}}{0.75} = 0.54\,\text{m/s}$

  $$N_2 = \frac{(v_p)_2}{\pi D} = \frac{(0.54)(60)}{(3.14)(4.2)} = 2.5\,\text{rpm}$$

* Third compartment

  $$(v_p)_3 = \left[\frac{2\times82}{(1.8)(30\times0.1)(1000)}\right]^{0.333}$$

  = 0.31 m/s

  Velocity of paddle $= \dfrac{0.31\,\text{m/s}}{0.75} = 0.42\,\text{m/s}$

  $$N_3 = \frac{(v_p)_3}{\pi D} = \frac{(0.42)(60)}{(3.14)(4.2)} = 1.9\,\text{rpm}.$$

Note :

1. Area of paddle board and velocity of paddle

   Area (Ap) refers to the combined area of the slots that are perpendicular to the cylinder of rotation.

   Ap shall not exceed 40% of the total area encompassed by the paddle.

   Velocity of the paddle tip (v_p) is the velocity relative to the water and is about 75% of the actual paddle tip.

   Paddle velocity should be less than 1 m/s and a minimum distance of 0.3 m should be maintained between paddle tips and all other structures in the flocculator to prevent local areas of excessive velocity gradients.

Example 6.82

Useful Power Dissipation–Shear Rate

1. Power is dissipated usefully in biological treatment system when it speeds up the transfer of nutrients to flocs and films by enhancing contact opportunity. The power–dissipation function is:

$$G^2 = \frac{P}{\mu V}$$

   where P is the useful power input, μ is the absolute viscosity of the liquid being treated, and V is the volume of liquid in which the power is usefully dissipated.

2. The possibility that not all of the actual power input, (Pa), is usefully employed can be re–expressed by:

$$k_G^2 = \frac{p\,P_a}{\mu\,V}$$

   where p is the fraction of the actual power input that enters usefully into the treatment proper and allows, too, for departures from nominal viscosity effect. The power dissipation in wastewater treatment processes is given as:

Table 46 : Power dissipation in wastewater treatment plant.

Treatment system	Power (P)	Volume (V)	k_G^2	Detention* time (t_d)
Activated sludge	$\rho gphV_aQ$	Qt_d	$\dfrac{gphV_a}{vt_d}$	$\dfrac{V}{Q}$
Trickling filter	$\rho gphQ$	Qt_d	$\dfrac{gph}{vt_d}$	*
Irrigation and Intermittent Filtration	ρgpQ_d	fhA	$\dfrac{gpQ_d}{vfA}$	$\dfrac{V}{Q_d}$

$$*t_d = Ch\left(\frac{v}{g}\right)^{1/3}\left(\frac{A/V}{Q}\right)^{1/3}$$

where h is the filter depth, n is the kinematic viscosity of fluid, g is the gravity constant, A/V is the specific area, Q is the hydraulic load (Q/A), and C is the constant.

In general, Q_d is the actual rate of dosing the area or bed (not average rate of flow), f is the porosity ratio of soils and sands, ρ is the mass density, V_a is the volume of air diffused into the aeration unit per unit volume of fluid treaded.

3. Determination of G, t_d and Gt_d for various systems.

   3.1   Sand filtration

       In a sand filter 4 ft deep with a porosity of 0.45 treating 180,000 gpad of clarified wastewater on three $-\frac{1}{4}$ acre beds in single daily doses, the applied water occupies a bed depth of:

$$\text{Sand bed depth} = \frac{60000\,\text{gallons}}{\left(\frac{1}{4}\text{acre}\times0.45\right)(7.48\,\text{gallons}/\text{ft}^3)(43560\,\text{ft}^2/\text{acre})}$$

$$= 1.6\text{ ft}$$

$$G^2 = \frac{gpQ_d}{vfA} = \frac{gph}{vf} = \frac{32.2\times1.6}{1.27\times10^{-5}\times0.45}, p = 1$$

$$[v = 1.27 \times 10^{-5}\text{ ft}^2 - \text{sec. at }59°\text{ F}]$$

$$G = 3000\text{ per sec}$$

$$t_d = \frac{V}{Q_d} = \frac{fAh}{Q_d} = \frac{0.45\times0.75\times43560\times1.6}{0.06\times1.548\times3}$$

$$= 8.44 \times 10^4\text{ sec}$$

$$Gt_d = 3000 \times 8.44 \times 10^4 = 2.53 \times 10^8.$$

   3.2   Trickling filter Q = 20 mgad, v = 1.27 × 10^{-5} ft² sec, h = 6 ft, and p = 1

$$G^2 = \frac{gph}{vt_d} = \frac{32.2\times6}{(1.27\times10^{-5})(10\times60)}$$

$$G = 1.6 \times 10^2\text{ per sec}$$

$[t_d$ = 20 to 60 min for filter depth of 6 ft and low loadings (3 to 6 mgad). For high rates of flow (20 to 30 mgad), t_d = 2 to 10 min (reduced in proportion to the two–thirds power of Q)]

$$t_d = 10\text{ min} = 600\text{ sec}$$

$$Gt_d = 1.6 \times 10^2 \times 6 \times 10^2 = 9.6 \times 10^4.$$

3.3 Activated sludge

Assume :

Detention time (t_d) : 8 hours

Aeration basin depth : 10 ft

Wastewater flow (Q) : 1 mgd

Air supply (V_a) : **0.8 ft³** of air/gal of wastewater at 59°F

$$G^2 = \frac{gph V_a}{v t_d} = \frac{32.2 \times 1 \times 10 \times 0.8}{1.27 \times 10^{-5} \times 8 \times 3600}$$

$$G = 26.54 \times (8 \times 3600)$$

$$t_d = 8 \times 3600 \text{ sec}$$

$$Gt_d = 26.54 \times (8 \times 3600)$$

$$= 7.64 \times 10^5$$

Proportionate usefulness of the applied power is small and yet it is important. Comparative values of G itself are largest for the long–time exposure generated in intermittent sand filters, smaller in TF or greatly reduced t_d, and smallest for ASP but for a relatively long time. The resulting product Gt_d are much the same for TF and ASP but considerably larger for intermittent sand filters. It is probable that the extent of useful power dissipation is least in intermittent sand filters and greatest in TF (the unavailable power dissipation by frictional resistance to flow is highest in sand filters.

3.4 Activated sludge

Assume:

Wastewater flow : 1 mgd

Basin depth   : 12 ft

Air supply   : 0.75 ft³/gal of wastewater treated

Detention time   : 6 hours

For V_a (ft³ air/gal of wastewater) in a tank of depth h, the power expended by the air approximates the vertical displacement of V_a (ft³) of water through a distance of h (ft) for each gallon of wastewater treated in time *t*.

$$P = 62.4 \frac{V_a h}{t}$$

$$= \frac{62.4 \times 0.75 \times 10^6 \times 12}{24 \times 60 \times 33 \times 10^3} = 12.5 \text{ hp}$$

$$G^2 = \frac{P}{\mu V} = \frac{12.5 \times 550 \times 7.48}{2.1 \times 10^{-5} \times 10^6} \quad \text{at } 60° \text{F}$$

$G = 49.5$ per sec

$Gt = 49.5 \times 60 \times 60 = 4.28 \times 10^6$.

3.5 Baffled channel (mixing)

Water wades through a baffled channel at a velocity of 0.5 fps and speeds upto 1.5 fps. There are 25 around the end baffles.

The head loss (h) incurred when the rate of flow is Q, the useful power input is given by :

$P = Q \, \rho g h$

$$\frac{Q}{V} = \frac{1}{t_d} = \frac{\sqrt{P/\mu V}}{G t_d}$$

$$= \frac{\sqrt{Qgh/vV}}{G t_d}$$

Each foot of lost head is $\dfrac{62.4 \times 1.547}{550} = 0.175 \dfrac{hp}{mgd}$

$$= 0.131 \frac{kw}{mgd}$$

Head losses commonly lie between 0.5 and 2 ft, velocities vary from 0.5 to 1.5 fps, and detention times run from 10 to 60 min.

For (n–1) equally spaced over and under or around the end baffles and for velocities, v_1 and v_2 in the channels and baffled slots, respectively, the head loss (h) is:

$$h = n\frac{v_1^2}{2g} + (n-1)\frac{v_2^2}{2g}$$

[In addition to normal channel friction, assumptions: necessary velocities must be redeveloped at each change in direction of flow and a substitutional estimate is normal channel friction increased by a reduction of Chezy or Hazen and William coefficient of discharge to 20 and 50, ASCE–Water Treatment Plant, Manual Eng. Pratc. 18, 32 (1940)]

$$h = n\frac{v_1^2}{2g} + (n-1)\frac{v_2^2}{2g}$$

$$= 26\frac{(0.5)^2}{2g} + 25\frac{(2)^2}{2g} = 0.1 + 1.55 = 1.65 \text{ ft}$$

Assume wastewater flow of 1 mgd (1.547 cfs) and $t_d = 40$ min

$P = 1.547 \times 62.4 \times 1.65 = 159.3$ lb–t/s

$$G^2 = \frac{P}{\mu V} = \frac{159.3}{2.74 \times 10^{-5} \times 1.547 \times 40 \times 60}$$

$$G = 39.57 \text{ per sec}$$

$$Gt_d = 39.57 \times 40 \times 60 = 9.5 \times 10^4$$

$$\text{Channel loading} \left(\frac{Q}{V}\right) = \frac{1.0 \times 10^6 \text{ gal/d}}{1.547 \times 40 \times 60} = 269 \text{ gpd/ft}^3$$

$$V = 3713 \text{ ft}^3$$

The baffled tank consists of 5 channels with the following dimensions:

$$37.13 \text{ ft (L)} \times 2.5 \text{ ft (W)} \times 8 \text{ ft (D)}.$$

3.6 **Paddle flocculators**

Assume the following data :

Wastewater flow : 10 mgd

Flocculator size   : 100 ft (L) × 40 ft (W) × 15 ft (D)

Paddle sizes   : 1 ft supported parallel to and moved by four horizontal shafts

Rotational speed : 2.0 rpm

Centre line of the paddles is 6.0 ft from the shaft which is at the mid–depth of the flocculation basin and two paddles are mounted on each shaft (one opposite the other)

Drag coefficient (C_D) : 1.8

Mean velocity of water (approximately) : 0.3 times the velocity of the paddle

Paddle velocity $v_i = 2\pi r n$

$$= 2 \times 3.14 \times 6 \times 2.0/60 = 1.257 \text{ ft/s}$$

Velocity differential (relative velocity) $v = v_i - k v_i = (1-k) v_i$

$$= 0.7 \times 1.257 = 0.88 \text{ ft/s}$$

Area of paddles $= 40 \times (2 \times 1) \times 4$

$$= 320 \text{ ft}^2$$

Useful power expended by the paddles

$$P = \frac{1}{2} C_D \rho A v^3$$

$$= 0.5 \times 1.8 \times \left(\frac{62.4}{32.2}\right) \times 320 \times (0.880)^3$$

$$= 380 \text{ lb.ft/s}$$

$$= \frac{380}{550} = 0.692\,\mathrm{hp}\left(\frac{380}{737.6} = 0.52\,\mathrm{kw}\right)$$

Energy utilized per mgd

$$= \frac{0.692 \times 24}{10} = 1.66\,\frac{\mathrm{hp.hr}}{\mathrm{mil.gal}} = 1.25\,\frac{\mathrm{kW.hr}}{\mathrm{mil.gal}}$$

[Add extra for mechanical + electrical losses: Normal values for paddle flocculators are 2 to 6 $\frac{\mathrm{kW.hr}}{\mathrm{mil.gal}}$]

Flocculator volume (V)

$$V = 100 \times 40 \times 15 = 6 \times 10^4 \ \mathrm{ft}^3$$

Detention time (t_d)

$$t_d = \frac{V}{Q} = \frac{6 \times 10^4 \times 7.481 \times 24 \times 60}{10 \times 10^6} = 65\,\mathrm{min}$$

Shear rate (G)

$$G^2 = \frac{P}{\mu V} = \frac{380}{2.74 \times 10^{-5} \times 6 \times 10^4}$$

$$G = 15.2 \ \mathrm{per\ sec}$$

$$Gt_d = 15.2 \times 6.5 \times 60$$

$$= 1.4 \times 10^4$$

Flocculator loading

$$\frac{Q}{V} = \frac{10 \times 10^6}{6 \times 10^4}\,\mathrm{gal/d}$$

$$= 16.7 \ \mathrm{gal/ft}^3.\mathrm{d}$$

3.6.1 Paddle flocculator

$$P = G^2\mu V_p$$

$$P = \frac{C_D \rho A V_p{}^3}{2}$$

where P is the power, C_D is the drag coefficient, V_p is the relative velocity of paddles with respect to the fluid (ft/s: Normally between 0.60 to 0.75 times the paddle tip velocity).

Determine power requirements and paddle area for the flocculator

Assume the following:

G = 50 per sec

Basin volume = $10^4 \ \mathrm{ft}^3$

$C_D = 1.8$ (rectangular paddles)

Paddle tip velocity : 2 ft/s

Relative velocity of the paddle : $0.75\ V_p$

$P = 50 \times (2.359 \times 10^{-5}) \times 10^4$

$\quad = 589.8$ ft lb/s

$$A = \frac{2P}{C_D \rho V_p^{\ 3}}$$

$$= \frac{2 \times 589.8}{1.8 \times (1.938\,\text{slug}\,/\,\text{ft}^3)(0.75 \times 2)}$$

$= 100$ ft^2.

3.7 Upflow clarification (hydraulics of upflow clarification)

The operations are aimed at the control of (1) floc growth (2) positioning of the flocculation–zone or floc–blanket surface, and (3) regulation of the intensity of floc shear. Hydraulic control is exerted by proper dissipation of hydraulic power and adjustment of residence time in the flocculation (contact zone). The power dissipated (P) is expressed as:

$P = \rho\ g\ h_f Q$ and head loss (h_f) is given by :

$$h_f = \left[\frac{(\rho_s - \rho)}{\rho}\right](1 - f_e)(h_2 - h_1)$$

$$= (S_s - 1)(1 - f_e)(h_2 - h_1)$$

where h_f is the head loss in passage through a zone of depth $(h_2 - h_1)$, ρ_s and S_s are respectively the mass density and specific gravity of the flocs, and f_e is the relative pore space of the flocculation zone

[The useful loss of head (h_f) is equal to the weight in water of the suspended flocs.]

$$\text{Zonal volume}(V) = \int A\,dh = 4\cot^2\theta \int h^2\,dh$$

$$= \frac{4}{3}\cot^2\theta(h_2^{\ 3} - h_1^{\ 3})\,[\text{Square pyramidal tank}]$$

$$= \frac{\pi}{3}\cot^2\theta(h_2^{\ 3} - h_1^{\ 3})\,[\text{Conical tank}]$$

$$\text{Detention time}(t_d) = \frac{f_e V}{Q}$$

$$f_e^{\ 5} = \frac{V}{V_s} = \left(\frac{V_h}{V_s}\right)^{5/4}$$

where v_s, v_h and v being the settling, hindered settling, and superficial velocities of the particles and fluid respectively.

$$G^2 = \frac{P}{\mu \times V}$$

$$= \left[\left(\frac{g}{v} \right) \frac{(S_s - 1)(1 - f_e)(h_2 - h_1)}{\dfrac{V}{Q}} \right]$$

$$Gt_d = f_e \left[\left(\frac{g}{v} \right)(S_s - 1)(1 - f_e)(h_2 - h_1)\frac{V}{Q} \right]^{0.5}$$

Cross–seational area is increased in the direction of flow by providing a wall angle (θ) of 40 to 65° with the horizontal (2 cot θ = 2.00 to 0.93) to create a diameter of circular tanks or width of square tank as large as 2.00 to 0.93 times the distance from the apex. At θ (wall angle) of 63°26', diameter D and width B equal the apical distance D = B = h.

Assume the following data:

Upward wastewater flow : 0.6 ft^3/s

Square pyramidal tank with wall angle (θ) : 45°

Flocculating zone between 2 ft above the apex (1 ft above the tank bottom) and 9 ft above the apex

Floc concentration : 40%, (f_e = 0.6)

Average specific gravity of floc : 1.1

Kinematic viscosity at 50°F : 1.41 × 10^{-5} ft^2/s

$$V = \frac{4}{3}\cot^2(45)(h_2{}^3 - h_1{}^3) \; [\text{Square pyramidal tank}]$$

$$= \frac{4}{3} \times 1 \times (9^3 - 2^3) = 961\,\text{ft}^3$$

$$t_d = \frac{fV}{Q} = \frac{0.6 \times 961}{0.6} = 961\,\text{sec}$$

$$G^2 = \frac{\left[\left(\dfrac{g}{v} \right)(S_s - 1)(1 - f_e)(h_2 - h_1) \right]}{\dfrac{V}{Q}}$$

$$= \frac{32.2 \times 0.1 \times 0.4\,(9 - 2)}{1.41 \times 10^{-5}\left(\dfrac{961}{0.6} \right)}$$

$$G = 20 \text{ per sec}$$

$Gt_d = 20 \times 961 = 1.9 \times 10^4$

G (upper 1 ft.), $(h_2 - h_1) = (9 - 8) = 1$

$$(h_2{}^3 + h_1{}^3) = (9^3 - 8^3) = (729 - 512) = 217$$

$$G = \left[\frac{(1/7)}{(217/721)} \right]^{0.5} \times 20 \quad \text{[On proportionality basis]}$$

$$= \left(\frac{721}{7 \times 217} \right)^{0.5} \times 20 = 13.8 \, \text{per sec}$$

G (lower 1 ft), $(h_2 - h_1) = (3 - 2)$

$$(h_2{}^3 - h_1{}^3) = (3^3 - 2^3) = 21$$

$$G = \left[\frac{(1/7)}{(19/721)} \right]^{0.5} \times 20$$

$$= \left(\frac{721}{7 \times 19} \right)^{0.5} \times 20 = 47 \, \text{per sec}$$

In summary, the upflow tank serves as a flocculation as well as settling unit, the values of G and Gt_d indicate:

- Floc should grow well at average values of G = 20 per sec and Gt_d = 1.9 × 10⁴.

- Floc formed is not likely to be destroyed by shear in the upper foot of the tank, where G = 47 per sec.

Note :

1. Static mixer

   The power consumed by static–mixing devices can be computed by:

   $$P = \gamma Q h$$

   where P is the power dissipated (lb.ft/s or kW), γ is the specific weight of water (lb/ft³ or kN/m³), Q is the flow rate (ft³/s or m³/s), and h is the headloss (ft or m).

2. Pneumatic mixing

   Power dissipated by rising bubbles in the estimated as:

   $$P = p_a V_a \ln (p_c/p_a)$$

   where P is the power dissipated (lb.ft/s or kw), p_a is the atmospheric pressure (lb/ft² or kN/m²), V_a is volume of air at atmospheric pressure (ft³/s or m³/s) and p_c is the air pressure at the point of discharge (lb/ft² or kN/m²), or alternatively,

$$P = KQ_a \ln\left(\frac{h+34}{34}\right) \quad \text{(US units)}$$

$$= KQ_a \ln\left(\frac{h+10.33}{10.33}\right) \quad \text{(S.I. units)}$$

where K is the constant [= 81.5 (US units) or = 1.689 (SI units)], Q_a is the overflow rate at atmospheric pressure (ft^3/min or m^3/min), and h is the air pressure at point of discharge expressed in height of water (ft or m).

3. Propeller and Turbine mixers

$$P = k\mu\ N^2D^3 \qquad \text{[Laminar : } R_e \text{ less than 10]}$$

$$P = k\mu\ N^3D^5 \qquad \text{[Turbulent : } R_e \text{ greater than 10,000]}$$

$$R_e = \frac{\rho ND^2}{\mu}$$

where P is the power requirement (lb.ft/s or W), k is the constant, μ is the dynamic viscosity(Ib.ft/ft^2 or N.s/m^2), ρ is the mass density of fluid (slug/ft^3 or kg/m^3), D is the diameter of the impeller (ft or m), and N is the revolutions per seconds.

Table 47 : Values of k (constant)

Impeller	Laminar	Turbulent
Propeller–square pitch, 3 blades	41.0	0.32
Propeller–pitch of two, 3 blades	43.5	1.00
Turbine–6 flat blades	71.0	6.30
Turbine–6 curved blades	70.0	4.80
Fan turbine–6 blades	70.0	1.65

Example 6.83

Mean Velocity Gradient

Derive an expression for mean velocity gradient as a function of power input.

Solution

1. Required expression for mean velocity gradient

    The mean velocity gradient is related to the power input through the sheer stress (t) on an element of fluid:

$$\tau = \mu\left(\frac{\overline{dv}}{dy}\right)$$

Power input (P) into the fluid element (ΔX. ΔY. ΔZ) can be expressed in terms of the mean velocity gradient $\left(\dfrac{\overline{dv}}{dy}\right)$:

$$P = \tau(\Delta X.\Delta Y.\Delta Z)\left(\frac{\overline{dv}}{dy}\right)$$

The power input per unit volume (P = p/V) is:

$$P = \frac{\tau(\Delta X.\Delta Y.\Delta Z)}{(\Delta X.\Delta Y.\Delta Z)}\left(\frac{\overline{dv}}{dy}\right) = \tau\left(\frac{\overline{dv}}{dy}\right)$$

P is defined as :

$$= \mu\left(\frac{\overline{dv}}{dy}\right)^2$$

$$= \left(\frac{\overline{dv}}{dy}\right) = G \text{ (mean shear gradient)} = \left(\frac{P}{\mu}\right)^{0.5}$$

Or, $P = \mu G^2$

$$G = \left(\frac{P}{\mu}\right)^{0.5} = \left(\frac{p}{\mu V}\right)^{0.5}$$

where τ = Shear stress [Pa : N/m²]

μ = Dynamic viscosity (Ns/m²)

v = Fluid velocity (m/s)

$\left(\dfrac{\overline{dv}}{dy}\right)$ = Mean velocity gradient (s⁻¹)

V = Volume.

2. Required sheer rate (G) for over and under baffle system, and mechanical paddle system

    Power input for an over and under baffle system is:

    $$P = \rho\, g\, Q\, h$$

    $$G = \left(\frac{\rho g\, Qh}{\mu V}\right)^{0.5} = \left(\frac{g h}{v\theta}\right)^{0.5}$$

    Power input in a mechanical paddle system is:

    $$F_D = C_D\, A\rho_w\, \frac{v^2}{2}$$

$$p = F_D \, v = C_D \, A \rho_w \, \frac{v^3}{2}$$

$$G = \left(\frac{C_D A \rho_w v^3}{2 \, V\mu} \right)^{0.5} = \left(\frac{E_f \, P_r}{\mu \, v} \right)^{0.5}$$

where ρ = Water density (kg/m³)

g = Gravity constant (9.81 m/s²)

Q = Waterflow rate (m³/s)

h = Head loss (m)

v = Kinematic viscosity (m²/s)

θ = Hydraulic detention time (s) [= V/Q]

F_D = Drag force (N)

C_D = Drag coefficient (1.8 for flat blades)

A = Total cross–sectional area of flocculator paddles (m²)

r_w = Water density (kg/m³)

E_f = Motor–flocculator efficiency

P_r = Motor power output

v = Fluid velocity

3. Determine the motor size [P$_r$] for flocculation process. Assume the following data :

E_f = 0.6

G = 30 s $^{-1}$

Q = 10,000 m³/d

θ = 30 min

3.1 Required motor size

$$G = \left(\frac{E_f \, P_r}{\mu \, V} \right)^{0.5}$$

$$P_r = \frac{V \mu G^2}{E_f}$$

$$V = (10,000 \, m^3/d) \left(\frac{30 \, min.}{1440 \, min./d} \right) = 208.5 \, m^3$$

$$P_r = \frac{(208.5)(10^{-3} \, kg/m.s)}{0.60} (40 \, s^{-1}) = 556 \, W$$

4. Power requirements for mixing

$$\text{Power number } (N_p) = \frac{P_p}{\rho n^3 Di^5}$$

Mixing operation is defined as:

$$\phi = \frac{N_p}{(F_r)^y} = K R_e^{x} \quad \text{[Unbaffled system]}$$

Power number for baffled system (y = 0)

$$\phi = N_p = K R_e^{x}$$

$$P = (\rho n^3 D_i^5) K \left(\frac{n\rho D_i^2}{\mu}\right)^{-1} \quad \text{[Laminar region, x = –1]}$$

$$P = K m n^2 D_i^3 \quad [R_e \leq 10]$$

$$P = K \rho n^3 D_i^5 \quad \text{[Turbulent region, x = 0]}$$

where   P = Power requirement (W)

ρ = Density of water (kg/m³)

n = Rotational speed (rps)

D_i = Diameter of mixer impeller (m)

φ = Power function

Fr = Froude number (n² D_i/g)

K = Constant

Re = Reynolds number (n ρ D_i^2/μ)

y, x = Constants.

4.1 Determine the power requirements for 2 m diameter, six blade (flat) turbine impeller mixer running at 20 rpm in a 10 m diameter mixing tank of standard configuration (Water temperature is 15°C).

4.2 Required power for mixing

$$R_e = \frac{n\rho D_i^2}{\mu} = \frac{\left(\frac{20}{60}\right)(1000 \text{ kg/m}^3)(2 \text{ m})^2}{(1.139 \text{ Ns}/\text{m}^2)}$$

$$= 1171$$

Power function (φ) at R_e = 1171 [From standard graphs]

$$\phi = 5.0$$

Power requirement (P):

$$P = \phi \rho n^3 D_i^5$$

$$= 5(1000 \text{ kg/m}^3)\left(\frac{20}{60}\text{rps}\right)^3 (2 \text{ m})^5$$

$$= 5(1000)(0.37)(32) = 5926 \text{ W}$$

Power requirement $\left(\dfrac{P}{V} = P\right)$:

$$V = \frac{\pi}{4}D^2H, \ [H = 3D_i, D_i = \text{Diameter of tank}]$$

$$= \frac{\pi}{4}(10 \text{ m})^2 (6 \text{ m})$$

$$= 417\text{m}^3 = 0.417 \times 10^3 \text{ m}^3$$

$$P = \frac{p}{v} = \frac{5926 \text{ W}}{0.417 \times 10^3 \text{ m}^3} = 14.21 \text{kW} / 10^3 \text{m}^3$$

Note :

1. Standard tank configuration for six–blade turbine mixer
    - **Impeller diameter** (D_i) = 0.33 (Tank diameter)
    - Impeller height from bottom (H_i) = 1.0 (Impeller diameter)
    - Impeller blade width (q) = 0.2 (Impeller diameter)
    - Impeller blade length(r) = 0.25 (Impeller diameter)
    - Length of impeller blade mounted on the central disk = 0.5 (r) = 0.125 (Impeller diameter)
    - Liquid height (H_L) = 1.0 (Tank diameter)
    - Number of baffles = 4 mounted vertically at tank wall and extending from the tank bottom to above the liquid surface
    - Baffle width (W_b) = 0.1 (Tank diameter)
    - Central disk diameter (S) = 0.25 (Tank diameter)

Example 6.84

Zeta Potential

1. The stability of a colloid is primarily due to electrostatic forces, neutralization of this charges is necessary to induce flocculation and precipitation. Although it is not possible to measure the psi potential, the zeta potential can be determined, and hence the magnitude of the charges and resulting degree of stability can be determined as well. The zeta potential is defined (i) as:

$$\tau = \frac{4 \pi \eta}{X E} \mu$$

where E is the dielectric constant of the medium, η is the viscosity of the medium, X is the applied potential, and μ is the electrophoretic mobility

2. Zeta potential can be re–expressed (practical consideration)

$$\tau \,(m\,V) = \frac{113000}{E}\,(EM)$$

where η is the viscosity of the medium (poise), and EM is the electrophoretic mobility [micron/(sec)(volt)(cm)]

After simplification,

$$\tau = 12.85 \; EM$$

Therefore, zeta potential is determined by measurement of the mobility of colloidal particles across a cell (as viewed from a microscope).

3. Consider an electrophoretic cell of 15 cm length (grid divisions are 200 μ at 8 power magnification. Determine zeta potential at an impressed voltage of 40 volts.

The time of travel between grid divisions is 50 seconds and the temperature is 20°C

$$EM = \frac{200 \; \mu \times 15 \; cm}{40 \; volts \times 50 \; sec} = 1.5\,\mu/(sec)(volt)(cm)$$

$$\text{Zeta potential} \; (\tau) = 1.5 \times 113,000 \times \frac{0.01}{80.36} = 21.1 \; m\,V$$

Note :

1. There will be a statistical variation in the mobility of individual particles, at least six values should be averaged for any one determination

2. The magnitude of the zeta potential for water and wastewater colloids has been found to average from –16 to –22 mV with a range of –12 to –40 mV

3. The zeta potential is uneffected by pH over a range of pH 5.5 to 9.5

4. The zeta potential is lowered by:

    Change in concentration of potential determining ions

    Addition of ions of opposite charge

    Contraction of the diffuse part of the double layer by increase in the concentration in solution.

5. Vast majority of colloids in industrial wastes possess a negative charge, the zeta potential is lowered and coagulation is induced by the addition of high-valence cations. The precipitating power of added ions increases geometrically with valence. The effectiveness of cation valence in precipitation of arsenims oxide is:

    $$Na^+ : Mg^{+2} : Al^{+3} = 1:63:570$$

6. Optimum coagulation will occur when zeta potential is zero, (iso-electric point) and the effective coagulation with usually occur over a zeta potential range of ± 0.5 mV.

Example 6.85

Flocculation in Water and Wastewater Systems

In a clarifier the processes of rapid mixing, flocculation, and solids/liquid separation are accomplished in a single unit. Estimate the products of collision (ΩGt) in an upflow clarifier. Use the following information:

Liquid detention time (t_w)	= 45 min
Solids concentration in the tank	= 1.5%
Influent solids concentration	= 150 mg/L
Specific gravity of solids in tank	= 1.01
Depth of the fluidized suspension	= 6 ft
Liquid depth of tank	= 12 ft
Water temperature	= 20°C

Solution

1. Residence time of the solids (t_s) in the tank

$$t_s = \frac{\text{Weight of solids in the tank}}{\text{Rate at which solids are added to the tank}}$$

$$= \frac{(15,000 \text{ mg/L})(6 \text{ ft})(A)}{150 \text{ mg/L}(Q)}$$

where Q is the flow rate to the tank, and A is the surface area of the tank

$$Q = \frac{V}{t_w} = \frac{(A)(10 \text{ ft})}{45} \times 60$$

$$t_w = \frac{(15,000)(6) A}{\left[\dfrac{150(12) A}{45} \times 60 \right]} = 37.5 \text{ hours}$$

$$= 1.35 \times 10^5 \text{ sec.}$$

2. Power dissipation (P)

$$P = Q\, \rho_l\, g\, h_f$$

where Q is the flow rate, ρ_1 is the liquid density, g is the gravitational constant, and h_f is the head loss

$$\text{Shear rate } (G) = \left(\frac{g\, h_f}{\overline{v}\, t_w} \right)^{0.5}$$

Head loss is encountered in the tank through the 6 ft bed of SS. This is equal to the buoyant weight of the solids

$$h_f = [1.01-1](15,000 \text{ mg/L})(6 \text{ ft})\left(\frac{L}{10^3 g}\right)\left(\frac{g}{10^3 mg}\right)$$

$$= 9 \times 10^{-4}$$

v at 20°C $= 1.01 \times 10^{-2}$ cm²/s

Therefore,

$$G = \left[\frac{980 \text{ cm/s}^2 \times 9 \times 10^{-4} \text{ ft} \times (2.54 \times 12 \text{ cm})/\text{ft}}{1.01 \times 10^{-2} \text{ cm}^2/\text{s} \times 45 \text{ min} \times 60 \text{ sec/min}}\right]^{0.5}$$

$$= 9.93 \times 10^{-1} = 0.993 \text{ s}^{-1}$$

The value of Ω in an upflow clarifier is in the order of 2×10^{-2}.

Therefore,

$$(\Omega G\ t_s) = (2 \times 10^{-2} \times 0.993 \times 37.5 \text{ hour} \times 60 \times 60)$$

$$= 2681$$

In a conventional flocculation, Ω is in the order of 10^{-4} (SS is 150), $t_s = t_w$ and ranges from 10^4 to 10^5

Therefore,

$$[\Omega(G\ t)] = (10^{-4})(10^5) = 10$$

An upflow clarifier provides $\left(\frac{2286}{10}\right)$ or 229 times more contact opportunities than a horizontal-flow system. Experimental determination of Ω is, therefore, very important.

Example 6.86

Flocculation for Water and Wastewater Treatment

In a jar test, flocculation is achieved by stirring a destabilized suspension at a velocity gradient of 15 sec⁻¹ for 45 min. The number of particles in suspension is reduced by 90%. Calculate the product of collision efficiency factor (η) and the volume fraction ($\eta\ \Omega$) for the wastewater being treated. Neglect floc breakup. Also calculate the detention period for a single CSTR, and for 4-CSTRs. What degree of aggregation is obtained if a single CSTR with 45 min detention period is used at a shear rate (G) of 10 sec⁻¹ and repeat it for 4 CSTRs with a total detention period of 60 min.

Solution

1. Orthokinetic flocculation (J_o)

The rate of changes in total concentration of particles with time due to Orthokinetic flocculation is expressed by a second order differential equation:

$$J_o = \frac{dN}{dt} = \frac{2\eta\, G\, d^3}{3} N^2 \tag{1}$$

where N is the total concentration (number) of particles in suspension at time (t), η is the collision efficiency factor, G is the shear rate (velocity gradient), and Ω is the volume fraction of colloidal particles (Ω, is the volume fraction of colloidal particles per unit volume of suspension)

Therefore,

$$\Omega = \left(\frac{\pi d}{6}\right)^3 N \tag{2}$$

where, d is the diameter of particle at any time, t

At time, t = 0, N = N_o and d = d_o; $\Omega = \left(\dfrac{\pi d_o^3}{6}\right) N_o$

Substituting equation (2) in equation (1):

$$\frac{dN}{dt} = -4\frac{\eta}{\pi} G\Omega N \tag{3}$$

Integrating equation (3) yields

$$\ln\left(\frac{N}{N_o}\right) = -\frac{4\eta}{\pi}\Omega G t$$

The rate of the orthokinetic flocculation is seen to be first order with respect to concentration of particles, the velocity gradient, and the floc volume fraction.

2. Product of collision efficiency factor (η) and volume fraction (Ω)

$$\eta\Omega = \frac{\pi}{4G\,t}\, ln\left[\frac{N_o}{N}\right]$$

$$= \frac{\pi}{4(15)(60)(45)}\, ln\left[\frac{N_o}{0.1N_o}\right] = \frac{\pi}{4(15)(60)(45)}\, ln[10]$$

$$= 4.46 \times 10^{-5}$$

3. Single-stage CSTR for flocculation

$$nt = \frac{\pi}{4\eta G\Omega}\left[\left(\frac{N_o}{N_m}\right)^{1/n} - 1\right],\ \text{for } n-\text{tanks in series} \tag{4}$$

For n = 1

$$t = \frac{\pi}{4\eta G\Omega}\left[\frac{N_o}{N} - 1\right]$$

$$= \frac{3.14}{4(4.5 \times 10^{-5})(15)} \left[\frac{N_o}{0.1 N_o} - 1\right]$$

$$= 174 \text{ min}$$

4. 3–stage CSTR

$$t = \frac{\pi}{4(4.5 \times 10^{-5})(15)} [(10)^{0.25} - 1] \tag{5}$$

$$= 15 \text{ min}$$

5. Considering single–stage CSTR

   Remaining Eqn. (4) for single–stage CSTR

   $$\frac{N_o}{N} = 1 + \frac{4(\eta \Omega) G t}{\pi}$$

   $$= 1 + \frac{4(4.5 \times 10^{-5}) 15 \times 60 \times 45}{3.14}$$

   $$= 1 + 2.32 = 3.32$$

   $$\frac{N}{N_o} = 0.30 \quad \text{or 70\% reduction in single-stage CSTR}$$

6. Considering plug flow

   $$\ln\left(\frac{N_o}{N}\right) = \frac{4}{\pi} (\eta \Omega) G t \tag{6}$$

   $$\ln\left(\frac{N_o}{N}\right) = \frac{4}{3.14} (4.5 \times 10^{-5}) 15 \times 60 \times 45$$

   $$\frac{N_o}{N} = \exp(2.32)$$

   or $\quad \dfrac{N}{N_o} = 0.098 \quad$ or $\quad 99.002\%$ reduction of particles

7. Considering 4-stage CSTR

   $$\frac{N_o}{N} = \left[1 + \frac{4(4.5 \times 10^{-5}) 15 \times 60 \times 45}{(3.14)(4)}\right]^4$$

   $$= [1 + 0.58]^4$$

   $$= 6.23$$

   or 84% reduction of particles

8. Summary of reaction in different types of reactors

   1. Plug flow   : 99.002% reduction of particles

2. 4–stage CSTR   : 84% of particles

3. Single–stage CSTR : 70% reduction of particles.

Note :

1. Perkinetic flocculation : Interparticle contacts are produced by Brownian motion and such random motion of colloidal particle results from the rapid and random bombardment of colloidal particles by molecules of the fluid

2. The rate of change in total concentration of particles with time due to **perkinetic flocculation** (J_p) follows the second order low of reaction

$$J_p = \frac{dN}{dt} = -\frac{4\eta kt}{3\mu}(N)^2$$

Integration yields:

$$\frac{N}{N_o} = \frac{1}{1 + \left(\dfrac{4\eta kTN_o}{3\mu}\right)t} \tag{1}$$

where N is the total number of particles in suspension at time t, η is the collision efficiency factor representing the fraction of the total number of collisions which are successful in producing aggregates, k is the Boltzmann's constant, T is the temperature ($^{\circ}$K), and m is the fluid velocity

If $\dfrac{3\mu}{[4\eta kTN_o]}$ is represented by $t_{0.5}$, the equation (1) can be expressed as:

$$N = \frac{N_o}{1 + \left(\dfrac{t}{t_{0.5}}\right)}$$

where $t_{0.5}$ represents the time necessary to halve the concentration of particles and at 25°C

$$t_{0.5} = \frac{1.6 \times 10^{11}}{\eta N_o} \quad [t_{0.5} \text{ in sec, } N_o \text{ in particle/mL and } \eta \text{ is dimension less}]$$

$t_{0.5}$ is inversely proportional to N_o and for low initial concentration (particle) or low destabilization (low η), $t_{0.5}$ will be quite large.

3. Orthokinetic flocculation

Agitation accelerates the aggregation of colloidal particles. In such systems, the velocity of the fluid varies both spatially and temporally. The spatial changes in velocity have been characterised by a velocity gradient (shear rate, G). Particles which follow the fluid motion will also have different velocities, so that opportunities exist for inter particle contacts. When contacts between particles are caused by fluid motion the process is termed orthokinetic flocculation.

4. Ratio of orthokinetic to perkinetic flocculation

$$J_p = \frac{dN}{dt} = -\frac{4\eta kT}{3\mu}(N)^2$$

$$J_o = \frac{dN}{dt} = -\frac{2\eta Gd^3 N^2}{3}$$

$$\text{Ratio of rates} = \frac{J_o}{J_p} = \frac{-\dfrac{2\eta Gd^3 N^2}{3}}{-\dfrac{4\eta kTN^2}{3\mu}}$$

5. In water at 25°C containing colloidal particles having a diameter of 1 μ, this ratio is unity when the velocity gradient is 10/sec. For colloidal particles having a diameter of 0.1 μ, it is apparent that a velocity gradient of 10,000/sec is necessary for orthokinetic flocculation to be as rapid as perkinetic flocculation. Similarly, for particles which are 10 μ in diameter, a velocity gradient of 0.01/sec will provide sufficient orthokinetic flocculation to equal the contacts resulting from diffusion. In water and wastewater treatment, mean velocity gradients of 10 to 100/sec are common.

Stirring, therefore, will not enhance the aggregation rate of small particles until they grow to a size of about 1 μ. Particle growth to a size larger than 1 μ requires fluid motion by agitation or other means. Since 1–μ particles do not settle well, tanks to provide orthokinetic flocculation must be included in treatment systems which use sedimentation tanks to separate solids from the water.

Such flocculation tanks will be ineffectual until the particles reach this 1 μ size through contacts produced by Brownian diffusion. They cannot, for example, aggregate viruses (0.1 μ or smaller in size) until they are adsorbed on or enmeshed in larger particles.

6. For example, a sample of water containing 10,000 viruses/mL and no other colloidal particles would require about 200 days before the concentration was reduced by one–half through coagulation, even if all of the virus particles were completely destabilized ($\eta = 1$). The removal of virus by coagulation in treatment systems must therefore require the presence of large numbers of other colloidal particles or enmeshment in a voluminous precipitate of metal hydroxide if effective removal is to be achieved within reasonable detention time.

7. An example for orthokinetic flocculation, consider a water sample containing 10,000 coliform organisms/ml and no other colloidal particles.

The floc volume fraction is calculated as 5.2×10^{-9}, assuming that the coliform organisms can be represented as spheres which are 1 μ in diameter. If the sample is treated with a coagulant so that the bacteria are completely destabilized ($\eta = 1$) and stirred to produce a velocity gradient of 10 sec^{-1}, a detention period of approximately 120 days would be required to reduce the concentration of coliform particles by one half through coagulation (equation (2)–(16)). The removal of bacteria and other micro-organism from water and wastewater by coagulation therefore usually required the presence of other colloidal particles, or enmeshment in a voluminous precipitate of metal hydroxide.

8. General consideration the flocculation time is inversely proportional to the floc volume fraction. Sludges in wastewater treatment plants (Ω in the order of 10^{-2}) could require a detention time about two orders of magnitude less than natural waters treated in conventional coagulation plants (Ω in the order of 10^{-4} or less). In other words, if 30 min of flocculation were adequate for a natural water, only 0.3 min would be required in a sludge–handling system, if other conditions (μ, η, $\bar{G}$, $\dfrac{N_0^o}{\bar{N}_m^o}$) were similar. Extended flocculation is unnecessary and even harmful in treating concentrated suspensions. This is particularly true in treating concentrated negative suspension destabilized by anionic polymers, where overstirring has been reported to produce restabilization (LaMer and Healy, 1963). A maximum value of the product $\bar{G}t$ is implied here.

9. Interparticle contacts for destabilization

   Three mechanisms have concerned the engineers, *viz*,

Perkinetic flocculation	: Contacts by thermal motion (Brownian motion or diffusion)
Orthokinetic flocculation	: Contacts resulting from bulk fluid motion (stirring)
Settling	: Contacts resulting from settling of particles (differential settling in which a rapidly settling particle overtakes and collides with a particle which is settling at a slower rate).

Example 6.87

Iron Salts: Coagulation

Determine the optimum pH for conducting coagulation operation using Ferrous sulphate (copperas) and ferric chloride as coagulants for wastewater having total dissolved solids of 140 mg/L at 25°C.

Solution

1. Required expressions for optimum pH using ferrous sulphate and ferric chloride

   1.1 Ferrous sulphate ($FeSO_4.7H_2O$) dissolved in water

   $$FeSO_4 \rightarrow Fe^{+2} + SO_4^{-2}$$

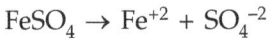

   * The equilibrium reactions are:

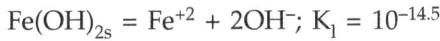

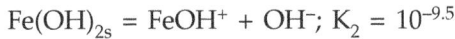

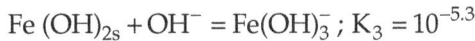

   $$Fe(OH)_{2s} = Fe^{+2} + 2OH^-; \ K_1 = 10^{-14.5}$$

   $$Fe(OH)_{2s} = FeOH^+ + OH^-; \ K_2 = 10^{-9.5}$$

   $$Fe(OH)_{2s} + OH^- = Fe(OH)_3^-; \ K_3 = 10^{-5.3}$$

   * Total moles of Fe^{+2} (SP_{Fe}) ion in water is given as:

   $$[SP_{Fe}] = [Fe^{+2}] + [FeOH^+] + [Fe(OH)_3^-] \tag{1}$$

   * Equilibrium values are:

   $$[Fe^{+2}] = \frac{\{Fe^{+2}\}}{\gamma_{Fe^{+2}}} = \frac{K_1}{\gamma_{Fe}\{OH^-\}^2} = \frac{K_1\{H^+\}^2}{\gamma_{Fe} K_w^2} = \frac{K_1 \gamma_H^2 [H^+]^2}{\gamma_{Fe} K_w^2} \tag{2}$$

   $$[FeOH^+] = \frac{\{FeOH^+\}}{\gamma_{Fe(OH)}} = \frac{K_2}{\gamma_{Fe(OH)}\{OH^-\}}$$

   $$= \frac{K_2[H^+]}{\gamma_{Fe(OH)} K_w} = \frac{K_2 \gamma_H[H^+]}{\gamma_{Fe(OH)} K_w} \tag{3}$$

   $$[Fe(OH)_3^-] = \frac{\{Fe(OH)_3^-\}}{\gamma_{Fe(OH)_3}} = \frac{K_3\{OH^-\}}{\gamma_{Fe(OH)_3}}$$

   $$= \frac{K_3 K_w}{\gamma_{Fe(OH)_3}\{H^+\}} = \frac{K_3 K_w}{\gamma_{Fe(OH)_3} \gamma_H[H^+]]} \tag{4}$$

   Substituting all three values in equation (1), and it yields

   $$[SP_{Fe}] = \frac{K_1 \gamma_H^2 [H^+]^2}{\gamma_{Fe} K_w^2} + \frac{K_2 \gamma_H[H^+]}{\gamma_{Fe(OH)} K_w} + \frac{K_3 K_w}{\gamma_{Fe(OH)_3} \gamma_H[H^+]} \tag{5}$$

   Differentiating equation (5) w.r.t. $[H^+]$, and equating it equal to zero resulting in the following equation, $\left[\dfrac{dSP_{Fe}}{d[H^+]} = 0 \right]$:

   $$\left[\frac{2K_1 \gamma_H^2}{\gamma_{Fe} K_w^2} \right][H_{opt}^+]^3 + \left\{ \frac{K_2 \gamma_H}{\gamma_{FeOH} K_w} \right\}[H_{opt}]^2 = \frac{K_3 K_w}{\gamma_{Fe(OH)_3} \gamma_H} \tag{6}$$

1.2 Ferric chloride ($FeCl_3$) and Ferric sulphate ($Fe_2(SO_4)_3$) dissociates in water as:

$$FeCl_3 \rightarrow Fe^{+3} + 3Cl^-$$

$$Fe(SO_4)_3 \rightarrow 2\ Fe^{+3} + 3SO_4^{-2}$$

* Equilibrium reactions are:

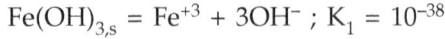

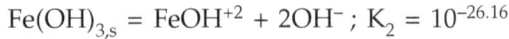

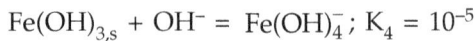

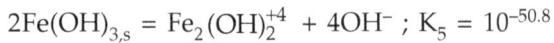

$$Fe(OH)_{3,s} = Fe^{+3} + 3OH^- \ ;\ K_1 = 10^{-38}$$

$$Fe(OH)_{3,s} = FeOH^{+2} + 2OH^- \ ;\ K_2 = 10^{-26.16}$$

$$Fe(OH)_{3,s} = OH^- + Fe(OH)_2^+ + OH^- \ ;\ K_3 = 10^{-16.74}$$

$$Fe(OH)_{3,s} + OH^- = Fe(OH)_4^- \ ;\ K_4 = 10^{-5}$$

$$2Fe(OH)_{3,s} = Fe_2(OH)_2^{+4} + 4OH^- \ ;\ K_5 = 10^{-50.8}$$

* Total molar concentrations of Fe^{+3} in solution is:

$$[SP_{Fe}] = [Fe^{+3}] + [FeOH^{+2}] + [Fe(OH)_2^+] + [Fe(OH)_4^-]$$

$$+2[Fe(OH)_2^{+4}] \tag{1}$$

* Equilibrium values are:

$$[Fe^{+3}] = \frac{\{Fe^{+3}\}}{\gamma_{Fe}} = \frac{K_1}{\gamma_{Fe}\{OH^-\}^3} = \frac{K_1\{H^+\}^3}{\gamma_{Fe}\,K_w^3} = \frac{K_1\,\gamma_H^3[H^+]^3}{\gamma_{Fe}\,K_w^3} \tag{2}$$

$$[FeOH^{+2}] = \frac{\{FeOH^{+2}\}}{\gamma_{FeOH}} = \frac{K_2}{\gamma_{FeOH}\{OH^-\}^2}$$

$$= \frac{K_2\{H^+\}^2}{\gamma_{FeOH}K_w^2} = \frac{K_2\,\gamma_H^2[H^+]^2}{\gamma_{FeOH}\,K_w^2} \tag{3}$$

$$[Fe(OH)_2^+] = \frac{\{Fe(OH)_2^+\}}{\gamma_{Fe(OH)_2}} = \frac{K_3}{\gamma_{Fe(OH)_2}\{OH^-\}}$$

$$= \frac{K_3\{H^+\}}{\gamma_{Fe(OH)_2}\,K_w} = \frac{K_3\,\gamma_H[H^+]}{\gamma_{Fe(OH)_3}\,K_w} \tag{4}$$

$$[Fe(OH)_4^-] = \frac{\{Fe(OH)_4^-\}}{\gamma_{Fe(OH)_4}} = \frac{K_4\{OH^-\}}{\gamma_{Fe(OH)_4}}$$

$$= \frac{K_3\,K_w}{\gamma_{Fe(OH)_4}\{H^+\}} = \frac{K_3\,K_w}{\gamma_{Fe(OH)_4}\,\gamma_H[H^+]} \tag{5}$$

$$[Fe_2(OH)_2^{+4}] = \frac{\{Fe_2(OH)_2^{+4}\}}{\gamma_{Fe_2(OH)_2}} = \frac{K_5}{\gamma_{Fe_2(OH)_2}\{OH^-\}^4}$$

$$= \frac{K_5\gamma_H^4[H^+]^4}{\gamma_{Fe(OH)_2}K_W^4} \tag{6}$$

Substituting all these values in equation (1) yields:

$$[SP_{Fe}] = \frac{K_1\gamma_H^3[H^+]^3}{\gamma_{Fe}K_W^3} + \frac{K_2\gamma_H^2[H^+]^2}{\gamma_{FeOH}K_W^2} + \frac{K_3\gamma_H[H^+]}{\gamma_{Fe(OH)_2}K_W} +$$

$$\frac{K_4K_W}{\gamma_{Fe(OH)_4}\gamma_H[H^+]} + \frac{2K_5\,\gamma_H^4[H^+]^4}{\gamma_{Fe_2(OH)_2}\,K_W^4} \tag{7}$$

Differentiating equation (4) with respect to [H$^+$] and equating it to zero yields $\left[\dfrac{dSP_{Fe}}{d[H^+]} = 0\right]$:

$$\left[\frac{8K_4\gamma_H^4}{\gamma_{Fe(OH)_2}K_W^4}\right][H_{opt}^+]^5 + \left[3\frac{K_1\,\gamma_H^2}{\gamma_{Fe}\,K_W^3}\right][H_{opt}^+]^4 +$$

$$\left[\frac{2K_2\gamma_H^2}{\gamma_{FeOH}\,K_W^2}\right][H_{opt}^+]^2 + \left[\frac{K_3\gamma_H}{\gamma_{Fe(OH)_2}\,K_W}\right][H_{opt}]^2 = \frac{K_4K_W}{\gamma_{Fe(OH)_4}\gamma_H} \tag{8}$$

Solving by trial and error for [H_{opt}^+] and knowing:

$$pH = -\log\{H_{opt}^+\} = -\log(\gamma_H[H_{opt}^+])$$

where γ = Activity coefficient = 10^{-x}

$\{\,\} $ = Activity

[ ] = Molar concentration

μ = Ionic strength [= 2.5 (TDS)]

TDS = Total dissolved solids (mg/L)

$\{\,\} = \gamma[\,]$

$$x = \frac{0.5Z_i^2(\mu^{0.5})}{1+1.41(\mu^{0.5})}$$

Z_i = Change in species i

K_W = [H$^+$][OH$^-$] = at 25°C

The equilibrium constants have been determined at 25°C, and Van't Hoff equation can be used to obtain values at other temperatures.

2. Required optimum pH for ferrous salt coagulation for wastewater having TDS of 140 mg/L

Activity coefficient

* $\mu = 2.5 \times 10^{-5}$ (TDS) $= 2.5 \times 10^{-5}(140) = 3.5 \times 10^{-3}$

* $\gamma_H = 10^{-x_H}$

* $x_H = \dfrac{0.5(1)^2 (3.5 \times 10^{-3})^{0.5}}{1 + 1.14(3.5 \times 10^{-3})^{0.5}}$

$\gamma_H = 0.94$

* $\gamma_{Fe} = 10^{-x\,Fe}$

$x_{Fe} = \dfrac{0.5(2)^2 (3.5 \times 10^{-3})^{0.5}}{1 + 1.41(3.5 \times 10^{-3})^{0.5}}$

$\gamma_{Fe} = 10^{-x_{Fe}} = 0.77$

$\gamma_{FeOH} = 10^{-x_{FeOH}}$

$x_{FeOH} = \dfrac{0.5(2)^2 (2.3 \times 10^{-3})^{0.5}}{1 + 1.14(2.3 \times 10^{-3})^{0.5}}$

$\gamma_{FeOH} = 10^{-x_{FeOH}} = 0.94$

$\gamma_{Fe(OH)_3} = \gamma_H = 0.94$

Therefore, substituting values in equation (6), section 1.1:

First member $= \dfrac{2K_1 \gamma_H^2}{\gamma_{Fe}\, K_W^2} = \left[\dfrac{2(10^{-14.5})\,(0.94)^2}{(0.77)(10^{-14})^2}\right] = 7.26 \times 10^{13}$

Second member $= \dfrac{K_2 \gamma_H}{\gamma_{FeOH} K_W} = \dfrac{10^{-9.4}\,(0.94)}{0.94\,(10^{-14})} = 3.98 \times 10^4$

Third member $= \dfrac{K_3 K_W}{\gamma_{Fe(OH)_3} \gamma_H} = \dfrac{10^{-5.1}\,(10^{-14})}{0.94\,(0.94)} = 8.99 \times 10^{-20}$

Final equation becomes as:

$$7.26 \times 10^{13} [H_{opt}^+]^3 + 3.98 \times 10^4 [H_{opt}^+]^2 = 8.99 \times 10^{-20}$$

Solving the equation by trial and error procedure yields:

$$[H_{opt}^+] = 1.11 \times 10^{-12}$$

Therefore, pH $= -\log(1.11 \times 10^{-12}) = 11.95$

3. Required optimum pH for ferric salt coagulation for waste water having TDS of 140 mg/L.

    Activity coefficient

    * $\mu = 2.5 \times 10^{-5}$ (TDS) $= 2.5 \times 10^{-5}(140) = 3.5 \times 10^{-3}$

    * $\gamma_H = 0.94$

    * $\gamma_{Fe_2(OH)_2} = 10^{-X_{Fe_2(OH)_2}}$

        $X_{Fe_2(OH)_2} = \dfrac{0.5(4)^2\,(3.5 \times 10^{-3})^{0.5}}{1 + 1.41(3.5 \times 10^{-3})^{0.5}}$

        $\gamma_{Fe_2(OH)_2} = 0.36$

    * $\gamma_{Fe} = 10^{-X_{Fe}}$

        $X_{Fe} = \dfrac{0.5(3)^2\,(3.5 \times 10^{-3})^{0.5}}{1 + 1.14(3.5 \times 10^{-3})^{0.5}}$

        $\gamma_{Fe} = 0.56$

    * $\gamma_{FeOH} = 10^{-X_{FeOH}}$

    * $X_{FeOH} = \dfrac{0.5(2)^2\,(3.5 \times 10^{-3})^{0.5}}{1 + 1.14(3.5 \times 10^{-3})^{0.5}}$

        $\gamma_{FeOH} = 0.77$

    * $\gamma_{Fe(OH)_2} = \gamma_H = 0.94$

    * $\gamma_{Fe(OH)_4} = \gamma_H = 0.94$

    Therefore, substituting values in equation (8), section 1.2:

    $$\text{First member} = \frac{8K_5\,\gamma_H^4}{\gamma_{Fe_2(OH)_2}\,K_W^4} = \frac{8(10^{-50.8})\,(0.94)^4}{(0.36)(10^{-14})^4} = 2.75 \times 10^6$$

    $$\text{Second member} = \frac{3K_1\,\gamma_H^3}{\gamma_{Fe}\,K_W^3} = \frac{3(10^{-38})\,(0.94)^3}{0.56\,(10^{-14})^3} = 4.45 \times 10^4$$

    $$\text{Third member} = \frac{2K_2\,\gamma_H^2}{\gamma_{FeOH}\,K_W^2} = \frac{2(10^{-26.16})(0.94)^2}{0.77(10^{-14})^2} = 158.78$$

    $$\text{Fourth member} = \frac{K_3\,\gamma_H}{\gamma_{Fe(OH)_2}\,K_W} = \frac{10^{-16.74}\,(0.94)}{0.94\,(10^{-14})} = 10^{-2.74}$$

$$\text{Fifth member} = \frac{K_4 K_W}{\gamma_{Fe(OH)_4} \gamma_H} = \frac{10^{-5}(10^{-14})}{0.94(0.94)} = 1.13 \times 10^{-19}, \text{ and}$$

the final expression becomes as:

$$2.75 \times 10^6 [H_{opt}^+]^5 + 4.45 \times 10^4 [H_{opt}^+]^4 + 158.78[H_{opt}^+]^3 + 10^{-2.74}[H_{opt}^+]^2$$

$$= 1.13 \times 10^{-19}$$

Solving this equation by trial and error procedure yields:

$$[H_{opt}^+] = 0.693 \times 10^{-8}$$

Therfore, $pH = -\log[H_{opt}^+] = -\log[0.693 \times 10^{-8}] = 8.2$

Example 6.88

Coagulation Treatment Plant Sludge

Determine the amount of sludge that must be disposed of daily from a coagulation treatment plant. Use the following data:

Treated wastewater flow   : $0.75 \text{ m}^3/\text{s}$

Alum dosage   : 25 mg/L (No other chemical added)

Influent SS   : 40 mg/L

Desired effluent SS   : 10 mg/L

Sludge solids concentration   : 0.9%

Specific gravity of sludge solids : 2.8

Solution

1. Required mass balance for the sedimentation basin

   [Accumulation] = [Input] – [Output]

   The pH range of 6 to 8 is where most wastewater treatment plants effect the coagulation, and an insoluble aluminium hydroxide complex of $Al(H_2O)_3(OH)_3$ probably predominates, resulting in production 0.44 kg of chemical sludge for each kg of alum added.

   Any suspended solids present in water will produce an equal amount of sludge. The amount of sludge produced per turbidity unit is not as obvious; however, in many waters a correlation does exit. Carbon, polymer, and clay will produce about 1 kg of sludge per kg of chemical addition.

   The sludge production for alum coagulation is approximated by:

   $$M_s = 86.40Q(0.44 \text{ A} + SS + M)$$

   where M_s is the dry sludge produced (kg/d), Q is the plant flow (m^3/s), A is the alum dose (mg/L), SS is the influent SS (mg/L), and M is the miscellaneous chemical addition [clay, polymer, and carbon, (mg/L)].

$$M_s = (86.40)(0.75 \text{ m}^3/\text{s})[0.44 (25 \text{ mg/L}) + 40 \text{ mg/L} + 0]$$

$$= 3304.8 \text{ kg/d}$$

2. Required weir output rate (W)

To estimate the mass flow (output rate) of solids leaving the clarifier through the weir an estimate of the concentration of solids and the flow rate is necessary. The mass flow through the weir is:

$$W = [\text{Concentration (mg/L)}] [(\text{Flow rate (m}^3/\text{s})] = \text{mg/s}$$

$$= [10 \text{ mg/L (or g/m}^3)][0.75 \text{ m}^3/\text{s}][86,400 \text{ s/d}][10^{-3} \text{ kg/g}]$$

$$= 648 \text{ kg/d}$$

3. Required accumulation rate

Accumulation = Input − Output

$$= 3304.8 - 648$$

$$= 2656.8 \text{ kg/d (dry mass)}$$

4. Required volume of water (W_o)

$$\%\text{Solids} = \frac{\text{Accumulated solids (kg/d)}}{\text{Accumulated solids (kg/d)} + W_o} \times 100$$

or, $\text{Acculmulated solids} + W_o = \dfrac{\text{Accumulated solids}}{\% \text{ solids}} \times 100$

$$W_o = \left[\frac{\text{Accumulated solids}}{\% \text{ solids}} \right] 100 - \text{Accumulated solids}$$

$$W_o = \left(\frac{2656.8}{0.9} \right) 100 - (2656.8)$$

$$= 2,92,543 \text{ kg/d.}$$

5. Required volume of sludge and water

$$\text{Volume} = \frac{\text{Mass}}{\text{Density}}$$

Total volume (V_T) = Volume of solids + Volume of water

$$= \frac{2656.8 \text{ kg/d}}{(2.8)(1000 \text{ kg/m}^3)} + \frac{2,92,543 \text{ kg/d}}{(1)(1000 \text{ kg/m}^3)}$$

$$= 0.95 + 292.543$$

$$= 293.493$$

[Sludge dewatering is essential as the solids form a small fraction of the total volume].

Note :

1. Softening sedimentation basin

   Softening process produces a sludge primarily $CaCO_3$ and $Mg(OH)_2$. Theoretically, each mg/L of calcium hardness removed produces 1 mg/L of $CaCO_3$ sludge; each mg/L of magnesium hardness removed produces 0.6 mg/L of sludge, and each mg/L of lime produces 1 mg/L of sludge. The theoretical sludge production is estimated as:

   $$M_s + (86,40)(Q)(C_{aR} + 0.58M_{gR} + L_A)$$

   where M_s is the dry sludge produced (kg/d), Q is the plant flow (m^3/s), C_{aR} is the calcium hardness removed as $CaCO_3$ (mg/L), M_{gR} is the magnesium hardness removed as $CaCO_3$ (mg/L), and L_A is the lime added as $CaCO_3$ (mg/L). When surface waters are softened, this equation is not valid. There will be additional sludge from coagulation of SS and precipitation of metal coagulants

2. Alum sludge leaving the sedimentation basin usually has a SS content of under 1%.

3. pH of alum sludge is normally in the range of 5.5 to 7.5.

4. Table 48, Range of cake concentrations obtainable

Table 48

Type of Thickening	Lime Sludge (%)	Coagulation Sludge (%)
Gravity thickening	15–20	3–4
Basket centrifuge	–	10–15
Scroll centrifuge	55–65	10–20
Belt filter press	–	10–15
Vacuum filter	45–65	NA
Pressure filter	55–70	30–45
Sand drying beds	50	20–25
Storage lagoons	50–60	7–15

Example 6.89

Alum–Coagulation and Sludge Production

A wastewater having a turbidity of 15 NTU is being treated by applying an alum dosage of 35 mg/L. Calculate the total sludge production, volume of the settled sludge, and quantity of filter wash water per mil–gal of treated wastewater. Assume the following data:

Settled sludge concentration : 1.4% solids

Backwash water with SS : 350 mg/L

Gravity thickened solids : 3%

Solution

1. Gravity thickening of waste in clarifier–thickener

    Total sludge solids (lb/mil–gal) = 2.75 × alum dosage (mg/L) + 8.34 × turbidity

    Sedimentation basin sludge (lb/mil–gal) = Total solids – Solids to filter

    1.1 Total sludge solids = 2.75 × 35 + 8.34 × 15

    $$= 96.25 + 125.1 = 221.36 \text{ lb/mil–gal}$$

    1.2 Solids to filter = 26 lb/mil–gal (Summer)

    (Assumed) = 60 lb/mil–gal (Winter)

    1.3 Sedimentation basin sludge = 221.36 – 26.00 = 195.36 lb/mil–gal

2. Volume of settled sludge

    $$= \frac{195.36}{0.014 \times 8.34 \times 1} = 1673 \text{ gal}$$

3. Volume of wash water

    $$= \frac{26}{\left(\dfrac{400}{10^6}\right) \times 8.34 \times 1} = 7793 \text{ gal}$$

4. Volume of thickened sludge

    $$= \frac{221.36}{0.030 \times 8.34 \times 1} = 885 \text{ gal}$$

Example 6.90

Centrifugal Separators

1. Derive the idealistic and maximum throughput expressions at a given particle size for removal of suspended matter from the liquid stream using centrifuge.

    1.1 Required general throughput expressions for centrifuges

    Centrifugal force (CF) acting on a particle is given as:

    $$CF = m \, \omega^2 \, r$$

    where   m = Particle mass (g)

    ω = Spinning of particles (rad/s; $2\pi N/60$)

    r = Radius from axis of rotation (cm)

    N = RPM

Relative centrifugal force (CF_r) can be defined as the force acting upon a particle in a centrifugal field in terms of its own weight in the gravitational field :

$$CF_r = \frac{\omega^2 r}{g}$$

$$= 1.11 \times 10^{-5} \, N^2 \, r$$

Effective mass of particle (m) is:

$$m = V(\rho_p - \rho_f)$$

where   V = Particle volume (spherical particle $= \pi d_p^3/6$)

$\quad\quad \rho_p$ = Particle density

$\quad\quad \rho_f$ = Fluid density

$\quad\quad d_p$ = Particle diameter

Force (F) producing motion is expressed as:

$$F = \frac{\pi d_p^3}{6} (\rho_p - \rho_f)\omega^2 r$$

Force (F) resisting particle motion is:

$$F = 3 \pi \mu d_p \, u$$

where μ = Fluid viscosity

$\quad\quad$ u = Particle velocity

Particle acceleration until it reaches a terminal velocity (u) where two forces are equal:

$$\frac{\pi}{6} d_p^3 (\rho_p - \rho_f)\omega^2 r = 3\pi\mu d_p u$$

$$u = \frac{d_p^2(\rho_p - \rho_f)\omega^2 r}{18\mu}$$

Terminal velocity (u) is proportional to radius (r)

If $\rho_p > \rho_f$ = Particle moves outwards from the axis of rotation (acceleration is considered as +ve).

If $\rho_p < \rho_f$ = Particle moves towards the axis of rotation (retardation is considered as +ve).

For centrifuges carrying relatively thin layers of liquid, the velocity of particles can be considered as approximately constant across such a layer. The radial distance travelled by a particle in time, t is X :

$$X = ut = \frac{d_p^2(\rho_p - \rho_f)\omega^2 \, r \, t}{18\mu}$$

To remove particles from liquid stream, the distance $X = (r_2 - r_1)$ for any given sample of the feed within the centrifugal bowl:

$$t = \frac{V}{Q}$$

Therefore, $(r_2 - r_1)X = \left[\dfrac{d_p^2(\rho_p - \rho_f)\omega^2 \, r}{18\mu}\right]\left(\dfrac{V}{Q}\right)$

Throughput at which particles of given size is eliminated from the liquid stream is given as:

$$Q = \left[\frac{d_p^2(\rho_p - \rho_f)}{18\mu}\right]\left(\frac{\omega^2 \, r \, V}{S}\right)$$

$$= k \, X \, Y$$

where S = Liquid layer thickness $(r_2 - r_1)$

         k = Constant

$$X = \frac{d_p^2(\rho_p - \rho_f)}{\mu}$$

$$Y = \frac{\omega^2 \, r \, V}{S}$$

      X = Distance measured normal to the disc (L)

      Y = Distance measured parallel to the disc (L)

Also, $\dfrac{\omega^2 \, r \, V}{S} = (CF)_r \dfrac{V}{S}$

Therefore, $Q = \dfrac{d_p^2(\rho_p - \rho_f)}{18\mu}(CF)_r \left[\dfrac{V}{S}\right]$

The equation predicts the maximum throughput (Q) at which particle of given size is eliminated from the liquid stream [Stoke's law is followed by the system].

2. Determine the sedimentation rate in the gravity separator and centrifugal separation for the limiting particle size d_m = 8 mm. Assume the following data:

Particle density (ρ_p) : 1050 kg/m^3

Liquid density (ρ_l) : 1000 kg/m^3

Viscosity of continuous phase (μ) : 1.0×10^{-3} N.s/m^2

2.1 Required ratio of sedimentation rate in centrifugal to gravity separation

Gravity separation:

$$\text{velocity } (u_g) = \frac{d_p^2(\rho_p - \rho_L)g}{18\mu}$$

$$= \frac{(8 \times 10^{-6})(1050 - 1000)(9.81)}{18(1 \times 10^{-3})}$$

$$= 1.76 \times 10^{-6} \text{ m/s}$$

Centrifugal separation:

$$u = \left(\frac{r\omega^2}{g}\right) u_g ; \text{ and } N = 5000 \text{ rpm}$$

$$\omega = \frac{2\pi N}{60} = \frac{2(3.14)(5000)}{60} = 523.6$$

$$u = \left[\frac{(2)(3.142)(5000)}{60}\right]^2 \left(\frac{0.2}{9.81}\right)(1.75 \times 10^{-6}), [r = 0.2 \text{ m}]$$

$$= 0.98 \times 10^{-2} \text{ m/s}$$

$$\text{Ratio of } \frac{u}{u_g} = \frac{0.98 \times 10^{-2}}{1.75 \times 10^{-6}} = 5600.$$

3. Estimate the separating power of a disc centrifuge rotating at 6000 rpm. Assume the following data:

Rotational speed (N)   : 6000 rpm

No. of spaces between discs (n)   : 65

Half angle of the conical discs (θ) : 35°

Radius of smaller end of disc stack (R_1) : 3.5 cm

Radius of larger end of disc stack (R_2) : 14.3 cm.

3.1 Required mathematical expressions for disc centrifuge

Assume: Cone has half angle, θ

               Spin is given as w (rad/s)

Volumetric throughput (Q) through the bowl is:

$$Q = \frac{\pi}{27} d_p^2 \frac{(\rho_p - \rho_f)}{\mu} \omega^2 n \cot\theta (R_2^3 - R_1^3)$$

Separating power of the disc centrifuge is:

$$= \frac{2\pi}{3} \frac{\omega^2}{g} n \cot \theta \, (R_2^3 - R_1^3)$$

Actual performance is around 50% of the theoretical value and, therefore,

$$P = \frac{2\pi}{3}(0.5)d_p^2 \frac{\omega^2}{g} n \cot \theta \, (R_2^3 - R_1^3)$$

$$= 0.0006 \, d_p^2 \left[\frac{(\rho_p - \rho_f)}{\mu} \right] N^2 . n \cot \theta \, (R_2^3 - R_1^3)$$

and separating power $= 13 \times 10^{-6} \, N^2 \, n \cot \theta \, (R_2^3 - R_1^3)$

3.2 Required magnitude of separating power

Speed of rotation (ω) $= \dfrac{2\pi N}{60}$

$$= \frac{2(3.142)(6000)}{60}$$

$$= 628.3 \text{ rad/s}$$

Separating power of a disc centrifuge is:

$$= 13 \times 10^{-6} \, N^2 \, n \cot \theta \, (R_2^3 - R_1^3)$$

$$= 13 \times 10^{-6}(6000)^2(65) \cot (35) \, [(14.5)^3 - (3.5)^3]$$

$$= 1.25.$$

Note :

1. Typical expressions for high speed tubular centrifuge

   For high speed tubular centrifuge, the velocity of the particles for size upto 10 mm does not perceptibility exceed 1 m/s. Reynolds number ranges between 10^{-4} to 10, and Stake's law is applicable. Therefore,

   $u = \dfrac{d_p^2 (\rho_p - \rho_f)}{18\mu} \omega^2 r$ can be applied for high speed tubular bowl centrifuge. The residence time in the centrifuge bowl (V/Q) introduces an error and, therefore, the performance evaluation on its basis may not be perfect. The most promising approach for the performance is to assume viscous flow in the bowl and use for its length some fraction of the geometric length estimates empirically.

   In a bowl of length L, having an inner wall radius R_B and the bowl is filled so that the liquid surface radius is R_L, the throughput (Q) and length (L) are given as :

$$Q = \frac{\pi}{18} d_p^2 \left(\frac{\rho_p - \rho_f}{\mu}\right) \omega^2 L \left[\frac{(R_B^2 - R_L^2)^2}{2R_B^2 \ln(R_B/R_L) - R_B^2 + R_L^2}\right]$$

$$L = \frac{18}{\pi} \frac{Q}{d_p^2} \left(\frac{\mu}{\rho_p - \rho_f}\right) \frac{[2R_B^2 \ln(R_B/R_L) - R_B^2 + R_L^2]}{\omega^2 (R_B^2 - R_i^2)^2}$$

For the fallacious block flow patter, the throughput (Q) is:

$$Q = \frac{\pi}{18} d_p^2 \left(\frac{\rho_p - \rho_f}{\mu}\right) \omega^2 L \left[\frac{R_2^3 - R_L^3}{\ln(R_B/R_L)}\right] \tag{1}$$

For a given centrifuge rotor with interior radius R_B, it is obvious that the depth of the liquid layer may be varied (different values of R_L can be used which will have a great influence upon the centrifuge performance as is clear from the above equation).

To consider the influence of the magnitude of R_L, the volumetric flow rate through the bowl is expressed as:

$$Q = \frac{\pi}{18} \frac{d_p^2 (\rho_p - \rho_L)}{\mu} \omega^2 L \nu, \ (cm^3/s) \tag{2}$$

where: ν = Velocity parallel to disc surface (cm/s), and can be calculated for various values of the ratio (R_B/R_L). Typical values of $R_B/R_L = 0.3$ to 0.5.

The value of ν can be calculated using equations (1) and (2). The ratio of $\nu_{13}/\nu_{14} = 0.75$ shows the magnitude of error caused by block flow condition. In general, the effective length of the bowl is 0.5 times the geometric length (L).

Practical value for the performance of a high speed tubular bowl centrifuge is :

$$Q = \frac{\pi}{18} \frac{d_p^2 (\rho_p - \rho_f)}{\mu} \omega^2 L (0.75 \times 0.5) \frac{(R_B^2 - R_L^2)}{\ln(R_B/R_L)}$$

$$\text{and } \omega = \frac{2\pi N}{60}; \ \ln\left(\frac{R_B}{R_L}\right) = 2.303 \log\left(\frac{R_B}{R_L}\right)$$

$$d_p = \frac{\eta}{10^4}$$

$$Q = 2.8 \times 10^{-12} \eta^2 \left(\frac{\rho_p - \rho_f}{\mu}\right) N^2 L \left[\frac{(R_B^2 - R_L^2)}{\log(R_B/R_L)}\right]$$

where η = Particle diameter (μm)

N = Rotor rpm

Example 6.91

Tube Settler

Design a tube settler for primary settling of domestic sewage placed before an activated sludge basin. Assume the following data:

Wastewater flow (Q)	: 2000 m³/h
Sludge setting velocity (u)	: 0.8 mm/s

Characteristics of tube settler:

A set of parallel pipes of a square cross–section of dimensions   :
(b) = 0.05 m × 0.05 m

Inclination (θ)	: 50°
Useful length of tubes	: 1m; width of wall common to the neighbouring pipes = 0.003 m
Shape coefficient (s)	: 11/8

Solution

1. Required water flow velocity (V) in the pipes

$$V = \frac{u\,(m/s)}{\alpha\,s}\left[\sin\theta + \frac{1}{b}\cos\theta\right]$$

Assume : a = Cofficient = 1.25

$$V = \frac{0.0008}{(1.25)\left(\frac{11}{8}\right)}[\sin 50° + \cos 50°]$$

2. Required net surface area of the pipe cross–section (excluding walls), and number of pipes

- Area $(A) = \frac{Q}{3600\,V} = \frac{2000 \text{ m}^3/\text{h}}{3600(0.0062 \text{ m/s})} = 89.6 \text{ m}^2$

- Number of pipes (n):

$$n = \frac{F}{b \times b} = \frac{89.6 \text{ m}^2}{0.05 \text{ m} \times 0.05 \text{ m}} = 35840$$

3. Required surface occupied by the settler (Ap)

   Assuming the no. of settlers (N) equal to 4, and the coefficient β relative to the increase of the surface due to the width of the walls separating the tubes (β) = 1.15

   $$Ap = \frac{A}{n \sin\theta}\,\beta = \frac{89.6 \text{ m}^2}{4(\sin 50°)}(1.15)$$

   $$= 33.67 \text{ m}^2$$

4. Required dimensions of the settler

   Assuming the width of each settler (B) = 3 m, the working length (L_w) of the settler filled with pipes is:

   $$L_w = \frac{Ap}{B} = \frac{33.67 \text{ m}^2}{3 \text{ m}} = 11.22 \text{ m} \cong 12 \text{ m}$$

   Assume the length of the inlet part of the settler (L_1) = 1.5 m; and the outlet length (L_2) of the settler = 1.5 m, than total length (L) is:

   $$L = L_1 + L_2 + L_w$$

   $$= (1.5 + 1.5 + 12.0) \text{ m} = 12 \text{ m}$$

5. Required flow dynamic conditions

   - Hydraulic load on the settler is:

   $$q = \frac{Q}{B(L)N} = \frac{2000 \text{ m}^3/\text{h}}{3 \text{ m} (12 \text{ m})(4)}$$

   $$= 13.9 \text{ m/h} \quad [\text{less than 18 m/h, OK}]$$

   - Hydraulic radius of each pipe (R_h) is:

   $$R_h = \frac{b^2}{4b} = \frac{(0.05)^2}{4(0.05)} = 0.0125 \text{ m}$$

   - Reynolds number (R_e) is:

   $$R_e = \frac{V R_h}{\upsilon} = \frac{(0.0062 \text{ m/s})(0.0125 \text{ m})}{(1.31 \times 10^{-6} \text{ m}^2/\text{s})}$$

   $$= 59 \quad [\text{less than 12500, OK}]$$

   - Froude number (F_e) is:

   $$F_e = \frac{V^2}{R_h\, g} = \frac{(0.0062)^2}{(0.0125 \text{ m})(9.81 \text{ m/s}^2)}$$

   $$= 3 \times 10^{-4} \text{ greater than } 10^{-6} \text{ [OK]}.$$

Example 6.92

Sedimentation Basin for Wastewater Treatment Facility

[A] Assume that a waste has a flow rate of 1000 ft³/hr and that the terminal velocity of the critical discrete non-flocculating particles is 4 ft/hr. The required surface area is calculated according:

$$A = \frac{Q}{v_t} = \frac{1000}{4} = 250 \text{ ft}^3$$

Table 49, The ratio of the depth to the detention time is 4 ft/hr and the following are acceptable designs:

Table 49

Depth (ft)	Detention Time (hr)	Surface Area (ft²)
8	2.0	250
10	2.5	250
12	3.0	250
14	3.5	250

The tank or tanks would now be chosen with the following in mind:

- Space required for scraping and skimming mechanisms;
- Required flexibility (*e.g.*, how would the system respond when part of it is shutdown for repairs?)
- Depth of adjacent tankage
- Allowance for future expansion
- Adequacy of safety factors.

[B] Consider a case in which a laboratory settling analysis of a water or waste influent gives the data presented, Table 50. The design of a settling tank then involves first plotting the data as illustrated in Figure 22. Assume that the desired depth for the sedimentation basin is 8 ft. Equation is then used to calculate the information presented, Table 51, and these results are used to construct graphs characterizing the percent removal as a function of surface overflow rate, and as a function of detention time, Figure 23. The final design overflow rate can be so selected as to compensate for losses in efficiency from scale–up by incorporating an appropriate safety factor, Figure 23. An overflow rate of 600 gal/(ft²)(day) is chosen for 68% removal with a safety factor of 2. The theoretical detention time for 68% removal is 70 min, or 140 min with a safety factor of 2, Figure 23.

Table 50 : Laboratory settling analysis data.

Time (min)	% Suspended Solids Removed at Indicated Depth and Time					
Depth (ft)	15 min	30 min	45 min	60 min	90 min	120 min
2	50	65	69	72	73	74
4	31	56	64	67	72	73
6	20	49	61	64	69	72
8	16	41	57	62	67	71
10	15	37	52	61	65	69

The required area to treat a flow of 1 mgd is calculated from Equation

$$v_t = \frac{\text{Tank depth}}{\text{Detention time}} = \frac{\text{Depth}}{\left(\dfrac{\text{Tank volume}}{\text{Flow rate}}\right)} = \frac{\text{Depth}}{\left(\dfrac{\text{Area depth}}{\text{Flow rate}}\right)} = \frac{Q}{A} \tag{1}$$

$$\text{Required area} = \frac{10^6 \text{ gpd}}{600 \text{ gal}/(\text{ft}^2)(\text{day})} = 1665 \text{ ft}^2$$

This area can be obtained with a circular tank a diameter of 46 ft.

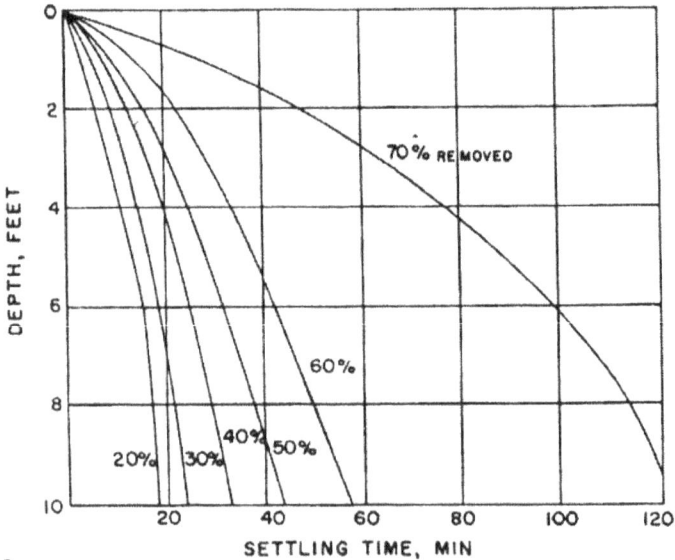

Figure 22 : Laboratory settling analysis.

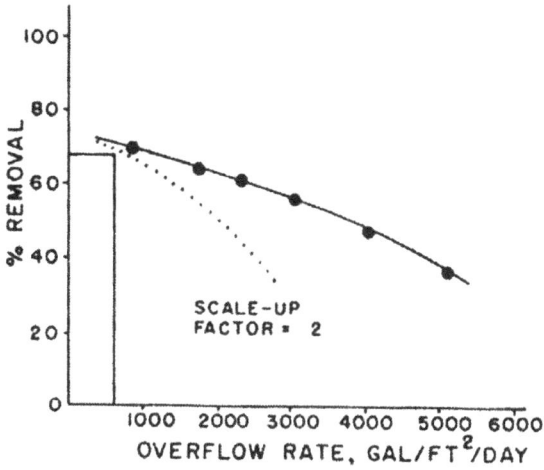

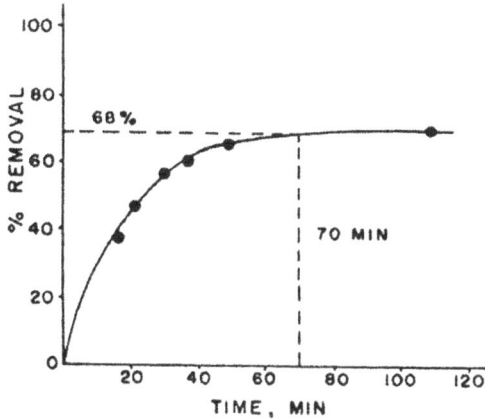

Figure 23 : Percent removal for 8-ft tank.

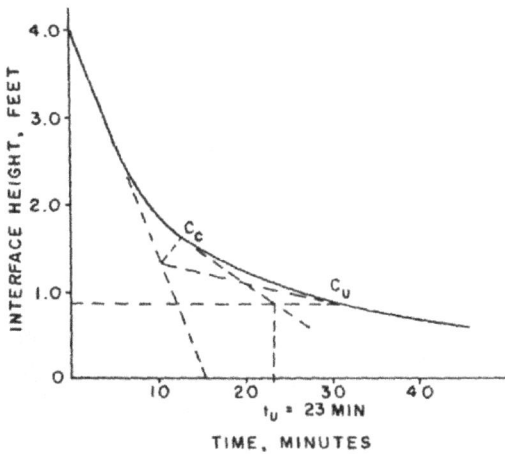

Figure 24 : Zone settling design.

Table 51 : Clarifier design calculation according to equation 2 for a tank 8 ft deep.

Time (min)	% SS Range	Avg. Depth of range (ft)	v_s (fpm)	v_o gal/ $(ft^2)(day)$	$\dfrac{v_s}{v_o}$	% SS (R) Removal
17.0	0–20	8.0	0.471	5080	1.0	20.0
	20–30	6.2	0.365		0.73	7.3
	30–40	4.0	0.236		0.50	5.0
	40–50	2.6	0.153		0.32	3.2
	50–60	1.7	0.100		0.21	2.1
(1)	(2)	(3)	(4)	(5)	(6)	(7)
	60–70	0.8	0.047		0.10	1.0
			% SS Removal		Total	38.6
21.5	0–30	8.0	0.372	4010	1.0	30.0
	30–40	6.0	0.280		0.75	7.5
	40–50	3.8	0.177		0.48	4.8
	50–60	2.4	0.112		0.30	3.0
	60–70	1.1	0.051		0.14	1.4
					Total	46.7
29.0	0–40	8.0	0.276	2980	1.0	40.0
	40–50	6.4	0.221		0.80	8.0
	50–60	4.1	0.141		0.51	5.1
	60–70	1.8	0.062		0.22	2.2
					Total	55.3
37.5	0–50	8.0	0.213	2300	1.0	50.0
	50–60	6.2	0.191		0.78	7.8
	60–70	2.8	0.075		0.35	3.5
					Total	61.3
50	0–60	8.0	0.160	1725	1.0	60.0
	60–70	4.4	0.088		0.55	5.5
					Total	65.5
115	0–70	8.0	0.070	755	1.0	70.0
					Total	70.0

$$v_o = v_s \ (\text{at depth})(\text{fpm}) \times (7.481 \ \text{gal/ft}^3) \times (24 \times 60 \ \text{min/day}) = \frac{\text{gal}}{\text{ft}^3.\text{d}}$$

$$R = (1 - f_1) + \int_0^{f_1} \frac{v_s}{v_o} \, df \qquad (2)$$

where f_1 is the fraction of particle with a subsiding velocity of v_o or less

[C] A waste has a solids concentration of 0.2 lb/ft³ and is to be thickened in a continuous–flow tank operated at a flow rate of 1.5 cfs. The desired under flow concentration is 0.9 lb/ ft³. Data obtained from a batch settling analysis performed on the waste are plotted. Figure 23. The height of the interface for the underflow concentration is calculated as

$$h_u = \frac{C_o h_o}{C_u}$$

Since the total weight of solids in the batch column must remain constant

Therefore,

$$h_u = \frac{0.2 \times 4}{0.9} = 0.889 \text{ ft}$$

The tangents to the hindered settling and compression portions are shown in Figure 23. The line bisecting the angle of these tangents located the critical concentration C_c. The tangent of the curve at C_c intersects a horizontal line constructed to pass through C_c at t_u. Figure 23 shows the construction and the determination of $t_u = 23$ min.

The area required for thickening is then determined according to

$$A = \frac{Q t_u}{h_o}$$

$$= \frac{1.5 \times 23 \times 60}{0.889} = 2330 \text{ ft}^2$$

The slope of the hindered settling portion of the curve establishes the subsidence velocity, v_s:

$$v_s = \frac{4 \text{ ft}}{15.5 \text{ min}} = 0.258 \text{ ft/min}$$

The area required for clarification

$$A = \frac{Q}{v_s}$$

where Q is the volume overflow rate

$$A = \frac{1.5 \times \dfrac{(4 - 0.889)}{4}}{\dfrac{0.258}{60}}$$

$$= 2610 \text{ ft}^2$$

The area requirement for clarification exceeds the area requirement for thickening and therefore controls the design.

Note :

1. A clarifier has at least three main functions to perform

   Settling tank must provide for effective removal of SS

   Sludge removal capacity must be adequate

   Thickening of the sludge may be important. Any failure in one of there functions will impair the performance of the tank and, if series may destroy the effectiveness of the process almost completely.

2. Tanks of any size are subject to eddies, density converts, unequal temperature effects, wind action, imperfect distribution of velocity vectors, resuspension of sludge through motion of the sludge rates, etc. These factors impair the rate of subsidence and cause a loss in effectiveness in tank performance. This is not neglected by any factor involved in the equations of quiescent laboratory system

3. Two paths available to the engineer for design of sedimentation basis

   Recognition that a loss in effectiveness is inherent in the projection of settling experiments

   Second alternative if the design will not permit the luxury of slight decline in efficiency is to reduce the experimental observed overflow rates by a safety factor.

4. Tanks may be small, well–baffled, and protected from wind by covers. Or tanks may be large, unbaffled, and unprotected from wind or other disturbances. Safety factors may vary from 1.5 in the former case to 3 in the later case.

Example 6.93

Horizontal Flow Tanks

If the inlet and outlet conditions are near ideal, a value of T_r of 0.95 can be used for primary settlement V can be set at 31/8 m/s and for a choice of v_0 of 0.5 mm/s, a ratio of L/D of approximately 8 is calculated. Calculate surface area of the tank and relative wall area (RWA).

Solution

Use the following equations:

$$Q = VBD$$

$$\frac{V}{v_0} = \frac{L}{D}$$

$$V_o = \frac{Q}{LB}$$

[Settling velocity of that group of particles which will be just completely removed is numerically equal to flow per unit surface area of the tank]

$$A = \frac{Q}{(T_r v_o)}, \text{ [Q is the maximum flow]}$$

$$RWA = \frac{\text{Wall area}}{\text{Surface area}} \times \frac{L}{D}$$

In order to reduce the effect of scouring, the following criteria are suggested:

$$\frac{L^2}{BD} \rangle 20$$

$$\frac{L}{D} \text{ between 5 and 10 (Primary tanks)}$$

$$\frac{L}{B} \text{ is likely to lie between 3 and 6}$$

where Q is the flow into the tank, B and D are breadth and depth respectively, V is the forward velocity, L is the length of the horizontal tank, v_o is the settling velocity, and T_r is a time ratio which expresses the extent to which the velocity profile of the tank, in plan, is ideal [variations in depth have been shown to have little effect unless sufficiently large to cause scour].

For a maximum (3 DWF) flow of 0.3 m^3/s the area (A) is given by:

$$A = \frac{0.3 \times 1000}{0.95 \times 0.5} = 631 \text{ m}^2$$

For n individual tanks of breadth B, and a total breadth B_o, the surface area required is

$$LB_o = A$$

Although a suitable value of B could be shown in order to ensure that L^2/BD > 20 from the point, Clement has suggested that since the cost of construction is a constraint, this ought to be included in the analysis.

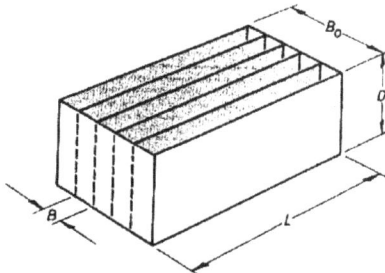

Figure 25 : Tank dimension.

RWA gives a relative estimate of the cost of the tank construction, since both L/D and surface area are constant.

For five tanks in parallel

$$RWA = \left(\frac{6LD + 2B_oD}{LB_o}\right)\frac{L}{D}$$

For any given number of tanks and for different values of L/B_o, plot can be constructed (number of tanks versus RWA at various L/B_o and number of tanks versus RWA at various L/B). Where L is the length of the tank, B_o is the width of each sub tank, and D is the depth of tanks.

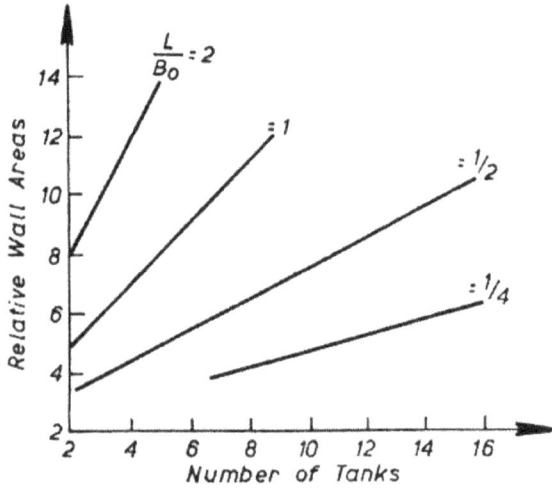

Figure 26 : L/B_o ratio versus relative wall area and number of tanks.

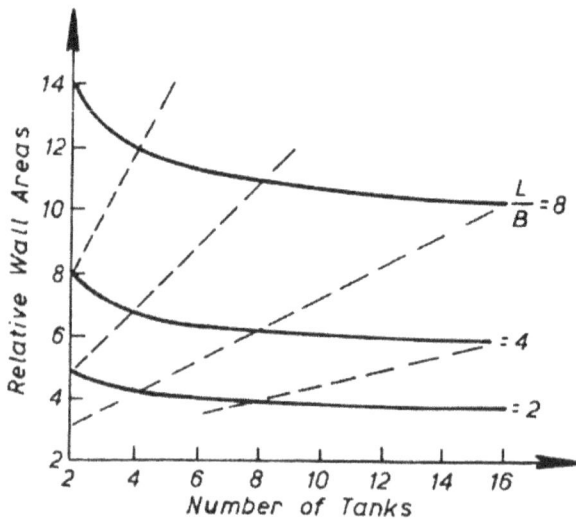

Figure 27 : Combination of L/B_o and L/B ratio.

To reduce the machinery costs n should be low and to reduce construction costs RWA should also be low. While values of L/B must be high enough to ensure $L^2/BD > 20$. A choice is therefore made within the range of $L/B_o = 1/2$ or $L/B_o = 1$. Figure 25 to 27 and 3 describe the system.

Example 6.94

Primary Sedimentation Basin (Flocculant Settling)

Determine the percent suspended solids (SS) removal versus detention time, the percent SS removal versus overflow rate, the percent SS remaining (fraction of particles with less than stated velocity) versus settling velocity for a residence time of 25 min, the primary clarifier cross–sectional area and effective depth to remove 50% of the SS (use a scale–up factor of 1.25), and the daily accumulation of sludge. Also determine the percent of SS removal if the flow rate is doubled. Assume the following data:

Wastewater flow (Q_o)   : 1 MGD

Influent SS_o concentration   : 430 mg/L

Underflow SS concentration : 10,000 mg/L (1% concentration)

Table 52 : Experimental settling column data (8 ft depth).

Time	SS Concentration at Various Depths (mg/L)		
	Tap – 1 (2 ft)	Tap – 2 (4 ft)	Tap – 3 (6 ft)
5	356.9	387.0	395.6
10	309.6	346.2	365.5
20	251.6	298.9	316.1
30	197.8	253.7	288.1
40	163.4	230.1	251.6
50	144.1	195.7	232.2
60	116.1	178.5	204.3
75	107.5	143.2	180.6

Solution

1. Required analysis of the flocculant settling in a primary sedimentation basin Types of settling

   * **Discrete settling :** No agglomeration of particles and the properties of the particles (size, slope, and specific gravity) remain unchanged during the operation (grit chamber).

   * **Flocculant settling :** Agglomeration of particles occurs accompanied by changes in density and settling velocity (particle size) [primary sedimentation].

   * **Zone settling :** Particles form lattice (blanket) and settle as a mass exhibiting a distinct interface with the liquid phase [secondary clarifier

biological activated sludge processes), and alum flocs in water treatment systems].

A practical design of settling column is 8 ft deep with sampling ports at depths of 2, 4, 6, and 8 ft [data to be taken at 2, 4, and 6 ft depth at various time intervals. Data from the 8 ft port are used for sludge concentration and compaction determination].

Solids remaining in suspension:

$$x = \frac{SS}{SS_o}$$

Fraction of solids removal $= 1 - x$

% of solids removed $= (1 - x)\,100$

Effective settling velocity (V_s):

$$V_s = \frac{\text{Effective depth (6 ft in this example)}}{\text{Time (t) required to travel the effection depth}}$$

$$= \frac{H}{t}$$

Overall suspended solids removal

$$\text{Overall removal} = (1 - x_o) + \int_0^{x_o} \left(\frac{V_1}{V_s}\right) dx \qquad (1)$$

where : x_o = Fraction of solids remaining at a given time and depth $\left(\dfrac{SS}{SS_o}\right)$

SS_o = Initial uniform SS concentration in the settling column

$(1 - x_o)$ = Fraction of suspended solids removed $\left(= 1 - \dfrac{SS}{SS_o}\right)$ at a given time and depth in a settling column.

V_1 = A variable $(0 \le V_1 \le V_s)$ with $V_1 = f(x)$.

The term $(1 - x_o)$ refers to the fraction completely removed corresponding to the particles with velocitties greater than V_s

and the term $\displaystyle\int_0^{x_o} (V_1/V_s)\,dx$ refers to the fraction of removal

corresponding to the particles with velocities less than V_s in the equation (1).

The equation (1) can be written as:

$$\text{Overall removal} = (1-x_o) + \int_0^{x_o} \left(\frac{V_1}{V_s}\right) dx$$

$$= (1-x_o) + \int_0^{x_o} \left(\frac{h}{H}\right) dx$$

$$= X_{tot} + \left[\frac{h_1}{H} + \frac{h_2}{H} + ----- + \right] \Delta x$$

$$= X_{tot} + \Sigma \left(\frac{h_{avg}}{H}\right) \Delta x$$

where $X_{tot} = (1 - x_o)$

Δx = percent interval (preferably 10%)

Material balance for a primary clarifier

Figure 28, The schematics of a primary clarifier is shown as:

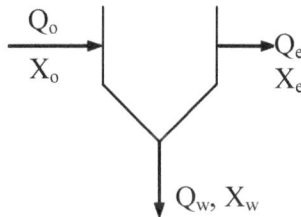

Figure 28

$Q_o = Q_e + Q_w$ (water balance)

$Q_o X_o = Q_e X_e + Q_u X_o$ (SS balance)

$$Q_e = \frac{Q_o(X_u - X_o)}{(X_u - X_e)}$$

$Q_u = Q_o - Q_e$

Over flow rate $(OR) = \dfrac{Q_e}{A}$

2. Table 53, Required fraction of solids remaining and removed at a 2 ft depth of the settling column SS_o = 430 mg/L

Similar table can be constructed at 4 and 6 ft for the settling column, Plot % SS removed versus time (Figure 29) at 2, 4, 6 ft for various time intervals (5,10,15 —— 75 min). Based on Figure 29, construct the settling profile at constant % SS removals. Figure 30.

3. Required overall suspended solids removal for settling time of 25 min

For settling time (t) = 25 min and settling depth (H) equal to 6 ft, 30% of SS are completely removed (Figure 30). Consider next the particles in each additional 10 % range. Particles in the range of 30–40% are removed in the proportion V_1/V_s or in the proportion of average settled depth (h_1) to the total depth of the settling column (H). The average settled depth (h_1) is estimated drawing (by interpolation) a curve corresponding to (30 + 40)/2, 35% constant removal and reading from if it the depth (h_1 = 4.2 ft) corresponding to settling time (t) = 25 min. Similarly, the curves for constant % SS removals of 45, 55, 65 and 75% area drawn and average settled depths of 2.4, 1.4, 0.84 and 0.28 ft are recorded for t = 25 min.

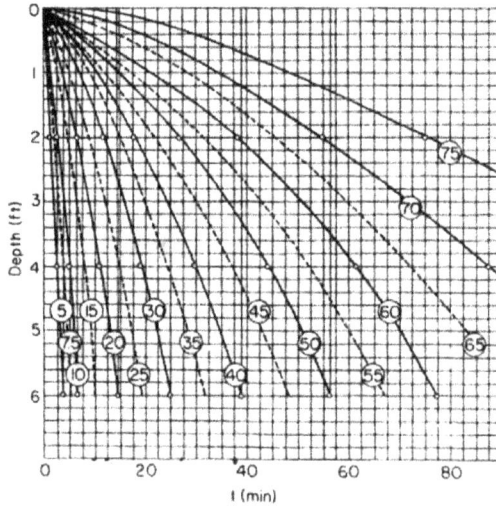

Figure 29 : Suspended solids removed versus time.

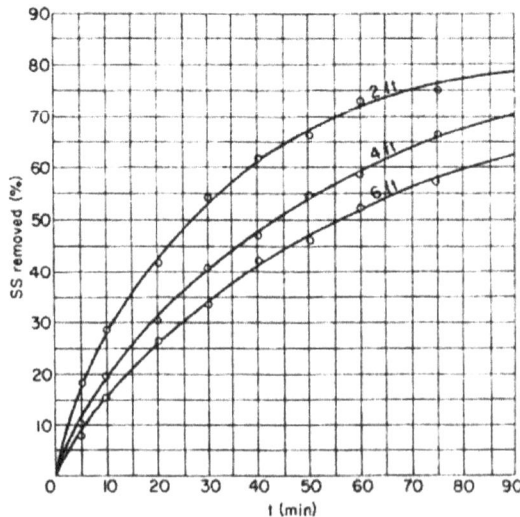

Figure 30 : Settling profile. Encircled numbers are percentage of SS removed.

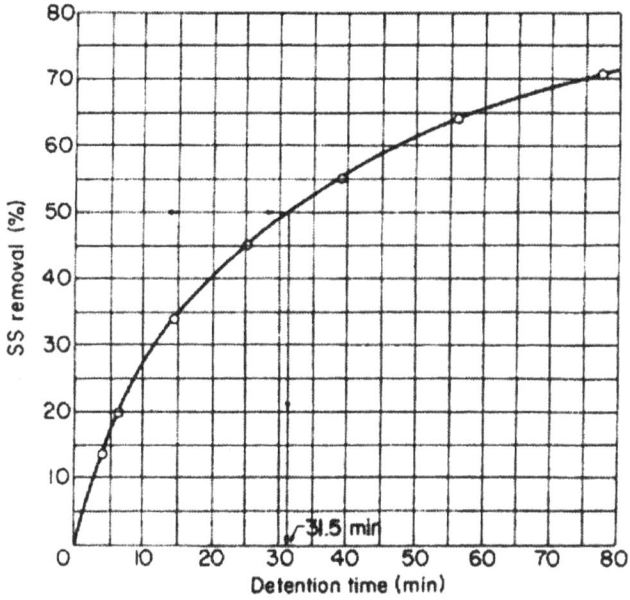

Figure 31 : Suspended solids removed versus detention time.

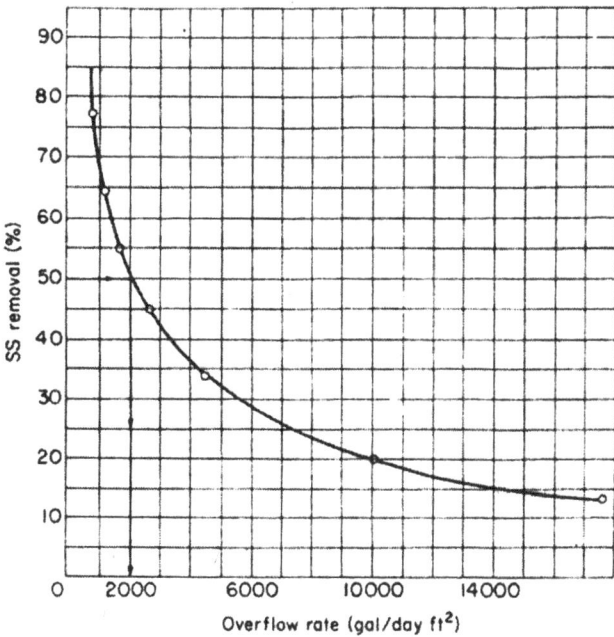

Figure 32 : Suspended solids removed versus overflow rate.

Table 53

Time (min)	SS remaining (mg/L)	% Solids remaining (SS/SS$_o$ × 100)	%Solids removal $100\left(1-\dfrac{SS}{SS_0}\right)$
5	356.9	83.0	17.0
10	309.6	72.0	28.0
20	251.6	58.5	41.5
30	197.8	46.0	54.0
40	163.4	38.0	62.0
50	144.1	33.5	66.5
60	116.1	27.0	73.0
75	107.5	25.0	75.0

The calculations for t = 25 min are shown as:

t = 25 min (0.417 h)

$$V_s = \frac{H}{t} = \frac{6 \text{ ft}}{(0.417 \text{ h})} = 14.38 \text{ ft/h } (14.4 \text{ft/h})$$

% SS removed during settling time (t) = 25 min (Figure 30):

Complete (100%) removal at 30% constant SS = 30.00%

Table 54

1st intervanl at 35% constant SS	$\left[\dfrac{4.2}{6.0}\times10\right] = 7.00\%2$
2nd interval at 45% constant SS	$\left[\dfrac{2.4}{6.0}\times10\right] = 4.00\%$
3rd interval at 55% constant SS	$\left[\dfrac{1.4}{6.0}\times10\right] = 2.33\%$
4th interval at 65% constant SS	$\left[\dfrac{0.84}{6.0}\times10\right] = 1.40\%$
5th interval at 75% cosntant SS	$\left[\dfrac{0.28}{6.0}\times10\right] = 0.46\%$
Remaining	= Negligible
Total SS removal during t = 25 min = 45.19%	

From Figure 30. The settling time for H = 6 ft at various % SS removals can be recorded and $V_s = \dfrac{H}{t}$ can be estimated as (Table 55) :

Table 55 : Estimated V_s versus SS and at various time.

% SS removal (Constant)	t (min) at H = 6 ft	Settling Velocity (V_s)
5	3.7	$\dfrac{360}{3.7} = 97.2$
10	6.5	55.2
20	14.5	24.8
30	25.0	14.4
40	39.0	9.2
50	56.5	6.35
60	77.5	4.64

Similar calculations are executed for other settling times as listed above [3.7, 6.5, 14.5, 39.0, 56.5, and 77.5 min], and the results are:

Table 56 : Total % SS removal V_s time.

t (min)	Total % SS removal
3.7	13.4
6.5	20.1
14.5	33.9
25.0	45.2
39.0	55.0
56.5	64.3
77.5	71.1

Plot % SS (total) removal versus settling time.

4. Required overflow rate (OR)

V_s in ft/h = ft^3/ft^2.h)

OR (gal/ft^2.d) = V_s[(ft/h)(24 h/d)(7.48 gal/ft^3)]

= 179.5 V_s

Table 57 : OR vs V_s and % SS.

Settling Time $(t_1$ min)	V_s	OR (gal/ft^2.d)	% Total SS removal
3.7	97.2	17450	13.4
6.5	55.2	9908	20.1
14.5	24.8	4452	33.9
25.0	14.4	2585	45.2
39.0	9.2	1651	55.2
56.5	6.35	1140	64.3
77.5	4.64	833	71.1

Plot % total SS removal versus overflow rate (OR).

5. Table 58, Required percent of particles with less than stated velocity versus settling velocity (t = 25 min)

Table 58

Solids Removal (%)	Particles with Less than Stated Velocity (%)	h (ft)	$V_1 = V_s (h/H)$ = 14.4 (h/H)(ft/h)
30	70	6.0	14.4
35	65	4.2	10.08
45	55	2.4	5.76
55	45	1.4	3.36
65	35	0.84	2.016
75	25	0.28	0.672

6. Required overflow rate (OR) and underflow rate (Q_u)

Figure 31. Read detention time at suspended solids (total) removal of 50%, and this is equal to 31.5 min. At 50% total SS removal, the overflow rate = 2000 gal/ft².d (Figure 32).

If flow rate (Q_o) is doubled, the detention is reduced to one–half (t = 31.5/2 = 15.8 min), the % SS (total) removal is 36% [Figure 31].

Note :

1. Design criteria for primary clarifiers

    Depth   : 7–12 ft

    Detention time   : 0.5–1.5 hour

    Flow through velocity (V)   : 1–5 ft/min

    Overflow rate (OR)   : 900–1200 gal/ft².h

    Efficiency (%): SS (BOD)   : 40–60% (30–50%)

    Scaling–up factor   : 1.25–1.75.

Example 6.95

Flocculant Settling of Particles

A flocculant suspension containing 400 mg/L suspended solids was subjected to a batch settling test in the laboratory. Samples were taken with time and depth from the settling column, and the percent removal of solids for each sample is given as follows:

Table 59 : Time V_s % SS removal at various time.

Time (min)	5	10	20	40	60	90	120
Percent removal at :							
2 ft	41	50	60	67	72	73	76
4 ft	19	33	45	58	62	70	74
6 ft	15	15	38	54	59	63	71

Determine the required size of the basin which reduce the suspended solids to 150 mg/L. Use scale–up factors of 1.25 for overflow rate and detention period, respectively.

Solution

1. Required iso–percent removal lines by interpolation/judgement. Plot depth as the ordinate, time as the abscissa, and parameters of percent removal.

2. Required percent removal in an ideal horizontal flow basin (for any depth and detention period)

$$R = R_o + \frac{h_1}{h_o} \Delta R_1 + \frac{h_2}{h_o} \Delta R_2 + \frac{h_3}{h_o} \Delta R_3 + - - - -$$

where
- R = Total percent removal
- R_o = Percent removal given by iso–percent line inter-secting height h_o
- h_o = Depth of the ideal basin
- t_o = Ideal basin detention time (time given by the intersection of h_o and the iso–percent time R_o)
- h_1, h_2, h_3, etc. = Average heights between the iso–percent lines directly above time t_o
- $\Delta R_1, \Delta R_2, \Delta R_3$, etc. = Differences between values of the iso–percent lines above and below h_1, h_2, h_3, etc

Ideal overflow rate (Q/A) corresponding to the percent removal at t_o is given by

$$\frac{Q}{A} = \frac{h_o}{t_o}$$

3. Required percent removal at t_o = 35 min and overflow rate of 6 ft/35 min = 0.171 (10770) = 1840 gal/ft^2.d

$$R = 50 + \frac{5}{6}(55-50) + \frac{3.4}{6}(60-55) + \frac{2.4}{6}(65-60) + \frac{1.6}{6}(70-65) + \frac{0.7}{6}(5)$$

= 60.9% [A sample case]

4. Required ideal detention period and overflow rate

$$\%\text{Reduction} = \left[\frac{400-150}{400}\right] \times 100 = 62.5\%$$

Figure 33 shows the results of calculations similar to section-3 for different values of detention time and overflow rate. Therefore, ideal detention time = 40 min, and overflow rate 1600 gal/ft^2.d.

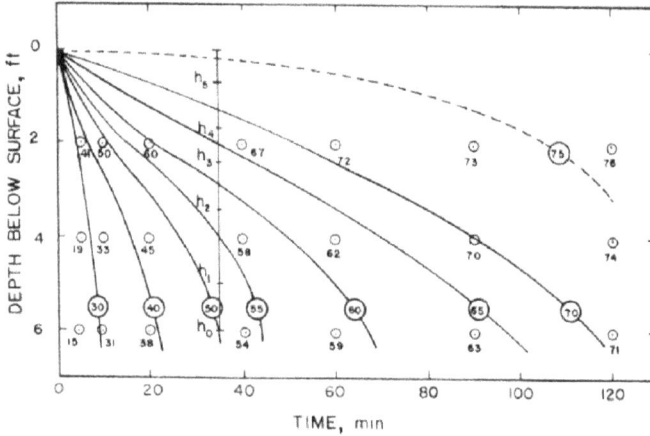

Fig. 32 : Flocculant suspensions batch settling data.

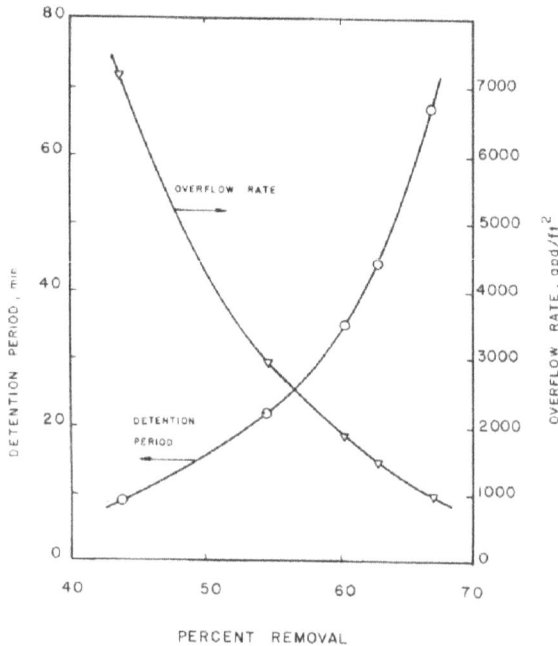

Figure 33 : Detention period and overflow rate vs percent removal.

5. Required under flow (Q_w) [Influent flow, $Q_0 = 1$ MGD]

   Scale–up factor   = 1.25

   Required detention time   = 31.5 × 1.25

              = 39.4 min

   Therefore, overflow rate OR = $\dfrac{2000}{1.25} = 1600$ gal/ft^2.d

   $$X_u = 10000 \text{ mg/L}$$
   $$X_e = 0.5(430) \text{ at } 50\% \text{ removal}$$
   $$Q_e = \frac{Q_0(X_u - X_o)}{(X_u - X_e)}$$
   $$= \frac{1.0(10,000 - 430)}{(10,000 - 215)}$$
   $$= 0.978 \text{ MGD}$$
   $$Q_u = Q_o - Q_e$$
   $$= (1.0 - 0.978) = 0.022 \text{ MGD}.$$

6. Required cross-sectional area of the clarifier

   $$A = \frac{Q_e}{\text{OR}} = \frac{1 \times 10^6 \text{ gal/d}}{1600 \text{ gal/ft}^2.\text{d}}$$

   $$= 625 \text{ ft}^2$$

   $$D = \left[\frac{4 \times 625}{3.14}\right]^{0.5} = 28.2 \text{ ft}$$

   $$H = \frac{\text{Volume of clarifier}}{\text{Cross-sectional area}} = \frac{Q_0(t)}{A}$$

   $$H = \frac{\left(\dfrac{1 \times 10^6 \text{ gal/d}}{1440 \text{ min/d}}\right)(39.4 \text{ min})\left(\dfrac{\text{ft}^3}{7.48 \text{ gal}}\right)}{625 \text{ ft}^2}$$

   $$= 5.8 \text{ ft}.$$

7. Required sludge flow rate

   $$Q_u = 0.022 \text{ MGD} = 22,000 \text{ gal/d of sludge}$$
   $$= (22,000 \text{ gal/d}) (8.34 \text{ lb/gal})$$
   $$= 1,83,480 \text{ lb/d of sludge}$$

   Amount of dry sludge = $(1,83,480 \text{ lb/d})\left(\dfrac{1}{100}\right)$

           = 1834.80 lb of dry solid [at 1% concentrate]

Example 6.96

Sedimentation Basin (Discrete Settling)

Prove that the settling velocity (V_s or V_1) is a function of the horizontal cross–sectional area (A), rather than of the depth (H) for a rectangular or a circular basin.

Solution

1. Required analysis for a rectangular basin

    The rectangular basin consists of four zones:

    - Inlet zone:

        Flow becomes quiescent and it is assumed that at the limit of this zone (along vertical line x t) particles are uniformly distributed across the influent cross–section

    - Sedimentation zone:

        Particles are assumed to be removed from suspension once it hits the bottom of this zone (horizontal line yt)

    - Outlet zone:

        Wastewater collected here prior to transfer to the next treatment

    - Sludge zone:

        It is provided for sludge removal.

        The schematics of sedimentation basin with discrete settling particle is shown as:

t = 0

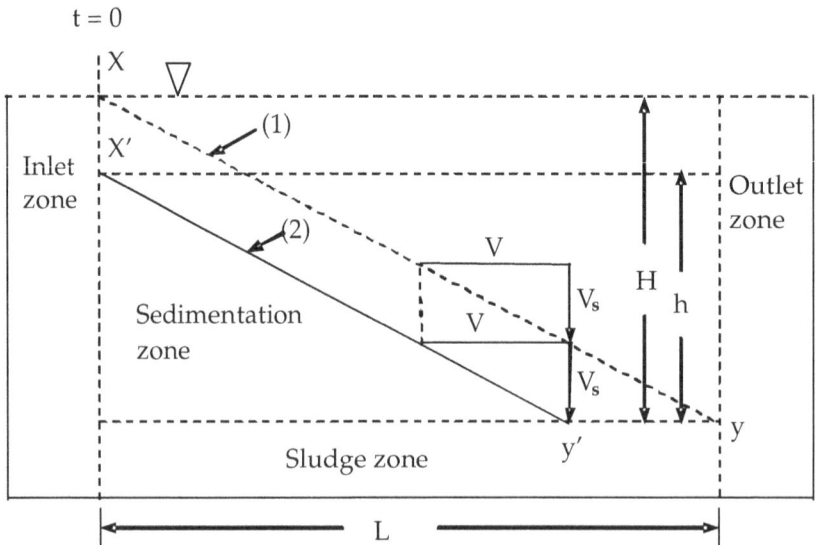

(a)

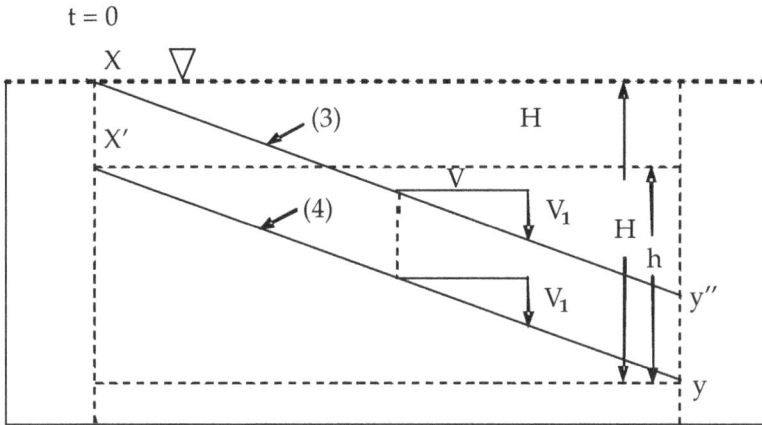

(b)

Figure 34 : Schematics of discrete settling particles in sedimentation basin.

The settling paths are indicated by lines x y (case 1) x' y' (case–2), x y'' (case 3), and x'y (case–4). The settling paths are resultant of two velocity vector components, *viz* :

- Flow though velocity (V) $= Q/A' = \dfrac{Q}{WH}$

  where   A' = Vertical cross–sectional area of sedimentation zone

  W = Width of the sedimentation zone

  H = Height of the sedimentation zone

- Settling velocity, indicated by either vectors V_s or V_1 (constant for discrete particles due to unhindered settling (no coalescence). For flocculent particles (coalescence occurs), the effective diameter of the particles increase resulting in increase of settling velocity (V_s) [thereby the settling paths are curved in comparison with discrete particles with straight line paths].

  Case 1:   The discrete particles from $(x, t) = (x, 0)$ reaches the bottom of the tank at y with a settling velocity V_s (Calculated from Newton's law).

  Case 2:   The discrete particles from $(x, t) = (x', 0)$ reaches the bottom of the tank at y´ with settling velocity V_s (or greater than V_s).

  Case 3:   The discrete particles having settling velocity V_1 but less than V_s and are not removed as they don't touch the bottom of the sedimentation basin.

  Case 4:   The discrete particles with settling velocity V_1 $(V_1 < V_s)$ and situated at x at t = 0 are removed in the sedimentation basin.

Using the property of similar triangular beds to:

$$\frac{V}{V_s} = \frac{L}{H} \qquad [\text{Figure } 34(a)]$$

$$V = V_s \left(\frac{L}{H}\right)$$

or $\qquad \dfrac{V}{L} = \dfrac{V_s}{H}$ $\qquad\qquad\qquad\qquad$ (1)

$$\frac{V}{V_1} = \frac{L}{h} \qquad [\text{Figure } 34(b)]$$

or $\qquad \dfrac{V}{L} = \dfrac{V_1}{h}$ $\qquad\qquad\qquad\qquad$ (2)

Comparing equations (1) and (2) yield:

$$\frac{V_s}{H} = \frac{V_1}{h}$$

or $\qquad \dfrac{V_s}{V_1} = \dfrac{h}{H}$

Flow through the rectangular basin is given by:

$$Q = VA' = VWH$$
$$= V_s\ (L/H)\ (WH)$$
$$= V_s\ LW = V_s\ A$$

where A (= LW) = Horizontal cross–sectional area of the sedimentation basin

or $V_s = \dfrac{Q}{LW} = \dfrac{Q}{A}$ [Independant of sedimentation basin depth]

Some considerations to be taken into account are:

Utilize sedimentation tank of large area (A) and low depths.

Reasonable depths are required to provide for removal of settled sludge (mechanical rakes).

Horizontal component of velocity (flow through velocity V) must be kept within permissible limits to prevent scouring the particles which have settled.

All particles with a settling velocity equal to or greater than V_s are removed.

All particles with settling velocity less than V_s (such as V_1) are removed in a proportion given by a ratio $\dfrac{V_1}{V_s}$.

2. Required analysis for a circular sedimentation basin

The schematics of discrete particle removal in a circular tank is shown as:

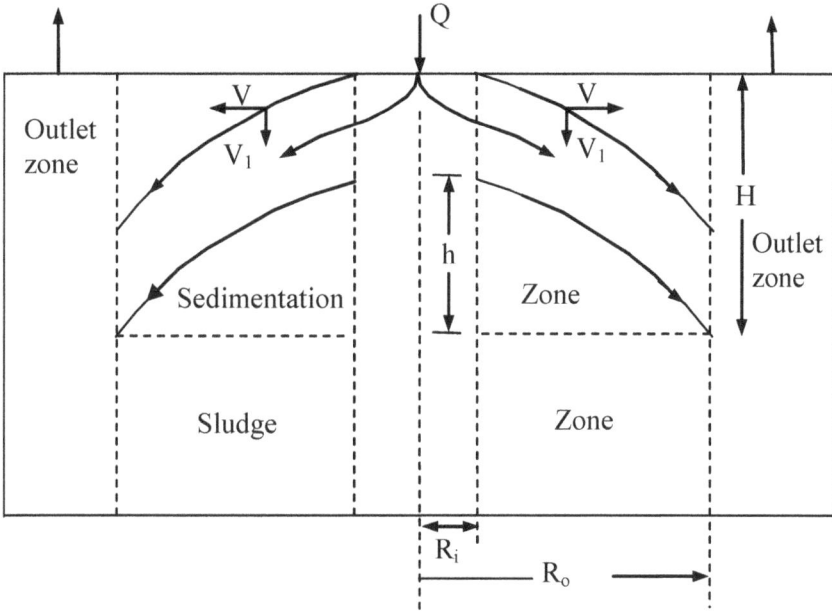

Figure 35 : Schematics of discrete particles removal in a circular basin.

The horizontal component of velocity (V) varies with the radius:

$$V = \frac{Q}{A} = \frac{Q}{2\pi r H} \tag{1}$$

$[R_o \geq R_i]$

Velocity (V) decreases with increasing radius whereas the vertical component (V_s or V_1) is constant for discrete particles, the paths of the particles are curved instead of straight lines. The slope of the settling curve at any radius is:

$$\frac{dh}{dr} = \frac{V_1}{V} \tag{2}$$

Multiplying the numerator and denominator of RHS of equation (2) by $2\pi rH$ yields:

$$\frac{dh}{dr} = \frac{2\pi rHV_1}{2\pi rHV} = \frac{2\pi rHV_1}{Q}$$

or $\quad \dfrac{dh}{H} = \dfrac{2\pi rV_1}{Q} dr$

Integrating, equation (3) :

$$\frac{h}{H} = \frac{2\,\pi\,V_1}{Q}\left[\frac{r^2}{2}\right]_{R_i}^{R_o}$$

$$= \frac{V_1}{Q}[\pi\,R_o^2 - \pi\,R_i^2] = \left(\frac{V_1}{Q}\right)A$$

or $\dfrac{h}{H} = \dfrac{V_1}{(Q/A)}$

Again, $\dfrac{V_1}{V_s} = \dfrac{h}{H}$

Overflow rate $= Q/A = V_s$ [Settling velocity of a particle settling through a distance exactly equal to the effective depth of the tank during the estimated retention time]

Retention time (t) $= \dfrac{HA}{Q}$, [A = LW]

Settling velocity $= \dfrac{H}{t} = \dfrac{H}{(HA/Q)} = Q/A$ [Independent of depth].

3. **Required scour velocity (V_{sc})**

The depth of the sedimentation basin should not be too low since V would rise above the scour velocity. Scour occurs when the flow through velocity (V) is sufficient to suspend the previously settled particles and is not usually a problem in large sedimentation basin [can be a problem in grit chambers and narrow channels].

The scanning velocity (V_{sc}) is the value of the flow–through velocity (V):

$$V = V_{sc} = \frac{Q}{A'} = \frac{Q}{WH}$$

The scour velocity is estimated by using the following empirical equation:

$$V_{sc} = \left[\frac{8\beta\,g\,d(s-1)}{f}\right]^{0.5}$$

where V_{sc} = Scour velocity (mm/s)

[Flow through velocity required to scour all particles of diameter (d) or smaller]

  β = Constant [0.04 (Unigranular sand) and 0.06 (non-uniform sticky material)

  f = Weisbach Darcy friction factor (0.03 for concrete)

  g = Gravity constant (9800 mm/s^2)

d = Particle diameter [mm, particles with diameter (d) less are scoured away]

s = Specific gravity of particles

4. Required settling velocity (V_s) for discrete particles

Newtons law is applicable for the settling velocity [terminal velocity (V_s)] for discrete spherical particles:

$$V_s = \left[\frac{4}{3} \left(\frac{g}{C_D} \right) \left[\frac{\rho_s - \rho_L}{\rho_L} \right] d \right]^{0.5}$$

This equation is derived using initial acceleration is considered to be zero:

Hydrostatic lift = Drag force

$$(\rho_s - \rho_L) g \, v = C_D A \left(\frac{\rho_L V_s^2}{2} \right)$$

where   v = Volume of a particle $\left[\frac{1}{6} \pi d^3 \right]$

A = Projected area of a particle $\left[\frac{1}{4} \pi d^2 \right]$

d = Particle diameter

ρ_s = Density of particle

ρ_L = Fluid density

C_D = Drag coefficient

Drag coefficient is approximately expressed as:

$$C_D = \frac{a}{R_e^n}$$

- Sloke's law $[R_e < 2]$; $C_D = \dfrac{24}{R_e}$

- Transition $[2 < R_e < 500]$; $C_D = \dfrac{18.5}{R_e^{0.6}}$
- Newton's law $[R_e > 500]$; $C_D = 0.4$ [constant]

where $R_e = \dfrac{\rho_L V_s d}{\mu_L}$

μ_L = Dynamic viscosity of fluid

Therefore, the expression for C_D is :

$$C_D = \frac{24}{R_e} + \frac{18.5}{R_e^{0.6}} + 0.4$$

Simplifying for the settling velocity with appropriate value of C_D results in the following expressions:

- Stoke's law $[C_D = 24/R_e]$

$$V_s = \left[\frac{1}{18} \frac{(\rho_s - \rho_L)}{\rho_L} g \right] d^2$$

$$= K\,d^2$$

- Transition $\left[C_D = \dfrac{18.5}{R_e^{0.6}} \right]$

$$V_s = \left[\frac{4g}{55.5} \left(\frac{\rho_L}{\mu_L} \right)^{0.6} \left(\frac{\rho_s - \rho_L}{\rho_L} \right) \right]^{0.7143} d^{1.143}$$

$$= K'd^{1.143}$$

- Newton's law $[C_D = 0.4]$

$$V_s = \frac{4}{3} \frac{g}{(0.4)} \left(\frac{\rho_s - \rho_L}{\rho_L} \right) d^{0.5}$$

$$= \left[3.33g \left(\frac{\rho_s - \rho_L}{\rho_L} \right) d^{0.5} \right]$$

$$= K''d^{0.5}$$

5. Determine the grit chamber surface (horizontal cross–sectional area] for removal of 70 percents of particles [uniform size (d) = 0.07 mm, ρ_s / ρ_L = s = 2.65 at 20°C], as also with another set with a uniformly larger particle size than 0.07 mm [where all particles will be removed]. Estimated the flow–through velocity (V_{sc}) so that all particles of lower settling velocity than those completely removed are scoured away.

5.1 Required horizontal cross–sectional area of the grit chamber when all particles are of 0.07 mm diameter

$$V_s = K'\, d^{1.143}$$

$$= 0.45 \text{ cm/s} = 53.1 \text{ ft/h} = 53.1 \text{ ft}^3/\text{ft}^2.\text{h}$$

Overflow rate $(Q/A) = V_s$

$$= (53.1 \text{ ft}^3/\text{ft}^2.\text{h})(7.48 \text{ gal/ft}^3)(24 \text{ h/d})$$

$$= 9533 \text{ gal/ft}^2.\text{d}$$

Assume wastewater flow (Q) = 1 MGD

$$A = \frac{10^6 \text{ gal/d}}{9533 \text{ gal/ft}^2.\text{d}} = 105 \text{ ft}^2 \text{ for } 100\% \text{ removed}$$

Retention time is reduced by 30% to obtain 70% removal:

$$A = (0.7)(105 \text{ ft}^2) = 73.5 \text{ ft}^2$$

The settling velocity with A = 73.5 ft² for 100% removal becomes:

$$V_s = \frac{Q}{A} = \frac{10^6 \text{ gal/d}}{73.5 \text{ ft}^2}$$

$$= 13605 \text{ gal/ft}^2.\text{d}$$

$$\text{Check: \% removal } = \frac{9533}{13605} = 0.70$$

Settling velocity (V_s) = 13605 gal/ft².d

(100% removed)

Settling velcity (V_l) = 9533 gal/ft².d

(70% removed)

Therefore, $\dfrac{h}{H} = 0.70$

5.2 Required diameter of particle size greater than 0.07 mm

The settling velocity of particle greater than d = 0.07 mm is:

$$\left(\frac{13605}{9533}\right)(0.45 \text{ cm/s}) = 0.642 \text{ cm/s}$$

Given the V_s = 0.642 cm/s, s = 2.65 at 20°C, and using the transition equation for V_s, d can be determined and is equal to 0.085 mm [100% removal].

5.3 Required scour velocity (V_{sc})

The flow through velocity (V or V_{sc}) is given as:

$$V_{sc} = \left[\frac{8\beta \text{ gd}(s-1)}{f}\right]^{0.5}$$

$$= \left[\frac{8(0.04)(9800)(0.07)(2.65-1)}{0.03}\right]^{0.5}$$

$$= 110 \text{ mm/s}$$

Assuming that sand contains only two particle sizes [0.07 and 0.085 mm], the scour velocity (V_{sc}) of 110 mm/s will scour all particles of d = 0.07 mm leaving behind those of d = 0.085 mm.

$$V = V_{sc} = \frac{(110 \text{ mm/s})}{(304.8 \text{ mm/ft})} = 0.36 \text{ ft/s}$$

Vertical cross–sectional area (A′) is given by:

$$V = \frac{Q}{A'} = \frac{Q}{WH}$$

or $\quad A' = \dfrac{Q}{V}$

$$= \left[\frac{(10^6 \text{ gal/d})}{(86400 \text{ s/d})(7.48 \text{ gal/ft}^3)}\right]\left(\frac{1}{0.36 \text{ ft/s}}\right)$$

$$= 4.3 \text{ ft}^2$$

A = 73.5 ft² = LW [Horizontal cross–sectional area]

A′ = 4.3 ft² = WH [Vertical cross–sectional area]

5.4   Determine the vertical (A′) and horizontal (A) cross–sectional areas for following distribution of particle size:

Table 60 : Weight of each particle size with particle size.

Sr. No.	Particle Size (d, mm)	Weight of each Particle Size (lb)
1	0.085	50
2	0.070	20
3	0.060	20
4	0.050	10
Total		100 lb

Wastewater flow (Q) = 1 MGD

Specific gravity (s) = 2.65

Temperature (°C) = 20°C

5.4.1   Required percent removal of various particle size

Table 61 : Particle size with removal (weight).

Sr. No.	Weight of each Particle Size (lb)	Particle Size (d, mm)	Settling Velocity $(V_{s,}$ cm/s)*	Fraction Removed	lb–Removed
1	50	0.085	0.642	$\left(\dfrac{0.642}{0.642}\right) = 1.0$	50 × 1 = 50
2	20	0.070	0.450	$\left(\dfrac{0.450}{0.642}\right) = 0.7$	20 × 0.7 = 14
3	20	0.060	0.350	$\left(\dfrac{0.350}{0.642}\right) = 0.545$	20 × 0.545 = 10.9
4	10	0.050	0.220	$\left(\dfrac{0.220}{0.642}\right) = 0.343$	10 × 0.343 = 3.43
Total	100 lb			Total	78.3 lb

*$V_s = K' \, d^{0.5}$.

5.4.2 Required vertical (A′) and horizontal (A) cross-sectional area of the grit chamber

If vertical cross–sectional area (A′) = 4.3 ft², then all particles of size (d) equal to or less than 0.070 mm will be scoured away. Only particles size d = 0.085 mm will be retained, thereby, the net removal will be 50 lb/100 lb [50% by weight]

If A′ is increased to 2 × 4.3 ft² = 8.6 ft², the value of V_{sc} = V will be equal to:

$$V = V_{Sc} = \frac{Q}{A'} = \frac{10^6 \text{ gal/d}}{8.6 \text{ ft}^2}$$

$$= 0.18 \text{ ft/s}$$

$$= 55 \text{ mm/s}$$

As, V_{sc} = (constant) $d^{0.5}$

Therefore, d = 0.0175 mm and all particles with d = 0.175 mm or less are removal by scouring, and the net removed in the grit chamber will be = 78.3 lb/100 lb = (78.3% by weight).

If   A = 73.5 ft² = LW [Horizontal cross–sectional area]

   A′ = 8.6 ft² = WH [Vertical cross–sectional area]

Assume H = 4 ft,

Therefore, $W = \frac{8.6}{4} = 2.15 \text{ ft}$

$$L = \frac{73.5}{2.15} = 34.2 \text{ ft}$$

Detention time (t) $= \frac{V}{Q}$

$$= \frac{A \times H}{Q}$$

$$= \frac{73.5 \text{ ft}^2 (4 \text{ ft})(7.48 \text{ gal/ft}^3)}{10^6 \text{ gal/d}}$$

$$= 0.0022 \text{ d}$$

$$= 3.17 \text{ min}$$

[Typical detention period is in the range of 20 s to 1 min and particle diameter ranges between 0.1 to 1.0 mm, and in the example the particle size is very small, therefore, larger detention time is required (this is not a typical example)].

Example 6.97

Sedimentation of Discrete Particle [Type I Settling]

Determine the percent removal of suspended solids in an ideal horizontal flow sedimentation basin operating at 1060 $gal/ft^2.d$. Use the laboratory settling data taken on a discrete particle, type – I suspension. Samples were taken 120 and 240 cm below the surface of the liquid in a batch sedimentation column:

Time (min)	0	15	30	45	60	90	180
C/C_0 at 120 cm	1	0.96	0.81	0.62	0.46	0.23	0.06
C/C_0 at 240 cm	1	0.99	0.97	0.93	0.86	0.70	0.32

where C = Suspended solids concentration at time–t taken from the respective sample ports, and C_0 is the initial suspended solids concentration.

Solution

1. Required distribution of settling velocities in the batch sedimentation column

Time (min)	0	15	30	45	60	90	180
h/t at 120 cm	–	8	4	2.67	2.0	1.33	0.67
h/t at 240 cm	–	16	8	5.33	4.0	2.67	1.33

The distribution of settling velocities. (Figure 36)

2. Required fractional removal of particles

$$R = (1 - p_o) + \frac{1}{v_o} \int_0^{P_o} v_s \, dp$$

where   R = Fractional removal of particles

p_o = Fraction of particles having a settling velocity (v_s) less than the basin surface loading rate (v_o)

p_o = Determined from the velocity distribution (Figure 36), and the integral in above equation is also evaluated graphically from Figure 36

$$v_o = \frac{[1060 \ gal/ft^2.d](60)}{(21200)} = 3.0 \ cm/min$$

p_o = 0.75, [Figure 36]

$$\int_0^{P_o} v_s \, dp = 1.19, \quad [\text{Figure 36, Area of the curve}]$$

$$R = (1 - 0.75) + \frac{1}{3.0} (1.19) = 0.647 \text{ or } 64.7\% \text{ removal}$$

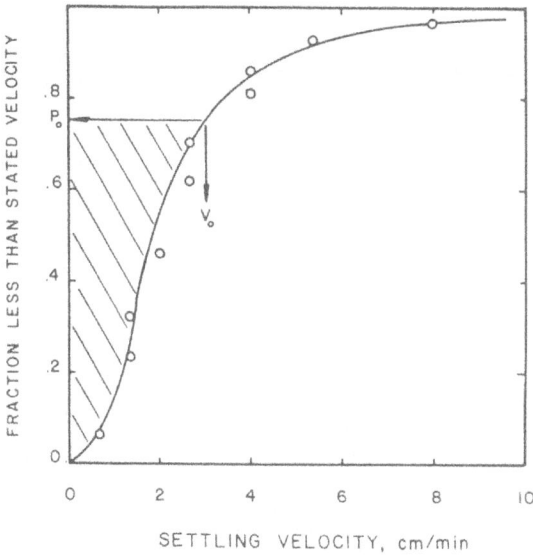

Figure 36 : Distribution of settling elocities

Example 6.98

Ideal Sedimentation

Determine the terminal velocities of single particles, groups of 8 particles, and groups of 64 such particles. Assume the following data:

Particle diameter (d_p) : 0.02 cm

Particle density (ρ_p) : 1.1 g/cm^3

Viscosity of water (μ) : 1 cp

Assume the packing arrangement is such that lines drawn from particle center to particle center are at right angles to each other.

Solution

1. Required force balance on the spherical particles

$$\frac{\pi d_p^3}{6} \rho_p \frac{dv_p}{dt} = \frac{\pi d_p^3}{6} g(\rho_p - \rho_l) - C_D \Delta_p \rho_l \frac{v_p^2}{2}$$

where d_p = Particle diameter

v_p = Particle velocity

ρ_p and ρ_l = Particle and water densities, respectively

A_p = Cross–sectional area of particle

C_D = Drag coefficient

CD is a function of R_e and is given as:

R_e	Less than 0.1	Between 0.1 and 1000	Greater than 1000
C_D	$24\ R_e^{-1}$	$18.5\ R_e^{-0.67}$	0.44

Assume that terminal velocity (v_p) is attained $\left(\dfrac{dv_p}{dt} = 0\right)$

Therefore,

$$v_p^2 = \left(\frac{\pi}{6}\right)\frac{d_p^3\, g(\rho_p - \rho_l)}{C_D A_p \rho_l}\ (2)$$

$$= \frac{\pi}{6}\frac{d_p^3(\rho_l\, v_p\, d_p)}{24\mu}\ \frac{g(\rho_p - \rho_l)(2)}{\dfrac{\pi d^2}{4}(\rho_l)}$$

$$= \frac{d^2 v_p(\rho_p - \rho_l)g}{\mu}$$

Therefore, $v_p = \dfrac{d^2(\rho_p - \rho_l)g}{18\mu}$

Important points are:

    Slowest settling particles are of interest rather than the faster ones.

    Critical particles have densities close to that of water and diameters (effective diameters in case of non–spheres) of less than 1 mm.

    Particle velocity is a function of d_p, ρ_l, ρ_p, and μ.

    ρ_l and μ cannot be changed in most treatment systems, but both the particle diameter and particle density can be changed through the aggregation of small particles into larger ones.

    Bulk–particle density of the aggregated particles decreases as the overall diameter increases due to inclusion of water in the floc. The decrease in density is extremely important as indicated in the above equation. A decrease in density from 1.1 to 1.09 is 0.9 percent decrease in density of the particle but a 10 percent decrease in $(\rho_p - \rho_l)$, and this is compensated by the fact that particle diameters are squared.

2.   Required particle settling velocities (effect of d_p and ρ_p)

    Single particle $(v_p 1)$

$$v_p 1 = \frac{d_p^2\, g(\rho_g - \rho_l)}{18(\mu)}$$

$$= \frac{(0.02)^2(1.1-1.0)}{18(0.01)} = 2.2 \times 10^{-4} \ cm/s$$

Aggregated of 8–particles with maximum packed volume

$$\rho_8 = \rho_p \frac{\pi}{6} + \left(1 - \frac{\pi}{6}\right)\rho_l$$

$$= 1.1\left(\frac{\pi}{6}\right) + \left(1 - \frac{\pi}{6}\right)$$

$$= 1 + \frac{0.1 \times 3.14}{6}$$

$$= 1 + 0.05 = 1.05 \ g/cm^3$$

$$d_8 = \left[\frac{6}{\pi} \ 8(d)^3\right]^{1/3} = 0.05 \ cm$$

$$v_p(8) = \frac{1}{18} \frac{0.05(1.05-1.0)}{(0.01)} = 6.9 \times 10^{-4} \ cm/s$$

Aggregated of 64 particles with maximum packed volume

$$\rho_{64} = 1.05$$

$$d_{64} = \left[\frac{\pi}{6}(64)d^3\right]^{1/3}$$

$$= 0.099 \ cm$$

$$v_p(64) = \frac{1}{18} \frac{(0.099)^2}{0.01} (1.05-1.0)$$

$$= 2.77 \times 10^{-3} \ cm/s.$$

Example 6.99

Dissolved Air Pressure Flotation

Determine the effluent recycle and dimensions of a flotation unit to local waste when the temperature of the latter is 25°C.

Assume the following data:

Waste flow rate	: 500 gal/min
Waste SS	: 800 mg/L
Air/Solids ratio (A/S)	: 0.03
Initial rise rate of the float interface at A/S = 0.03	: 0.5 ft/min

Air released after the clarified effluent is saturated with air at 50 psig : 6.0×10^{-4} lb of air/gal of effluent, the pressure is reduced to 1 atm

Air required at A/S = 0.03 : 0.08 lb/min

Solution

1. Mass rate of suspended solids

   $500 \times 8.34 \times 8 \times 10^{-4} = 3.336$ lb/min

2. Required rate of saturated effluent recycle

   $$= \frac{8 \times 10^{-2}}{6.0 \times 10^{-4}} = 133 \text{ gal/min}.$$

   However, for the retention times ordinarily employed in the retention tank, the air concentration is less than saturation. Assuming that the recycle is only 50% saturated as it flows to the flotation unit, the rate of recycle must be 2 × 133 gal/min = 266 gal/min. Therefore, total flow through the flotation unit will be = 500 + 266 = 766 gal/min.

3. Required surface area

   At a float concentration of 4%, only 6 gal/min of the liquid will remain with the solids. The overflow rate of the clarified liquid from the unit will be (766 – 6) = 760 gal/min. At a rise rate of 0.5 ft/min, the required area will be:

   $$\frac{760}{(0.5)(7.481)} = 203 \text{ ft}^2.$$

4. Required detention time

   Assume a minimum depth of 6 ft, the retention time in the flotation unit will be:

   $$\frac{203 \times 6 \times 7.48}{760} = 12 \text{ min}.$$

Note :

1. Vacuum flotation is not economical because limited quantity of oxygen can be released out of the solution saturated at one atmosphere.

2. The factors of greatest importance in the design of a pressure flotation operation are:

   Feed solids concentration

   Quantity of air used

   Over flow rates

   Retention time.

3. Over flow rates very from 1 to 4 $gal/ft^2.min$.

4. As a rule, a depth of 6 ft is required in the flotation unit to minimize turbulence and short circuiting. For this depth and the overflow rates indicated above, retention time will vary from 10 to 40 min.

5. Air is introduced to the influent at the suction end of a centrifugal pump discharging into a retention tank at a pressure ranging from 25 to 50 psig. Air has a retention time of 30 to 60 seconds and is provided with bleed times for removing any undissolved air that separates in the unit. From the retention tank the waste flows through a back–pressure regulating device to the flotation unit.

Example 6.100

Air Flotation

1. Develop the design equations for air flotation units with and without recycle.

    1.1   Design equations of dissolved air flotation system without recycle

        The schematics of the air flotation unit is shown below:

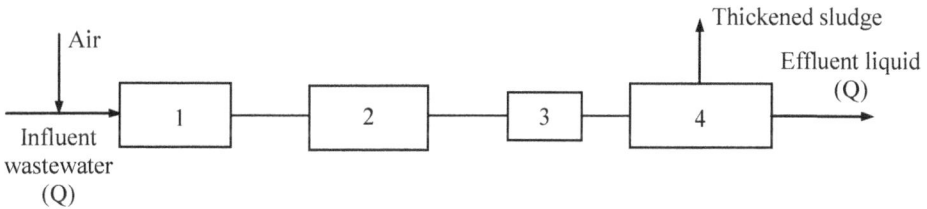

Figure 37 : Schematics of the air flotation system.

    where   1 = Pressuring unit

               2 = Retention tank

               3 = Pressure reducing value

               4 = Flotation unit.

Mass flow rate of dissolved air entering the flotation unit (W_{in}):

$$W_{in} = QC_i$$

Mass flow rate of dissolved air in liquid leaving the flotation unit (W_{out}):

$$W_{out} = QC_e$$

Mass rate of air released for flotation of solids (W_{air}):

$$W_{air} = W_{in} - W_{out} = Q(C_i - C_e)$$

Relationship of dissolved air with pressure:

At low dissolved air concentration where Henry's law is

applicable, the dissolved air concentration at saturation are proportional to air pressures:

$$\frac{C_i}{C_e} = \frac{p_i}{p_e} = \frac{p_i}{1}$$

where p_i is in atmospheres and p_e is taken as 1 atm

Mass rate of air released expressed in pressure is:

$$W_{air} = Q(C_i - C_e) = QC_e(p - 1)$$

Since complete saturation of liquid is often not achieved in the retention tank, a correction factor (f) is applied to the pressure:

$$W_{air} = QC_e(fp - 1)$$

Mass flow rate of suspended solids (X^o) entering with sludge feed (W_{ss}):

$$W_{ss} = QX^o$$

Ratio of air to solid :

$$\frac{W_{air}}{W_{ss}} = \frac{C_e(fp-1)}{X^o}$$

1.2   Design equation of dissolved air flotation unit with recycle

The schematics of the air flotation unit is given below :

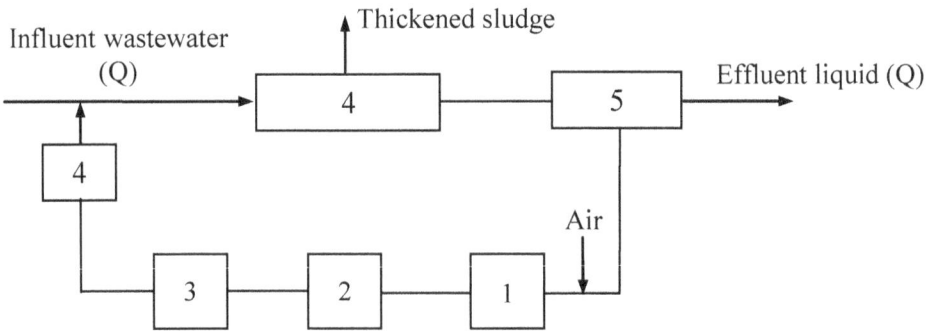

Figure 38 : Schematics of the air flotation unit.

Mass rate of the dissolved air entering the flotation tank (W_{in}):

$$W_{in} = QRC_t + QC_f$$

where C_t = Dissolved air concentration in the retention tank

        C_f = Dissolved air concentration in the fresh feed sludge

Mass rate of dissolved air out of the flotation unit (W_{out}) :

$$W_{out} = Q(1+R)C_e$$

Mass rate of the air released to float solids:

$C_f = C_e$ if both fresh feed and flotation tank effluent are saturated with air at 1 atm

$$W_{air} = W_{in} - W_{out} = QR\,(C_f - C_e)$$

Assuming dissolved air concentrations at saturation are proportional to pressure and applying the correction factor (f), the mass rate of air released is:

$$W_{air} = QRC_e(fp - 1)$$

Ratio of air to solid

$$\frac{W_{air}}{W_{ss}} = \frac{RC_e(fp-1)}{X^o}$$

2. Determine the retention tank pressure, area and solids loading for flotation units, with and without recycle. Assume the following data:

Influent solid concentration   : 0.2%

Effluent solids concentration   : 4.5%

Wastewater flow (Q)   : 0.022 m^3/s

Air to solids ratio   : 0.025

Overflow rate   : 0.5 $L/m^2.s$

Correction factor (f)   : 0.6

Recycle ratio (R)   : 1.5

2.1 Without recycle

Retention tank pressure (p):

$$\frac{W_{air}}{W_{solids}} = \frac{C_e(fp-1)}{X^o}$$

Assume C_e (solubility of air in water) at 20°C = 22.4 mg/L

$$0.025 = \frac{22.4(0.6\,p-1)}{2000}$$

$$p = \frac{1}{0.6}\left[\frac{0.025(2000)}{22.4}+1\right]$$

$$= 5.39 \text{ atm}$$

Flotation tank area (A)

$$A = \frac{0.022 \text{ m}^3/s}{0.5 \text{ L/m}^2.s\,(10^{-3} \text{ m}^3/L)}$$

$$= 44 \text{ m}^2$$

Solids loading (SL)

$$SL = \frac{(2 \text{ kg/m}^3)(0.022 \text{ m}^3/\text{s})(3600 \text{ s/hr})}{44 \text{ m}^2}$$

$$= 3.6 \text{ kg/m}^2.\text{hr}$$

With recycle

$$\frac{W_{air}}{W_{solids}} = \frac{RC_e(fp-1)}{X^o}$$

$$0.025 = \frac{1.5(22.4)(0.6p-1)}{2000}$$

$$p = \frac{1}{0.6}\left[\frac{0.025(2000)}{1.5(22.4)}+1\right]$$

$$= 4.15 \text{ atm}$$

Flotation tank area (A)

$$A = \frac{(1+1.5)(0.022 \text{ m}^3/\text{s})}{0.5 \text{ L/m}^2.\text{s} \, (10^{-3} \text{ m}^3/\text{L})}$$

$$= 110 \text{ m}^2$$

[Total flow rate entering the flotation unit is the fresh feed flows the recycle]

Solids loading (SL)

Assume solids concentration in the recycle stream is negligible

$$SL = \frac{(2 \text{ kg/m}^3)(0.022 \text{ m}^3/\text{s})(3600 \text{ s/hr})}{110 \text{ m}^2}$$

$$= 1.44 \text{ kg/m}^2.\text{hr}.$$

Example 6.101

Flotation Systems in Wastewater Treatment

Derive mathematical expressions for separation of solid–liquid flotation technique.

Solution

Flotation has been applied in separating particles from liquids in grease removal and in ore or coal separation. In recent years, many have advocated the treatment of sewage and industrial wastes and the purification of water by flotation. The process of flotation has previously been developed on an empirical basis, although it has been well-known that there are theories, which explain the phenomenon of flotation. These theories are the following: "electrical theory," "contact angle theory," "gas theory," and "adsorption theory." Mechanically, flotation is a reverse process of sedimentation. This chapter discusses the mathematical aspect of the mechanism of flotation.

1. Mathematical Derivation

The mathematical explanation of the phenomenon of flotation will be limited in this chapter to discrete particles without the interference of surface–active foaming agents. Let F be the constant of proportionality of the retarding force against the floating of the particle, X the accelerating force of flotation, m the mass of the particle, ρ_1 the density of particle, ρ the density of liquid having a viscosity coefficient μ, V_g the velocity of falling, V_r the velocity of rising, x the displacement of the particle rising under the force of flotation, and t the time. Then it is possible to write the equation of motion for the rising particle having D as its diameter:

$$m\frac{d^2x}{dt^2} = F\frac{dx}{dt} + X \tag{1}$$

This equation is well-known to physicists studying the flotation of a charged particle under the effect of an electrical force and a gravitational force such as in the Wilson's Cloud Chamber. It is also familiar to methamaticians in the problems of a damping body under a retarding or an accelerating force. At the steady state of equilibrium, when the particle reaches the top surface of the liquid, equation (1) becomes:

$$0 = -F\frac{dx}{dt} + X, \text{ or } \frac{dx}{dt} = \frac{X}{F} \tag{2}$$

The falling velocity of the particle therefore is

$$V_g = \frac{X}{F} = \frac{m'g}{F} \tag{3}$$

where m'g is the gravity force minus the buoyant force acting on the particle without other external force and is equal to

$$m = \left(\frac{\rho_1 - \rho}{\rho_1}\right)g$$

when the particle is rising under an upward mechanical force $P\left(\dfrac{\pi D^2}{4}\right)$

(where P is the effective pressure lifting the particle), the average rising velocity will be:

$$V_r = \frac{\left(\dfrac{P\pi D^2}{4} - m'g\right)}{F} \tag{4}$$

Combining equations (3) and (4), it is found that:

$$V_r = \frac{\left(\frac{P\pi D^2}{4} - m'g\right)}{V_g/m'g}$$

$$= \left(\frac{P\pi D^2 \cdot \rho_1}{4m(\rho_1 - \rho)g} - 1\right) V_g \tag{5}$$

Applying Stock' law to the settling of particle in the liquid as in equation (3),

$$V_g = \frac{g}{18\mu}((\rho_1 - \rho).D^2 \tag{6}$$

Then

$$V_r = \left[\frac{P\pi D^2 \cdot \rho_1}{4m\,(\rho_1 - \rho)g} - 1\right]\frac{g}{18\mu}((\rho_1 - \rho).D^2 \tag{7}$$

Since　　$m = \dfrac{\pi D^3 \rho_1}{6}$

$$V_r = \left[\frac{3P}{2D(\rho_1 - \rho)} - 1\right]\frac{g}{18\mu}((\rho_1 - \rho).D^2$$

$$= \frac{PDg}{12\mu} - \frac{g}{18\mu}((\rho_1 - \rho).D^2 \tag{8}$$

Equation (8) indicates that the velocity of rising of the particle increases with the size of the particle and the applied uplift pressure, but will be decrease as μ increases or when ρ_1 increases. In other words, V_r is large when P and D are reasonably large and when m and ρ_1 are reasonably small. Or, it can be said that light particles of large sizes will ascend more rapidly than heavy particles of smaller sizes.

Let f be the time for a particle to ascend per unit length. Then f equals $1/V_r$, or

$$\frac{18\mu}{\left[\dfrac{3P}{2D(\rho_1 - \rho)} - 1\right]g(\rho_1 - \rho).D^2} = \frac{18\mu}{\left[\dfrac{3PDg}{2} - g(\rho_1 - \rho).D^2\right]}$$

This shows that f decreases as P and D increase and that it increases as μ and ρ_1 increase.

2. In the case of flotation applied to a liquid flowing through a tank (say a rectangular one), the relation between the velocity of horizontal flow of the liquid and the velocity of rising of the particles may be explained by reference to Figure 39.

Let V_h be the velocity of fluid flow, Q the quantity of flow per unit time, X_1 the depth of the tank, A_h the surface area of the tank, L its horizontal length, ϕ its effective volume, and t the designed maximum rising time for a particle to reach the top of the fluid in the tank under the condition explained in equation (5). Therefore t is equal to X_1/V_r. The detention time for the particle to rise from the bottom of the tank (when the fluid enters the tank) to a point when it leaves, is T which depends upon Q and is equal to ϕ/Q.

The ratio of rising particles to total particles therefore equals x/X_1 or T/t. It must be noted that for the particles to reach the top of the tank T is ϕ/Q or X_1A_h/Q and that t is X_1/V_r. Then the ratio of T/t can written as follows:

$$\frac{T}{t} = \frac{X_1A_h/Q}{X_1/V_r} = \frac{V_r}{Q/A_h} \tag{9}$$

The ratio T/t thus represents the ratio of maximum removal of solid by floatation.

It also can be shown that

$$X = \int_0^t V_r\, dt = \left[\frac{PDg}{12\mu} - \frac{g}{18\mu}((\rho_1 - \rho).D^2\right].t \tag{10}$$

$$X \sim \left[\frac{PDg}{12\mu} - \frac{g}{18\mu}((\rho_1 - \rho).D^2\right]\frac{L}{V_h}$$

$$\frac{X}{L} \sim \frac{\left[\frac{PDg}{12\mu} - \frac{g}{18\mu}((\rho_1 - \rho).D^2\right]}{V_h} \tag{11}$$

$$\frac{X}{L} \sim \frac{V_r}{V_h} \tag{12}$$

and $L = V_h t$, neglecting the horizontal resistance offered by fluid.

3. Let C be the solids concentration in the fluid, and dC the change of C due to the ascent of solid particles during the time element dt of flotation. A small column of liquid of an area dA_h extending from the bottom of the tank to the top (the depth of the tank is X_1 feet) will then have a total amount of solids $C X_1. dA_h$. The ratio of change of solids in the liquid due to the change of solids concentration dC caused by the rising of solid particles will be :

$$\frac{dC.X_1.dA_h}{C.X_1.dA_h} = \frac{dC}{C}$$

and

$$\frac{dC}{C} = -\frac{V_r.dt}{X_1} \tag{13}$$

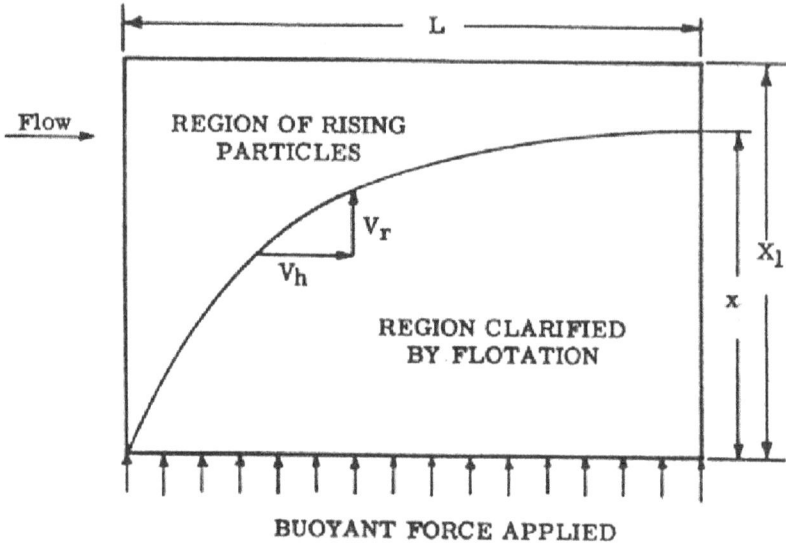

Figure 39 : Theoretical clarification in an ideal flotation tank.

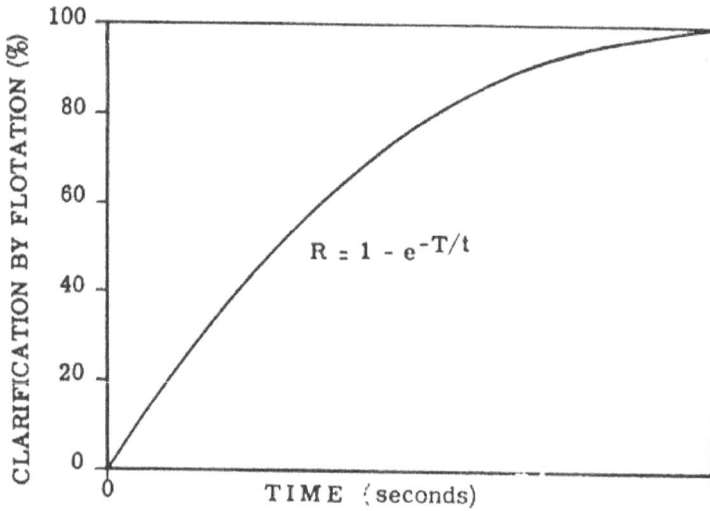

Figure 40 : Flotation efficiency as influenced by time.

because the bottom is clarified by flotation through a distance Vr.dt which has the amount of solids reduced equal to the amount of solids removed at the top of the tank. The negative sign indicates the decrease of solids concentration as time proceeds. Assuming that the initial solids concentration at the time of entering the flotation tank is C_i and that at the time of leaving the tank it is C_0 then for a detention of T it can be shown that :

$$\int_{C_i}^{C_0} \frac{dC}{C} = -\int_0^T \frac{V_r}{X_1} . dt$$

or $$\ln \frac{C_0}{C_i} = -\frac{V_r . dt}{X_1}$$

and $$\ln \frac{C_0}{C_i} = \exp\left[-\frac{V_r . dt}{X_1}\right]^*$$

If C_i is large, T must be reasonably large to have a minimum $\frac{C_0}{C_i}$ required for best efficiency. However, if C_i is small (low turbidity), T must be extremely large in order to obtain a high efficiency of solids removal. The ratio of total removal of solids after flotation is $1 - \frac{C_0}{C_i}$, and this ratio become Figure 40 :

$$R = 1 - \exp\left[-\frac{V_r T}{X_1}\right]$$

$$R = 1 - \exp\left[-\frac{T}{t}\right]$$

$$R = 1 - \exp\left[-\frac{V_r}{Q/A_h}\right]$$

$$R = 1 - \exp\left[-\left(\frac{X_1}{Q/A_h} . \frac{1}{X_1/V_r}\right)\right]$$

$$R = 1 - \exp\left[-\frac{X_1}{Q/A_h} . \frac{1}{t}\right] \qquad (14)$$

which shows that the ratio of solids removal is governed by the effective depth of the tank, the overflow rate, and the time provided for the particles to reach the surface of the fluid from the bottom of the tank.

When X_1 = 80 ft and Q/A_h = 2.0 gpm per sq ft, equation (14) can be solved for the values of R vs. t plotted in Figure 41.

If r is substituted for X_1 and t′for r/V_r (designating r for the hydraulic radius of the tank which can be assumed to be in the form of an open channel), then the ratio of solids removal can be written:

$$R = 1 - \exp\left[-k\left(\frac{T}{t'}\right)\right]$$

$$R = 1 - \exp\left[-k\left(\frac{r}{Q/A_h} \cdot \frac{1}{r/V_r} \right) \right] \tag{15}$$

which indicates the importance of the hydraulic radius of the tank in the efficiency of solids removal by flotation. In this expression k is a constant of the ratio r/X_1.

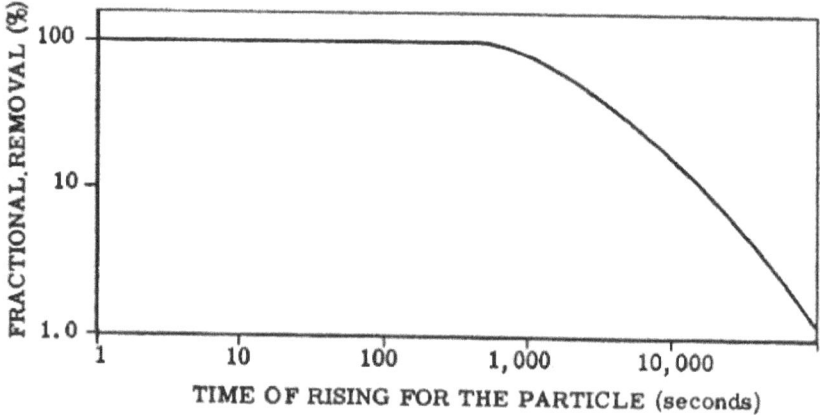

Figure 41 : Example of theoretical fractional removal of solid particles having various required rising times in an ideal floatation tank

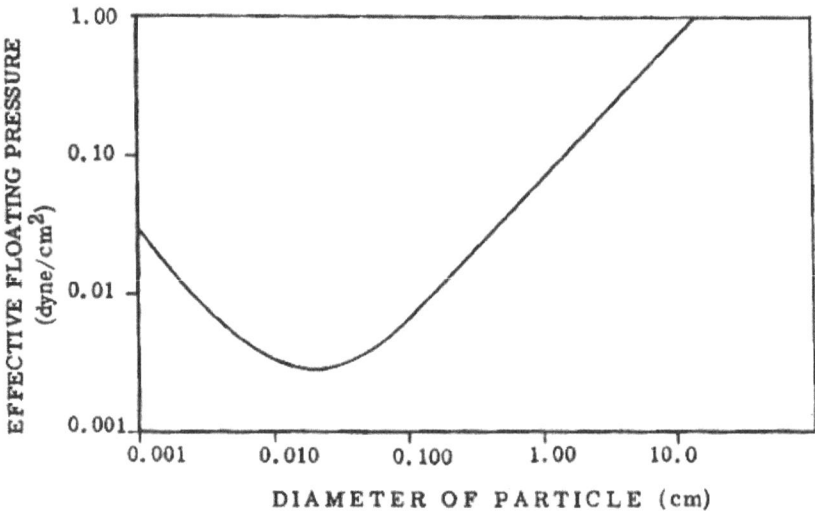

Figure 42 : Example of relation of effective floating pressure and particle size for same fractional removal $(r_1 > r)$

Equation (14) may also be applied to flotation in circular tanks, except that A_h must be the surface area of the particular circular tank and X_1 its effective depth. However, equation (15) is applied strictly to a channel–shaped flotation tank.

The design of an efficient flotation tank must be based on the possible maximum removal of solids, which can only be obtained from data after carefully planned experiments to be interpreted by equations (14) or (15).

In the application of flotation with the aid of surface–active agents, the reduction of solids may be due to the decrease of surface tension or to the increase in the size of the particles; the latter may result from the combination of smaller particles into larger ones either through an electrical process or adsorption. Since the physical properties of the active agents in the process of flotation are not fully known, the mathematical expression will not be considered here.

4. It must be remembered that the removal of solids by flotation is governed by P, D, m, μ, ρ_1, ρ, Q, A_h and X_1. From equation (14) the entire flotation mechanism can be written to describe the total fraction of removal:

$$R = 1 - \exp\left[-\left(\frac{X_1 V_r}{QX_1/A_h}\right)\right]$$

$$R = 1 - \exp\left[-\left(\frac{X_1}{QX_1/A_h}\right)\left(\frac{PDg}{12\mu} - \frac{g}{18\mu}C(\rho_1 - \rho).D^2\right)\right] \tag{16}$$

$$R = 1 - \exp\left[-\left(\frac{X_1}{QX_1/A_h}\right)\left(\frac{PDg}{12\mu}\right) + \left(\frac{X_1 V_r}{QX_1/A_h}\right)\left(\frac{g}{18\mu}(\rho_1 - \rho).D^2\right)\right] \tag{17}$$

For a liquid containing particles of uniform sizes, the total fraction of removal

$$R = 1 - \exp\left[-\frac{C_1}{\mu}\left(\frac{X_1}{Q/A_h}\right)\left(\frac{P}{X_1}\right) + \frac{C_2}{\mu}\frac{(\rho_1 - \rho)}{Q/A_h}\right]$$

$$R = 1 - \exp\left[-\frac{1}{\mu}\left(\frac{X_1}{Q/A_h}\right)(C_1 P - C_2(\rho_1 - \rho))\right] \tag{18}$$

When the densities of the solid particles and liquid do not change, then the total fraction of removal

$$R = 1 - \exp\left[-\text{Constant}\left(\frac{1}{\mu}\right)\left(\frac{X_1}{Q/A_h}\right)\left(\frac{P}{X_1}\right)\right] \tag{19}$$

The use of equation (16) can be illustrated by the following example:

$$g = 980 \text{ cm/sec}^2$$

μ (at 20°C) = 0.01009 poise

$$\frac{A_h}{Q} = 7.2 \text{ sec/cm}$$

$$R = 0.80$$

$$X_1 = 244 \text{ cm}$$

Required P = Effective $P + X_1\rho$

$$\rho = 1.00 \text{ gm/cm}^3$$

Figure 42 illustrates the solution of equation (16) when ρ_1= 1.10 gm/cm³ ($\rho_1 > \rho$); Figure 43 illustrates the solution when ρ_1= 0.90 gm/cm³ ($\rho_1 < \rho$).

Naturally, the second term of the last expression of equation (16) must be as small as possible in order to achieve a high efficiency of removal of the original suspended solids in the raw water or raw waste liquid. equation (18) also shows that the efficiency of flotation depends greatly on the quantity ($\rho_1 < \rho$) and μ which in turn may show the effect of temperature on flotation, for the density and viscosity vary at different temperatures. If the properties of water or liquid and its contents such as m, D, ρ_1, ρ and μ do not vary too much in a specific liquid being treated, then the factors governing the efficiency of removal will be only X_1, P, Q, and A_h. In other words, the efficiency of flotation depends on the depth of the tank, the average effective applied pressure of diffusion per unit length of depth, and the overflow rate of the tank. This is indicated in equation (19).

However, in the case of industrial wastes which may have a varying concentration of solids, the concentration of solids in the liquid is indirectly explained and considered in terms of in equation (19). For a liquid–solids mixture having particles of different sizes, the average efficiency of solids removal can be evaluated from equation (16), and it can be shown that

$$R = \frac{a_1 R_1 + a_2 R_2 + \ldots + a_n R_n}{a_1 + a_2 + \ldots + a_n} \tag{20}$$

where $a_1, a_2, \ldots a_n$ are the respective percent of solid of different size and $R_1, R_2, \ldots R_n$ are the respective specific efficiencies of removal. However, since $a_1 + a_2 + \ldots + a_n = 1$, equation (20) can be written as :

$$R = \sum_{n=1}^{m} a_n R_n$$

5. Practical Applications

As previously stated, flotation has been used to separate particles from liquids in grease removal and in the separation of ores and coal. For flotation of substances heavier than water, mechanical means or compressed air must be used to produce small air bubbles, which buoy the particles to the surface of the liquid. Special chemical agents such as sulfur or nitrogen compounds and foaming agents such as oil, resin or glue are used to increase the efficiency of flotation.

In the case of water purification by flotation, Hopper et al. reported successful results using certain quaternary ammonium compounds. They reported 100 percent removal of cysts and 99 percent removal of bacteria in raw water samples containing 15 to 300 ppm turbidity. The removal of turbidity was about 70 percent. In their experiments, 10 ppm of certain ammonium compounds were used, and the clear water contained less than 1 ppm. The mechanical part of flotation was accomplished by the application of air to the liquid at a rate of 1.5 litres per min through fine air diffusers (1 to 2 mm diameter) in 3 to 5 minutes.

Gibbs reported that 96 percent of the 300 ppm undissolved fatty acids in a liquid was removed by flotation. The overflow rate of the flotation process was 3 gal per sq ft per minute.

The removal of microorganisms by flotation, without the addition of active chemical compounds is most probably similar to the removal of solids by the same method. However, when the surface–active agents (particularly those which are toxic to microorganisms) are used, the removal is not only mechanical but also bactericidal. Physical phenomena such as adsorption or adhesion may also be involved.

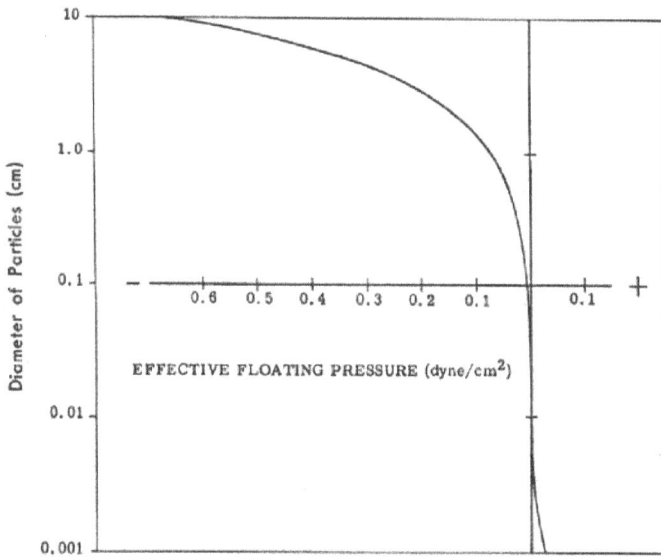

Figure 43 : Example of relation of effective floating pressure and particle size for same fractional removal ($r_1 < r$).

Example 6.102

Gravity Separator (Decanter)

In the manufacture of 2–ethyl hexanol by low pressure oxoprocess, n–butyraldehyde is reacted with 2% by weight of sodium hydroxide solution.

Approximately 90% of the butyradehyde is converted to 2–ethyl hexanol. The mixture of the organic phase and aqueous phase is separated in a decanter. Assume the following data :

Rated production capacity of 2–ethyl hexanol	: 60 tonne/d
Organic phase flow rate	: 2883.5 kg/h
Aqueous phase flow rate	: 425.4 kg/h
Density of organic phase	: 830 kg/m^3
Density of aqueous phase	: 1050 kg/m^3
Viscosity :	
Organic phase	: 6.5×10^{-3} N.s/m^2
Aqueous phase	: 1.1×10^{-3} N.s/m^2

Solution

1. Required schematics of the decanter

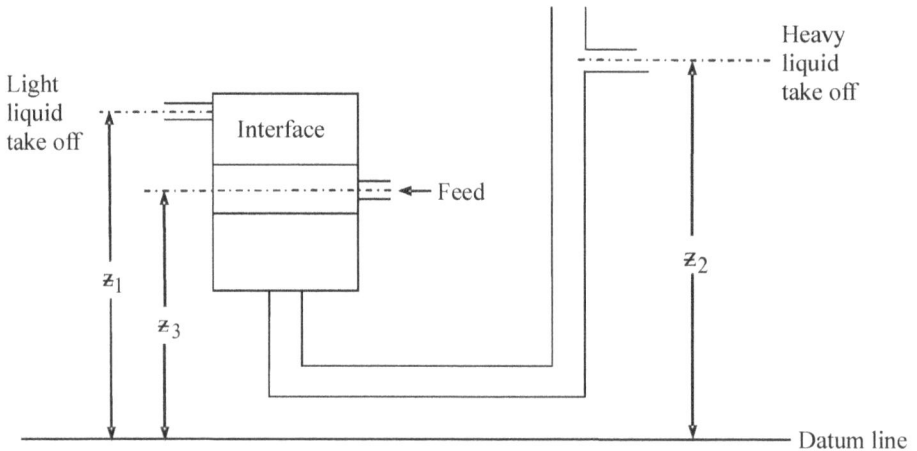

Figure 44 : Schematics of the decanter

2. Required velocity of the dispersed phase (V_d)

   Aqueous phase flow rate is much smaller, and can be considered as dispersed phase

   Assume a droplet diameter of 120 mm

   Dispersed phase velocity is :

   $$V_d = \frac{d^2(\rho_d - \rho_c)g}{18\mu_c}$$

$$= \frac{(120 \times 10^{-6})^2 (1050 - 830)(9.81)}{18(6.47 \times 10^{-3})}$$

$$= 2.67 \times 10^{-4} \text{ m/s}$$

3. Required area of interface (A_i)

   Mass rate of the continuous phase = 2883.5 kg/h

   $$\text{Volumetric flow rate } (Q_c) = \frac{2883.5 \text{ kg/h}}{(830 \text{ kg/m}^3)(3600 \text{ s/h})}$$

   $$= 9.65 \times 10^{-4} \text{ m}^3/\text{s}$$

   Area of interface (A_i) is :

   $$A_i = \frac{Q_c}{V_d}$$

   $$A_i = \frac{9.65 \times 10^{-4} \text{ m}^3/\text{s}}{2.67 \times 10^{-4} \text{ m/s}} = 3.614 \text{ m}^2$$

4. Required sizing of the decanter

   $$A_i = \frac{\pi D^2}{4}, \text{ or}$$

   $$D = \left[\frac{4(3.614)}{3.14}\right]^{0.5} = 2.145 \text{ m} \cong 2.15 \text{ m}$$

   Assume Ratio of $\frac{H}{D} = 2.0$

   Height of decanter (H) = 2(D)

   $$= 2(2.15) = 4.30 \text{ m}$$

   Height of the dispersion band can be considered as 5% of the total height of the limit:

   Dispersion band height = 0.05 (4.30 m)

   $$= 0.215 \text{ m}$$

   Residence time (t) of the droplets in dispersion band is :

   $$t = \frac{0.215 \text{ m}}{V_d} = \frac{0.215 \text{ m}}{2.67 \times 10^{-4} \text{ m/s}}$$

   Size of the aqueous phase droplet (d) that can be entrained with the organic phase is given as:

   $$d = \left[\frac{18 V_d \mu_d}{g(\rho_d - \rho_c)}\right]^{0.5}$$

$$= \left[\frac{18(3.11\times10^{-5})(1.1\times10^{-3})}{9.81(1050-830)}\right]^{0.5}$$

$= 16.9 \times 10^{-6}$ m $=16.9$ μm

[It is much below the assumed value and is, therefore, satisfactory]

Position of the interface can be fixed at the central portion of the unit

Organic phase removal can be arranged at 90% of the decanter height.

Therefore, $z_1 = 0.9(4.3$ m$) = 3.87$ m

$z_3 = 0.5(4.3$ m$) = 2.15$ m

$$z_2 = \left[\frac{3.87-2.15}{1050}\right]+2.15$$

$= 3.15$ m

Example 6.103

Dissolved Air Flotation

Determine the recycle rate, the surface area of the flotation unit and the total sludge produced (emulsified oil and alum sludge) for treating refinery wastewater. It is necessary to reduce the oil and grease content to 15 mg/L.

Assume the following data:

Influent oil and grease concentration	: 200 mg/L
Required alum does	: 100 mg/L
Required pressure for the flotation unit	: 70 psig (84.7 psia)
Required sludge production	: 1.0 mg/mg Alum
Sludge concentration	: 4% (by weight)
Wastewater flow (Q)	: 100 gal/min
Air to solid ratio (A/S)	: 0.065 lb air reduced/lb solids applied

Solution

1. Required recycle rate (R)

$$\frac{A}{S} = \frac{S_a\,R}{S_a\,Q}\left(\frac{fP}{P_a} - 1\right)$$

or $\quad R = \dfrac{(A/S)\,(Q)\,S_a}{S_a\left(f\dfrac{P}{P_a} - 1\right)}$

where P is the absolute pressure, P_a is the atmosphere pressure, f is the fraction of saturation in the retention tank, Q is the wastewater flow, S_a is the influent oil and/or suspended solids, A/S is the air to solid ratio, and s_a is the air saturation at atmosphere pressure

$$R(\text{at} = 30^{\circ}C) = \frac{(0.065)\,(100)\,(200)}{(21)\left[0.8\left(\frac{84.7}{14.7}\right) - 1\right]}, \quad f = 0.8$$

$$= 17.15 \text{ gal/min}$$

2. Required surface area (A)

$$A = \frac{\text{Flow rate}}{\text{Loading rate}}$$

$$= \frac{(Q + R)}{\text{Loading rate}}$$

$$= \frac{100 + 17.15}{3.0}$$

$$= 39 \text{ ft}^2$$

[Loading is determined through experiment, plotting effluent oil and grease concentration versus surface loading rate and is equal to 3.0 gal/ft².min].

3. Required sludge production

Alum sludge = (100 mg/L)(1.0 mg/mg Alum)

$$= 100 \text{ mg/L of sludge}$$

$$\text{Alum sludge production rate} = \frac{(100 \text{ mg/L})(100 \text{ gal/min})(1440)(8.34)}{10^6}$$

$$= 120 \text{ lb} - \text{sludge/d}$$

$$\text{Oil and grease sludge} = \frac{(200 - 15)(100)(1440)(8.34)}{10^6}$$

$$= 222 \text{ lb/d.}$$

Total sludge = (222 + 120) = 342 lb/d.

$$\text{Total sludge volume} = \frac{342 \text{ lb/d}}{0.04 \text{ lb/lb}}$$

$$= \left(\frac{342}{0.04}\right)\left(\frac{\text{gal}}{8.34 \text{ lb}}\right)$$

$$= 1025 \text{ gal/d}$$

Note : 1.

Table 62 : Solubility of air*

Temperature		Volume Solubility		Weight Solubility		Density	
°C	°F	mL/L	ft³/1000 gal	mg/L	lb/1000 gal	g/L	lb/ft³
0	32	28.8	3.86	37.2	0.311	1.293	0.0808
10	50	23.5	3.15	29.3	0.245	1.249	0.0779
20	68	20.1	2.70	24.3	0.203	1.206	0.0752
30	86	17.9	2.40	20.9	0.175	1.166	0.0727
40	104	16.4	2.20	18.5	0.155	1.130	0.0704
50	122	15.6	2.09	17.0	0.142	1.093	0.0682
60	140	15.0	2.01	15.9	0.133	1.061	0.0662
70	158	14.9	2.00	15.3	0.128	1.030	0.0643
80	176	15.0	2.01	15.0	0.125	1.000	0.0625
90	194	15.3	2.05	14.9	0.124	0.974	0.0607
100	212	15.9	2.13	15.0	0.125	0.949	0.0591

* In absence of waster vapour and at 14.7 psia (1 atm).

Example 6.104

Settling Basin for Type–2 Settling

Determine the size of the long–rectangular settling basin, as also the circular settling basin. Assume the following data:

Water flow   : 10,000 m³/d

Overflow rate : 20 m³/m².d

Tank depth   : 3 m

Solution

1. Required size of the long–rectangular settling basin

    Assume two basins with a flow of 5000 m³/d

    $$\text{Surface area} = \frac{5000\,\text{m}^3/\text{d}}{20\,\text{m}/\text{d}} = 250\,\text{m}^2$$

    Assume length to width ratio of 3:1

    W × 3W = 250 m²

    $$W = \left(\frac{250}{3}\right)^{0.5} = 9\,\text{m}$$

    L = 27 m

Check the detention time (t)

$$t = \frac{\text{Volume}}{\text{Flow rate}} = \frac{9 \times 27 \times 3 \, \text{m}^3}{5000 \, \text{m}^3/\text{d}} = 3.5 \, \text{h}$$

Check for horizontal velocity (v_h)

$$v_h = \frac{(5000 \, \text{m}^3/\text{d})\left(\dfrac{d}{24h}\right)}{(9\text{m} \times 3\text{m})} = 7.7 \, \text{h} \, [\text{Greater than 3.5h}]$$

Check for weir loading rate [Simple weir is located across the end of the basin]

$$= \frac{5000 \, \text{m}^3/\text{d}\left(\dfrac{d}{24h}\right)}{9\text{m}} = 23 \, \text{m}^3/\text{m.h}$$

Normally it ranges between 6 $\text{m}^3/\text{m.h}$ [light flocs to 14 $\text{m}^3/\text{m.h}$ [heavier flocs]

Therefore, four times this length will be required (weir length)

Schematics of the operation is :

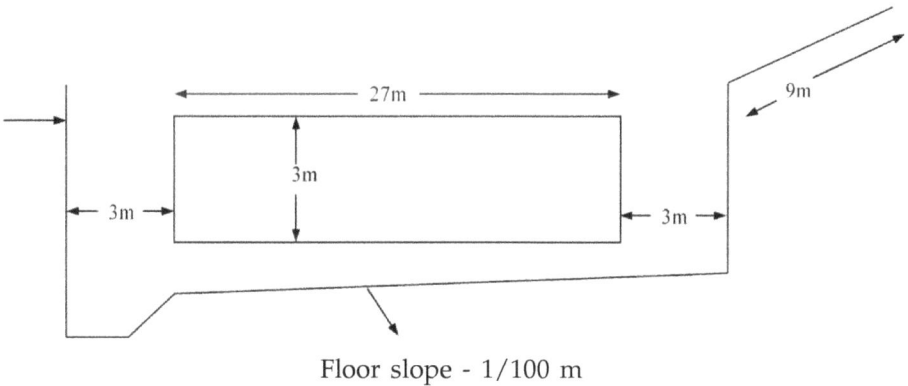

Floor slope - 1/100 m

Figure 45 : Schematics of system.

Total depth = 3 m (settling zone) + 0.5 m (free board) + 0.5 m (sludge zone) = 4.0 m

2. Required circular settling basin dimension

Two basins are required

Surface area of each basin = 250 m²

Diameter of the basin is :

$$d = \left[\frac{250(4)}{3.14}\right]^{0.5} = 18\text{m}$$

Schematics of the circular settling basin

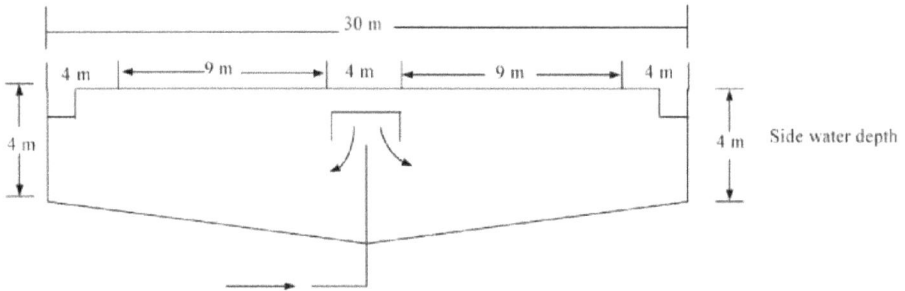

Figure 46 : Schematics of circular settling basin.

Note :

1.   Rectangular settling basin

Depth : 2.5 to 3 m (discrete particles)

: 3 to 4 m (flocculating particles)

: > 12 m, creates problems with sludge removal equipment

Lengths to be kept less than 48 m

Overflow rates : 1.0 to 2.5 m/h [discrete particles]

: 0.6 to 1.0 m/h [flocculating particles]

Detention time : 2 to 4 hours [discrete particles]

: 4 to 6 hours [flocculating particles]

Horizontal velocity (v_h) and the weir flow rate (qw) are equally important

Motion of the sludge removal mechanism may momentarily respond lighter particles, and flocs a few centimeters above the scraper blades, since excessive horizontal velocities would move this material progressively towards outlet zone where it would be lost in the overflow, horizontal flow velocity should not exceed 9.0 m/h for light flocculent suspensions or about 36 m/h for heavier discrete–particle suspension

Large weir overflow rates results in excessive velocities at the outlet. These velocities extend backward into the settling zone, causing particles and flocs which would otherwise be removed as sludge to be drawn into the outlet. Overflow rates ranging from 6 m^3/.m.h for light flocs to about 14 m^3/m.h for heavier discrete–particle suspensions are commonly used. It may be necessary to provide special in–board weir design to accommodate the lower weir over flow rates.

2. Circular settling basin

   Horizontal velocity of the wastewater is continually decreasing as the distance from the centre increases [Central inflow for radical type settler]. A discrete particle with a settling velocity (V_o) is continually undergoing a change in its absolute velocity due to decrease in horizontal velocity. The particle path in a circular basin is a parabola as opposed to the straight particle path line in the long rectangular basin.

   Sludge removal mechanisms are simpler and require less maintenance. Excessive weir overflow should never be a problem because the entire circumference is used for overflow. In fact, to prevent extremely thin sheets of water from being drawn off, overflow weirs on circular basins usually consist of V–notched metal plates, which reduce the effective overflow area. These strips are bottled on to the collection trough and can be adjusted to correct for differential settling of the basin after construction.

   Weir plates should be precisely level, since a very slight difference in elevation will result in considerable short circuiting (direct channeling from influent to effluent). Uneven distribution and wind currents can also cause short circuiting (flow control more difficult in circular basin than in long–rectangular ones). Therefore, diameters of the circular basin should be less than 30 m.

   Design is based on overflow rates and detention time (neither horizontal velocity nor weir overflow rates are a consideration in the design of circular settling basin).

Example 6.105

Chemostat Study : Rate Equation for Growth

Determine the rate equation for cell growth

A Chemostat study was performed with yeast. The medium flow rate was varied and the steady state concentration of cells and glucose in the fermenter was measured and recorded. The inlet concentration of glucose was set of 100 g/L. The volume of the fermenter was 500 mL.

Table 63 : Flow rate V_s cell conc. and substrates conc.

Flow rate, F(mL/hr)	Cell concentration, C_X (g/L)	Substrate concentration, C_s (g/L)
31	5.97	0.5
50	5.94	1.0
71	5.88	2.0
91	5.76	4.0
200	0	100.0

Solution

1. Required CSTR equations

    Mass balance for micro–organism in CSTR

    $$FC_{Xi} - FC_X + Vr_X = \frac{VdC_X}{dt}$$

    $$\tau_m = \frac{V}{F} = \frac{C_X - C_{Xi}}{r_X} \text{ at steady state}$$

    If $C_{Xi} = 0$ (Input stream stenile), and $r_X = \mu C_X$

    $$\tau_m = \frac{1}{\mu} = \frac{1}{D}$$

    where: D = Dilution rate and is equal to the reciprocal of the residence time (τ_m)

    F, C_{Si}, C_{Xi}, C_{Pi}

    F, C_S, C_X, C_P

    Figure 47 : Schematics of Chemo-reactor.

    $$D = \mu = \frac{1}{\tau_m} = \frac{\mu_{max}C_s}{K_s + C_s}$$

    Or, $C_s = \dfrac{K_s}{\tau_m\,\mu_{max} - 1}$

    Cellmass :

    $$C_X = Y_{X/S}\,(C_{Si} - C_S)$$

    Product mass :

    Similarly, $C_P = C_{Pi} + Y_{P/S}\left(C_{Si} - \dfrac{K_s}{\tau_m\mu_{max} - 1}\right)$

    Valid only when $\tau_m\,\mu_{max} > 1$.

2. Required rate expression (r_X)

    Calculated dilution rate (D) $= \dfrac{F}{V}$

    Plot D^{-1} verses C_s^{-1} resulting in

    Intercept : $\mu_{max}^{-1} = 3.8$

    Slope : $K_s\,\mu_{max}^{-1} = 5.2$

    Therefore, $\mu_{max} = 0.26$ h^{-1}, and $K_s = 1.37$ g/L

$$\text{Rate expression } (r_X) = \frac{0.26 C_S C_X}{1.37 + C_S}$$

3. Required washout considerations

$$C_X = Y_{X/S} \left(C_{Si} - \frac{K_s}{\tau_m \, \mu_{max} - 1} \right) > 0, \text{ and}$$

$$\tau_m = \frac{V}{F} > \frac{K_s + C_{Si}}{C_{Si} \, \mu_{max}}$$

$$\text{Therefore, } F < \frac{V C_{Si} \, \mu_{max}}{K_s + C_{Si}} = \frac{0.5(100)(0.26)}{1.37 + 100}$$

$$= 0.128 \text{ L/h}$$

Example 6.106

Dead Volume for CSTR System

Determine the dead volume (ineffective) for the CSTR system using impulse dose tracer technique.

Assume the following data :

Flow rate : 0.10 m³/s

Non reactive tracer (impulse input) study data :

Time(s)	10	50	100	150	200	300	400
Tracer concentration ratio (C/C$_i$)	0.95	0.78	0.61	0.47	0.37	0.22	0.14

Volume of CSTR system : 30 m³

Solution

1. Required hydraulic detention time (θ)

$$\theta = \frac{30 \text{ m}^3}{0.10 \text{ m}^3/s} = 300 \text{ s}$$

2. Required detention time (θ)

Impulse dose tracer (non–reactive) added at $t = 0$ and the response is expressed as:

$$C = C_i \exp\left[-\frac{t}{\theta} \right] \tag{1}$$

$$\text{Hydraulic detention time } (\theta) = \frac{t}{\ln(C/C_i)} \tag{2}$$

Using equation (2), the hydraulic detention (q) is at various time (t) of the tracer is presented as:

Time(s)	10	50	100	200	300	400
Hydraulic detention time $(q)^*$	205	202	202	201	198	203

Calculations of hydraulic detention time (θ) at various time (t) are:

$$\theta = \frac{10}{ln(1.05)} = \frac{10}{0.0488} = 205 \text{ s (at t} = 10 \text{ s)}$$

$$\theta = \frac{50}{ln(1.28)} = \frac{50}{0.248} = 202 \text{ s (at t} = 50 \text{ s)}$$

$$\theta = \frac{100}{ln(1.64)} = \frac{100}{0.494} = 202 \text{ s (at t} = 100 \text{ s)}$$

$$\theta = \frac{200}{ln(2.70)} = \frac{200}{0.99} = 202 \text{ s } \text{ (at t} = 200 \text{ s)}$$

$$\theta = \frac{300}{ln(4.55)} = \frac{300}{1.514} = 198 \text{ s (at t} = 300 \text{ s)}$$

$$\theta = \frac{400}{ln(7.14)} = \frac{400}{1.966} = 203 \text{ s (at t} = 400 \text{ s)}$$

Average hydraulic detention time (θ)

$$= \frac{1}{6} [205 + 202 + 202 + 202 + 198 + 203]$$

$$= 202 \text{ s}$$

3. Required dead volume

$$\text{Active volume} = (0.1 \text{ m}^3/\text{s})(202) = 20.2 \text{ m}^3$$

$$\text{Dead volume} = \text{Actual volume} - \text{Active volume}$$

$$= (30 - 20.2) \text{ m}^3$$

$$= 19.8 \text{ m}^3.$$

[Increase in turbulence as a result of increased flow rate (Q) may result in reduction in decreased volume reactor].

Example 6.107

Two–Step Phase Treatment (Slurry Reactor)

1. A laboratory test evaluated a two step batch process to treat creoste waste: (1) suspending creosote waste with surfactant for seven days at 20% solids in a roughing vessel, and (2) transferring the supernatant to a polishing

reactor for 14 days of additional biological treatment. Assuming the above bench–scale data are representative of a larger operation and starting at the same concentrations, how many days of treatment in a pilot–scale operation would be necessary to degrade each of the constituents to 100 ppm or less.

1.1   Required mathematical expressions

Specific growth rate (μ) :

$$\mu = \frac{\mu_{max} S}{K_s + S}$$

Biomass growth $\left(\dfrac{dX}{dt}\right)$:

$$\frac{dX}{dt} = \mu X = \left(\frac{\mu_{max} S}{K_s + S}\right) X$$

Yield coefficient (Y) :

$$Y = \frac{dX/dt}{ds/dt}$$

Substrate reduction rate (dS/dt) :

$$-\frac{dS}{dt} = \frac{1}{Y}\left(\frac{\mu_{max}S}{K_s + S}\right) X$$

or $\quad -\dfrac{dS}{dt} = \dfrac{kSX}{K_s + S}; \left[k = \dfrac{\mu_{max}}{Y} \right]$

or in terms of biomass production (dX/dt) :

$$\frac{dX}{dt} = \left(\frac{Y K S}{K_s + S}\right) X - k_d X$$

For the typical hazardous waste application in which target organic concentration is at a very low concentration ($S \ll K_s$), the substrate reduction equation becomes:

$$-\frac{dS}{dt} = \left(\frac{kX}{K_s}\right) S$$

or $\quad \ln[S/S_0] = -\left[\dfrac{k}{k_s} X\right] t$

$$S = S_0 \exp[-k' t]$$

$$k' = \left[\frac{k}{k_s} X\right]$$

Half life time $(t_{0.5})$ is:

$$t_{0.5} = \frac{0.693}{k'}$$

1.2   Required treatment period for degradation of organics to 100 ppm

$$S = S_0 e^{-k't}$$

$$\frac{500}{6000} = \exp[-k'(21)]$$

[It takes 21 days to degrade organics from 6000 ppm to 500 ppm]

Therefore, $k' = 0.12\ d^{-1}$

Time required to degrade to 100 ppm is:

$$100 = 6000 \exp[-0.12(t)]$$

$$t = 34\ \text{days}.$$

2.   The sludge from an inactive pit at petroleum refinery were mixed with dilution water and aerated in two tanks operated in a slurry phase, batch mode. One tank was operated at 5% solids, the other tank at 12%. Which of the two tests resulted in faster degradation? Assume the following data:

[A]

Table 64 : Slurry–phase treatment of petroleum refinery sludge at different concentration:

Parameters	Pit Sludge Test	Pond Sludge Test
Tank diameter (ft)	26	40
Tank depth (ft)	10	10 ft @ perimeter 12 ft @ center
Actual operating volume	17,300 gal	70,000 gal
Percent solids (day)	12%	5%
Surfactant addition	Yes	Yes
Nutrient addition	Yes	Yes
Aeration power	10 hp	40 hp+
Batch operating period	57 days	92 days

+40 hp was disrupting mixing and was replaced by a 7.5 hp mixer and 10 hp mixer on 65th day.

[B]

Table 65 : Reduction in sludge constituents

Parameters	Pit Sludge Test [12% solids]	Pond Sludge Test [5% solids]
Oil and grease		
Initial concentration	27.4%	51.4%
Reduction	50%	63%
Total poly aromatic hydrocarbons [PAHs]		
Initial concentration	1445 mg/kg	1904 mg/kg
Reduction	81%	76%
Carcinogenic PAHs		
Initial concentration	335 mg /kg	532 mg/kg
Reduction	60%	25%

2.1   Required rate constant (k') and treatment time

Total PAHs:

$$\frac{S}{S_o} = \exp(-k't)$$

$$S = (1 - 0.81)S_o$$

$$= 0.19(1445) \text{ mg/kg}$$

$$t = 57 \text{ days}$$

$$k' = -\frac{\ln(S/S_o)}{t}$$

$$= -\frac{\ln(0.19S/S_o)}{57}$$

$$= 0.029 \text{ d}^{-1}$$

Table 66 : Similarly k' can be calculated for oil and grease, and carcinogenic PAHs:

Parameters	Rate Constant (k', d⁻¹)	
	12% solids	5% solids
Oil and grease	0.012	0.011
Total PAHs	0.029	0.016
Carcinogenic PAHs	0.016	0.003

12% solids degraded PAHs at a faster rate, and oil and grease at essentially same rate as with 5% solids.

3.   The slurry phase treatment in two impoundments of petroleum refinery was carried out. Six mixers [5 × 25 hp surface aerators and 1 × 15 hp mixer] were installed in each impoundment to suspend the sludge with

supernatant, yielding a mixed solids concentration of 15% as a minimum surfactant, nutrients, pH control chemicals, and an adapted microbial culture were added to enhance degradation. What are the degradation rates for oil and grease? Assume the following data :

Table 67 : Various parameters for two impoundments.

Parameters	Impoundment	
	1	2
Sludge volume (cubic yard)	4000	2600
Sludge volume (%)	66	62
Dry sludge oil and grease concentration (%)	32	41
Operating period (design)	21	61
Operating temperature (°C)	18	14
Reduction in sludge volume (%)	68	61
Reduction in oil and grease (%)	62	87

3.1 Required rate constant (k') in the two impoundment at two temperatures

Rate constants (k') for oil and grease :

$$k' = \frac{-ln\,(S/S_o)}{t}$$

$$= \frac{-ln\,(0.38)}{21} \qquad \text{[Impoundment – 1]}$$

$$= 0.046 \text{ d}^{-1} \text{ at } 18° \text{ C}$$

Temperature coefficient (θ) :

$$k' = \frac{-ln\,(S/S_o)}{t}$$

$$= \frac{-\ln\,(0.13)}{61} \qquad \text{[Impoundment – 2]}$$

$$= 0.033 \text{ d}^{-1} \text{ at } 14° \text{ C.}$$

Temperature coefficient (θ) :

$$\frac{r(T)}{r(20)} = \theta^{T-20}$$

$$\frac{[r(T)/r(20)]_1}{[r(T)/r(20)]_2} = \frac{[\theta^{T-20}]_1}{[\theta^{T-20}]_2}$$

$$\frac{[r(T)]_1}{[r(T)]_2} = \frac{[\theta^T]_1}{[\theta^T]_2}$$

$$\frac{r(18)}{r(14)} = \theta^{18-14} = \frac{0.046}{0.033}$$

$$\theta^4 = 1.086$$

Difference in reduction is because of functional temperature in the impoundments.

3.2 Required ratio of mass of oil and grease removed per unit energy consumption

Impoundment–1 :

Power = $(5 \times 25 + 1 \times 15)$ hp

$= (125 + 15)$ hp

$= 140$ hp

Energy = $(140$ hp$)(21$ days$)(24$ hours/day$)$

$= 71,000$ hp–hr

Initial mass of oil and grease is:

$=$ (Sludge volume)(density)$(1 -$ moisture content)(Oil and Grease content)

$= (4000$ yd$^3)(1800$ lb/yd$^3)(1 - 0.62)(0.32)$

$= 783,360$ lb

$$\text{Ratio} = \frac{(783,360 \text{ lb Oil and Grease})(0.62)}{71,000 \text{ hp} - \text{hr}}$$

$= 6.84$ Ib Oil and Grease removal per hp – hr

Impoundment–2

Energy = $(140$ hp$)$ $(61$ days$)$ $(24$ hours /day$)$

$= 2,00000$ hp–hr

Initial oil and grease is :

$= (2600$ yd$^3)(1800$ lb/yd$^3)(1 - 0.62)$ (0.41)

$= 730,000$ yd^3

$$\text{Ratio} = \frac{730,000 \text{ yd}^3(0.87)}{200,000 \text{ hp} - \text{hr}} = 3.2 \text{ lb Oil and Grease removal per hp} - \text{hr}$$

Higher temperature impoundment is more than twice as energy efficient as impoundment – 2.

Example 6.108

Kinetic Constants Development and Kinetic Constants from Batch Reactors and CSTR–ASP

1. Kinetic constant development

   The growth curve of micro–organisms has a log phase, an acceleration phase, a log phase, stationary phase and finally an endogenous decay phase. In a crude approximation of the first order reaction, the log phase of the growth rate is expressed as:

   $$\left(\frac{dX}{dt}\right)_g = \mu X$$

   where　　　X = Biomass

   　　　　　μ = Specific growth rate

   $$\left(\frac{dS}{dt}\right)_u = qX$$

   where q = Specific substrate utilization

   $$-Y\left(\frac{dS}{dt}\right)_u = \left(\frac{dX}{dt}\right)_g$$

   Yield factor $(Y) = \left(\frac{dX}{dS}\right)$

   $$Y = -\frac{\dfrac{dX}{dt}}{dS/dt} = -\frac{\mu X}{qX}$$

   Therefore, $\mu = -Yq$

   Again,

   $$\left(\frac{dX}{dt}\right)_g = \left(\frac{dX}{dt}\right)_T - \left(\frac{dX}{dt}\right)_E$$

   where subscripts g, T, E and μ, respectively refer to growth, total, decay, and utilization

   $$\left(\frac{dX}{dt}\right)_E = k_d X$$

   $$-Y\left(\frac{dS}{dt}\right)_u = -Y_T\left(\frac{dS}{dt}\right)_u - k_d X$$

   $$Yq = Y_T q + k_d$$

   $$Y = \frac{k_d}{Y} + Y_T$$

$$= \frac{Y_T}{1 + \dfrac{k_d}{\mu}}$$

or $\qquad Y = \dfrac{Y_T}{1 + k_d \theta_c} \qquad [\mu = \theta_c^{-1}]$

[If $k_d = 0; Y = Y_{ob}$]

Using Monod's from of expression for q and μ :

$$q = q_m \frac{S}{K_m + S}$$

$$\mu = \mu_m \frac{S}{K_m + S}$$

Then, $\qquad \dfrac{dX}{dt} = \dfrac{\mu_m XS}{K_m + S}$

$$\frac{dS}{dt} = \frac{-\mu_m XS}{Y(K_m + S)}$$

2. Kinetic constants from batch reactors.

    Case–1 : $[S_o \ll K_m]$

    $$\frac{dS}{dt} = \frac{-\mu_m X_o S}{Y K_m}$$

    Integrating, keeping X_o constant :

    $$\ln \left[\frac{S_o}{S} \right] = \frac{\mu_m}{K_m Y} X_o t$$

    Plot $\ln \left[\dfrac{S_o}{S} \right]$ Verses t, slope $= \dfrac{-\mu_m X_o}{K_m Y}$

    Case–2 : $[S_o \ll K_m]$

    $$\frac{dX}{dt} = \frac{\mu_m XS}{K_m + S} = \frac{\mu_m X}{1 + \dfrac{K_S}{S}} = \mu_m X$$

    Integrating

    $$\ln \left[\frac{X}{X_o} \right] = \mu_m t$$

    $$X = X_o \exp(\mu_m t)$$

    $$-\frac{dS}{dt} = \frac{\mu_m X}{Y} = \frac{\mu_m (X_o e^{\mu_m t})}{Y}$$

or $\quad \dfrac{S-S_o}{X_o} = \dfrac{1}{Y}(e^{\mu_m t} - 1)$

Plot $\dfrac{S-S_o}{X_o}$ Vs $(e^{\mu_m t} - 1)$, assuming different values of μ_m, the one that becomes more or less a straight line through the origins will have a slope of $1/Y$ [Both μ_m and Y can be determined].

Case–3 : $[S \approx K_m]$

$$\dfrac{dS}{dt} = -\dfrac{\mu_m X S}{Y(K_m + S)}$$

Integrating

$$-\int_{S_o}^{S} \left(\dfrac{K_m + S}{S}\right) dS = \int_{o}^{t} \dfrac{\mu_m X}{Y} X dt$$

Gates and Marlar's solution:

$$\dfrac{1}{t} \ln\left[\dfrac{S}{S_o}\right] = C\left[\dfrac{\ln(1+ad)}{t}\right] - b$$

where $\qquad a = \dfrac{Y}{X_o}$

$$b = \dfrac{\mu_m}{(YK_m)(X_o + YS_o)}$$

$$c = 1 + \dfrac{(X_o + YS_o)}{(YK_m)}$$

$$d = (S_o - S)$$

If $\dfrac{1}{t} \ln\left[\dfrac{S}{S_o}\right]$ is plotted against $\dfrac{\ln[1+ad]}{t}$, both c and b can be determined. Since the value of a cannot be determined directly, a values are assumed, until the best straight time is obtained:

$$\mu_m = \dfrac{b}{c-1}$$

$$K_m = \dfrac{\left(\dfrac{1}{a+S_o}\right)}{(c-1)}$$

$$Y = a\,X_o$$

Case–4 : $[X \approx X_o]$

Integrating

$$\frac{dS}{dt} = -\frac{\mu_m XS}{Y(K_m + S)}$$

$$\frac{1}{t} \ln\left[\frac{S_o}{S}\right] = \frac{S - S_o}{K_m t} + \frac{\mu_m X_o}{YK_m} \qquad \text{[Henry equation]}$$

Plot of $\frac{1}{t} \ln\left[\frac{S_o}{S}\right]$ Vs $\frac{(S_o - S)}{t}$, yields :

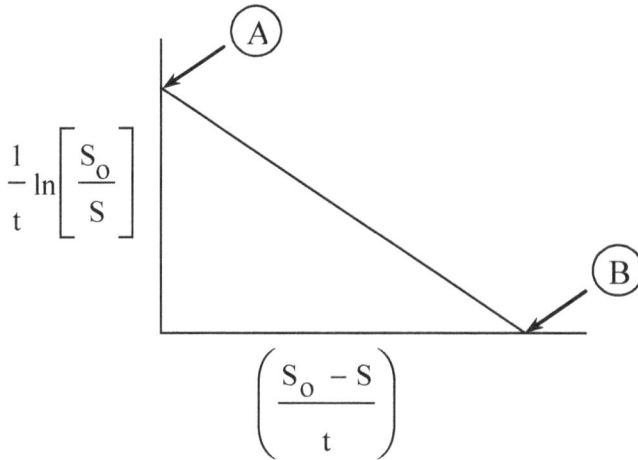

Figure 47 : Plot of $\frac{1}{t} \ln\left[\frac{S_o}{S}\right]$ Vs $\left[\frac{S_o - S}{t}\right]$

$$\text{Slope} = -\frac{1}{K_m}$$

$$A = \frac{k_o X_o}{YK_m}, \qquad B = \frac{k_o X_o}{Y}$$

$$K_o = Y \mu_m$$

3. Assume the following data obtained from bench scale experimentation for a completely mixed activated sludge process reactor (without recycle):

Table 68 : Influent and Effluent COD abd MLSS.

Time (hr), q	Influent COD (mg/L), S_o	Effluent COD (mg/L), S	MLSS (mg/L)
0	–	–	–
2.8	1043	303	324
3.5	1040	220	359
3.8	986	162	358
6.0	987	80	388
9.0	1010	58	401
11.25	1037	47	409

Determine the following kinetic coefficients Y, k_d, μ_m and K_m.

3.1   Determine $\left(\dfrac{S_o - S}{X}\right)$, $\dfrac{\theta X}{(S_o - S)}$ and $\dfrac{1}{S}$

Table 69: Values of q, S_0, S, X, $\left[\dfrac{S_0 - S}{X}\right]$, $\left[\dfrac{\theta X}{S_0 - S}\right]$ and S^{-L}.

q (hr)	S_o (mg/L)	S (mg/L)	X (mg/L)	$\left(\dfrac{S_o - S}{X}\right)$	$\left(\dfrac{\theta X}{S_o - S}\right)$	$S^{-1} \times 10^2$
2.8	1043	303	324	2.28	1.23	0.33
3.5	1040	220	359	2.28	1.53	0.45
3.8	986	162	358	2.30	1.65	0.62
6.0	987	80	388	2.34	2.56	1.25
9.0	1010	58	401	2.37	3.80	1.72
11.25	1037	47	409	2.42	4.65	2.13

Basic equation is :

$$\frac{S_o - S}{X} = \frac{1}{Y} + \frac{k_d \theta}{Y}$$

Slope = 0.0163 $\left(\dfrac{k_d}{Y}\right)$ and Intercept = 2.233 $\left[\dfrac{1}{Y}\right]$

Therefore, Y = 0.45, k_d = 7.34 × 10^{-3} hr^{-1}

Basic equation is:

$$\frac{\theta X}{S_o - S} = \frac{Y K_m}{\mu_m}\left(\frac{1}{S}\right) + \frac{Y}{\mu_m}$$

Slope = 186 = $\dfrac{K_m Y}{\mu_m}$ or K_m = 337 mg/L

Intercept = 0.552 = $\dfrac{Y}{\mu_m}$ or μ_m = 0.82 hr^{-1}

Example 6.109

River Sanitation

The outfall from a municipal wastewater treatment plant is located on a river 10 km above a bathing beach. A plant malfunction causes the coliform bacteria count in the river to rise almost instantaneously to 400 Cells/100 mL. The river velocity is 10 km/d and the diffusion coefficient is 10 km^2/d. The die–off rate of bacteria is first order with respect of bacterial count, with a rate constant k = 0.5 d^{-1}. The bathing standard is 200 Cells/100 mL. Determine if the bathing standards at the beach will be violated.

Solution

1. Required advection–dispersion–reaction (ADR) equation (One dimensional)

$$D\frac{\partial^2 C_i}{\partial x^2} - v_x \frac{\partial C_i}{\partial x} + \gamma_i = \frac{\partial C_i}{\partial t}$$

disperssion number $= \dfrac{D}{v_x L}$

With first order reaction r = – kC and steady state condition:

$$D\frac{d^2 C}{dx^2} - v_x \frac{dC}{\partial X} - kC = 0$$

Integration of this equation yields:

$$\frac{C}{C_{in}} = \frac{4a \exp\left[0.5 \frac{v_x L}{D}\right]}{(1+a)^2 \exp\left[0.5 \frac{v_x L}{D}\right] - (1-a)^2 \exp\left[-0.5a \frac{v_x L}{D}\right]}$$

$$a = \left[1 + 4\left(\frac{kL}{v_x}\right)\left(\frac{D}{v_x L}\right)\right]^{0.5}$$

where C_{in} is the bacterial concentration in the river at the outfall. Dispersion coefficient:

$$= \frac{D}{v_x L} = \frac{10 \text{ km}^2/\text{d}}{(10 \text{ km/d})(10 \text{ km})} = 0.1$$

$$a = \left[1 + 4\left(\frac{0.5 \times 10}{10}\right)(0.1)\right]^{0.5} = 1.095$$

$$\frac{C}{C_{in}} = \frac{4(1.095)\exp(0.5/0.1)}{(1+1.095)^2 \exp\left[\frac{0.5 \times 1.095}{0.1}\right] - (1-1.095)^2 \exp\left[\frac{-0.5 \times 1.095}{0.1}\right]}$$

$$= \frac{4(1.095)(148.4)}{(4.389)(238.65) - (9.025 \times 10^{-3})(238.65)}$$

$$= 0.620$$

$$C = (0.620)(C_{in}) = (0.620)(400) = 248 \text{ Cells}/100 \text{ mL}$$

[It violates the stipulated standards].

2. Required dispersion effect

Neglecting the dispersion term $\left(D\dfrac{\partial^2 c}{\partial x^2} \right)$, ADR equation becomes

$-v_x \dfrac{\partial C}{\partial x} - kC = 0$, and it integrates to

$$\dfrac{C}{C_{in}} = \exp\left(\dfrac{-kx}{v_x} \right)$$

$k = 0.5$ d^{-1}, $x = 10$ km; $C/C_{in} = 0.606$ [Therefore, the dispersion effect is minimal].

3. Required effect of higher levels of dispersion

Assume $\dfrac{D}{v_x L} = 1.0$

$\dfrac{C}{C_{in}} = 0.645$ [All other values same as given in section 1]

An increase in $\dfrac{D}{v_x L}$ means that more elements of the fluid reach the bathing area in a shorter time than would be true if transport were by advection only. The fractional concentration, therefore, increases because less time is available for bacterial die–off.

Example 6.110

Stream Flow

Develop the mass balance differential equation and solve (integrate) for the concentration as a function of time for the following cases:

Steady–state, increasing flow and cross–sectional area with distance (x) and first order reaction decay

Steady–state, exponentially decreasing rate constant as a function of distance in the stream (easiest–to–degrade material is the most rapidly degraded near the point of discharge leaving recalcitrant compounds).

[Exponentially increasing flow and area in a stream at steady–state]

Solution

1. Required differential equation

$$\dfrac{\partial C}{\partial t} = -\dfrac{1}{A}\dfrac{\partial QC}{\partial X} - r$$

This is a general equation for one–dimensional transport in a stream or river (neglecting dispersion). All coefficients (Q, A, r) may be functions of distance and time.

2. Required solution at steady–state, increasing flow and cross–sectional area with distance (x), and first order reaction decay

$$0 = \frac{1}{A(x)} \frac{d[Q(x)C(x)]}{dx} - kC$$

$$0 = -\frac{1}{A_o e^{ax}} \frac{d[Q_o e^{qx} C(x)]}{dx} - kC$$

$$0 = -\frac{Q_o}{A_o e^{ax}} \left[e^{qx} \frac{dC}{dx} + Cq e^{qx} \right] - kC$$

$$0 = -\frac{Q_o}{A_o} e^{(q-a)x} \frac{dC}{dx} - \left[\frac{qQ_o}{A_o} e^{(q-a)x} \right] C - kC$$

$$\frac{Q_o}{A_o} e^{(q-a)x} \frac{dC}{dx} = -\left[\frac{qQ_o}{A_o} e^{(q-a)x} + k \right] C$$

$$\int_{C_o}^{C} \frac{dC}{C} = \int_0^x -\left[q + \frac{k A_o e^{(a-q)x}}{Q_o} \right] dx$$

$$\ln \frac{C}{C_o} = \left[-qx - \frac{k A_o}{(a-q)Q_o} e^{(a-q)x} \right]_0^x$$

$$\ln \frac{C}{C_o} = -qx - \frac{k A_o}{(a-q)Q_o} e^{(a-q)x} + \frac{k A_o}{(a-q)Q_o}$$

$$C = C_o \exp \left[-qx - \frac{k A_o}{(a-q)Q_o} e^{(a-q)x} + \frac{k A_o}{(a-q)Q_o} \right]$$

3. Required solution at steady state, exponentially decreasing rate constant as a function of distance (x) in the stream

$$0 = -\frac{Q}{A} \frac{dC}{dx} - (k_o e^{-rx})C$$

$$\bar{u} = \frac{Q}{A}$$

$$\int_{C_o}^{C} \frac{dC}{C} = -\frac{k_o}{\bar{u}} \int_0^x e^{-rx} dx$$

$$\ln \frac{C}{C_o} = -\frac{k_o}{\bar{u}} \left[-\frac{1}{r} e^{-rx} \right]_0^x = \frac{k_o}{\bar{u} r} [e^{-rx}]_0^x$$

$$\ln \left(\frac{C}{C_o} \right) = \frac{k_o}{\bar{u} r} (e^{-rx} - 1)$$

$$C = C_o \exp \left[\frac{-k_o (1 - e^{-rx})}{\bar{u} r} \right]$$

Example 6.111

Stream Re–aeration

A stream can be defined as a body of flowing water in which the velocity is the only significant component of the flux through a cross section normal to the direction of flow and in which longitudinal dispersion may be neglected without serious error. Consequently, the time rate change in concentration of a substance being transported in a stream can be described mathematically.

$$\frac{\partial c}{\partial t} = -\frac{1}{A(x,t)}\frac{\partial[Q(x,t)c]}{\partial x} - S(c,x,t) \tag{1}$$

Both flow and cross–sectional area may vary with distance and time. Eqn. (1) can be used to describe the transport of conservative and non–conservative materials in streams. For many applications of the Equation (1) to the transport of dissolved oxygen, both volumetric flow rate and cross–sectional area can be considered to remain constant. For such cases, the equation can be written as

$$\frac{\partial c}{\partial t} = -\frac{Q}{A}\frac{\partial c}{\partial x} + k_2(c^* - c) - k_1 L(x,t) - S_R(x,t) \tag{2}$$

where c = dissolved–oxygen concentration (ML^{-3})

t = time (t)

Q = volumetric rate of flow (M^3t^{-1})

A = cross–sectional area of stream (M^2)

x = distance (L)

k_2 = re–aeration coefficient. (t^{-1})

c^* = dissolved–oxygen saturation concentration (ML^{-3})

k_t = deoxygenation constant (t^{-1})

L = first–stage BOD (ML^{-3})

S_R = rate at which dissolved–oxygen concentration changes as result of remaining sources and sinks in stream ($ML^{-3}t^{-1}$)

In Eqn. (2) the term S_R includes the combined effects of photosynthesis and algal respiration. The sign convention is such that when S_R is positive the rate of removal is greater than the rate of addition.

The dissolved–oxygen concentration is often expressed in terms of the dissolved–oxygen saturation deficit D:

$$D = c^* - c \qquad (3)$$

Through substitution, Eqn. (2) becomes

$$\frac{\partial D}{\partial t} = -\frac{Q}{A}\frac{\partial D}{\partial x} - k_2 D + k_1 L(x,t) + S_R(x,t) \tag{4}$$

The assumption that the spatial variation of flow Q and area A are negligible applies most correctly to reaches of streams with contributory drainage basins varying in size from 500 to 5,000 km².

Flow in most stream drops to low, relatively stable values during late summer and early fall, a time of year when temperatures are the highest. Consequently, stream conditions are most critical at such times, and steady–state flow conditions are frequently assumed in the determination of the spatial distribution of dissolved–oxygen concentrations. The time required to reach steady state in a stream where dispersion is insignificant is simply the time of travel from the point of pollution to the location under considerations.

At steady state, $\partial D/\partial t = 0$, and Eqn. (4) becomes

$$U\frac{dD}{dx} = -k_2D + k_1L(x) + S_R(x) \tag{5}$$

where $\quad U = \dfrac{Q}{A}$

By applying Equation (2) to the transport of the organic material expressed in terms of its biochemical oxygen demand, one has

$$\frac{\partial L}{\partial t} = -\frac{Q}{A}\frac{\partial L}{\partial x} - k_1L(x,t) - k_3L(x,t) + L_a(x,t) \tag{6}$$

where $\quad k_3$ = rate constant for BOD removal through sedimentation and/or adsorption (t^{-1})

$\quad L_a$ = rate of addition of BOD by local runoff or by resuspension of organics from Bottom sludge deposits. $(ML^{-3}t^{-1})$

At steady state, $\partial L/\partial t = 0$, and Eqn. (6) becomes

$$U\frac{\partial L}{\partial x} = -(k_1 + k_3)L(x) + L_a(x) \tag{7}$$

By assuming that U, k_1, k_3, and L_a remain constant for a given reach of the stream. Eqn. (7) can be integrated over the length of the reach to yield

$$L(x) = L_oF_1 + \frac{L_a}{k_1 + k_3}(1 - F_1) \tag{8}$$

where $\quad F_1 = \exp\left[-(k_1 + k_3)\frac{x}{U}\right] \tag{9}$

$\quad L_o$ = first–stage BOD at $x = 0$. (ML^{-3})

Eqn. (8) is the steady–state equation for the spatial distribution of BOD over the length of the stream reach under consideration. By substituting this expression into Eqn. (5), one has

$$\frac{dD}{dx} = -D\frac{k_2}{U} + \left[L_oF_1\frac{L_a}{k_1 + k_3}(1 - F_1)\right]\frac{k_1}{U} + \frac{S_R}{U} \tag{10}$$

By assuming that k_1, k_2, k_3, U, L_a and S_R remain constant over the reach of stream being considered, Eqn. (10) can be integrated to give:

$$D(x) = D_o F_2 + \frac{k_1}{k_2 - (k_1 + k_3)}\left(L_o - \frac{L_a}{k_1 + k_3}\right)(F_1 - F_2)$$

$$+ \left[\frac{S_R}{k_2} + \frac{k_1 L_a}{k_2(k_1 + k_2)}\right](1 - F_2) \qquad (11)$$

where $\quad F_2 = \exp\left(-k_2 \dfrac{x}{U}\right)$ \qquad (12)

D_o = dissolved oxygen saturation deficit at $x = 0$. (ML^{-3})

A typical plot of Eqn. (11) yields a oxygen–sag–curve shown in Figure 48. Oxygen–sag–curve provides a visual representation of the spatial distribution of the dissolved–oxygen concentration in a polluted stream. Of particular interest is the low point along the curve located at a distance x_c from the origin at $x = 0$. The deficit at this point, called the critical deficit D_c, is the maximum deficit that will occur under the given conditions of loading and stream phenomena.

By differentiating Eqn. (11), equating the derivative to 0, and solving for x, an expression is obtained for the critical distance x_c:

$$x_c = \frac{U}{k_2 - (k_1 + k_3)} \ln\left[\frac{k_2}{k_1 + k_3} + \frac{k_2 - (k_1 + k_3)}{(k_1 + k_3)L_o - L_a} \times \left(\frac{L_a}{k_1 + k_3} - \frac{k_2 D_o - S_R}{k_1}\right)\right] \qquad (13)$$

The solution of Eqn. (13) requires knowledge of the BOD at $x = 0$, L_o. The value of L_o can be calculated from a rearrangement of Eqn. (11):

$$L_o = \frac{D - [S_R/k_2 + k_1 L_a/k_2(k_1 + k_3)](1 - F_2) - D_o F_2}{\{k_1/[k_2 - (k_1 + k_3)]\}(F_1 + F_2)} + \frac{L_a}{k_1 + k_3} \qquad (14)$$

The cross–sectional area, flow, and re–aeration coefficient may vary considerably over a relatively short distance. Furthermore, tributaries, dams, and wastewater inputs may result in significant discontinuities in these and other characteristics of the stream. Consequently, it is often desirable to segment streams in order to achieve a more realistic correspondence between the stream and its mathematical model. In such cases, the stream is divided so that within each segment the geometric and hydraulic characteristics are reasonably uniform. Segmentation is arranged so that points of discontinuity coincide with boundaries between segments.

The continuous model can also be applied to a segmented stream. Eqns. (8) and (11) can be solved for points along the first reach upstream. At the boundary between the first and second reaches, the necessary changes are made in the parameters, and the BOD and dissolved oxygen deficit computed at the downstream end of the first segment become input variables at the upper end of the second segment. This procedure is continued throughout the length of the stream.

Solution Procedure

The term k_1, k_3 and L_a are independent of each other. The term k_1 is a specific reaction rate for a biologic process, whereas k_3 and L_a are measures of two independent physical processes. In many cases, k_3 and L_a are negligible. The term k_1 should be determined in the laboratory using samples from the river reach under consideration. For waters polluted by domestic sewage, k_1 will normally have a value in the neighborhood of 0.25 d^{-1}.

The values of k_3 and L_a for any particular situation can be determined from stream field data. The procedure for determining these terms will be illustrated by three cases, in each of which it is assumed that k_1 has been determined from a laboratory analysis.

Case I : The BOD L decreases along the reach, and L is less than is predicted by

$$L = L_o e^{-k_1(x/u)} \tag{15}$$

This indicates that BOD is decreasing faster than the rate of removal due to oxidation. Therefore, it can be concluded that k_3 is positive and that the reduction in BOD due to sedimentation and/or adsorption is greater than any addition of BOD to the flowing load from the benthic deposit or other sources. In such cases, assume that $L_a = 0$, and compute the effective value of k_3, from Eqns. (8) and (9) using stream data.

Case II: The BOD L decreases along the reach, but L is greater than is predicted by Equation 15. This indicates that the rate of addition of BOD along the reach exceeds the rate of removal by sedimentation and/or adsorption. In such cases, assume that $k_3 = 0$, and compute the value of L_a from Eqns. (8) and (9) using stream data.

Case III : The BOD L remains constant or increases along the reach. Such cases as this are handled in the same manner as Case II was handled.

The procedure outlined in three cases above treats any net reduction in BOD in excess of that predicted by Equation (15) as being proportional to the concentration present L, whereas a net increase is treated as being due to a uniform addition along the reach. Actually, both removal and addition can be occurring simultaneously in a grossly polluted stream that contains sludge deposits.

As was pointed out previously, the term S_R can be positive or negative, depending on the relative magnitudes of photosynthesis, algal respiration, and the oxygen demand of any aerobic layer at the top of a benthic deposit. Compared with photosynthesis and respiration, the oxygen demand exerted by benthic deposits is considered to be large. More than likely, the principal effect of benthic deposits is to contribute BOD to the flowing load, thereby exerting its demand in this manner.

If k_1, k_3 and L_a are known (or can be calculated), then only S_R and k_2 remain unknown for the solution of Equation (11). Either one can be calculated from

Equation (11) provided the other can be established by an independent procedure. In as much as the value of the re–aeration coefficient k_2 depends upon a physical process, it is more readily subjected to theoretical estimation.

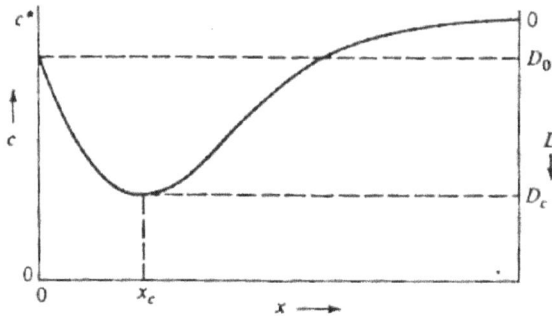

Figure 48 : Oxygen sag curve.

Example 6.112

Stream Re–aeration : Calculation of Re–aeration Coefficient

The rate of surface removal (r) has been defined by O' Conner and Dobbins, by the fluid turbulence [mixing length $(\overline{l})$ and vertical velocity fluctuations $(\overline{v})$] as follows

$$r = \frac{\overline{v}}{\overline{l}}$$

Two cases are considered :

1. Non–isotropic turbulence where a velocity gradient $\left[\dfrac{dU}{dy}\right]$ and shearing stress exists (Shallow steams). The ratio of the vertical velocity gradient fluctuation and mixing length is equal to the velocity gradient at the surface and the liquid film coefficient is

$$k_L = \left(D_L \frac{dU}{dy}\right)^{0.5}$$

Therefore, re–aeration coefficient (k_2) is:

$$k_2 = \frac{480\ D_L^{0.5}\ s^{0.25}}{H^{1.25}}, d^{-1}$$

where D_L is the molecular diffusion coefficient (ft²/d), s is the slope of the river channel (ft/ft), H is the average stream depth (ft)

2. Isotropic turbulence where neither a significant velocity gradient non shearing stress exist (deep steams). The vertical velocity fluctuation and

length are approximately equal to one–tenth of the forward flow velocity and the average depth, respectively. The liquid film coefficient (k_L) is:

$$k_L = \left(\frac{D_L U}{H}\right)^{0.5}$$ and the re – aeration coefficient (k_2) is

$$k_2 = \frac{(D_L U)^{0.5}}{2.3 H^{1.5}}$$

where U is the average stream velocity (ft/s).

3. The distinction between the two types of turbulence is established by the roughness coefficient of the channel (Chezy coefficient, B)

$$U = B\sqrt{H\,s}$$

If the value of B is less than 17, turbulence is considered as non–isotropic and if B is greater than 18, turbulence is considered as isotropic

4. Consider the following for a drought flow condition of a river at 20°C:

Average temperature : 20°C

Average depth   : 1.0 ft

Average velocity   : 0.5 ft/s

Slope   : 0.003 ft/ft

D_L at 20°C   : 0.00195 ft²/day

Calculate the re–aeration coefficient

$$\text{Chezy coefficient (B)} = \frac{U}{\sqrt{H\,s}} = \frac{0.5}{(1.2 \times 0.003)^{0.5}}$$

B is less than 17, therefore, the flow is non–isotropic

$$\text{and } k_2 = \frac{480\,D_L^{0.5}\,s^{0.25}}{H^{1.25}} = \frac{480\,(0.00195)^{0.5}\,(0.003)^{0.25}}{(1.0)^{1.25}}$$

$$= 4.96 \text{ d}^{-1}$$

5. Consider the following where the river was following at a higher velocity at 20°C

Average depth   : 3 ft

Average velocity : 3 ft/s

Average slope   : 0.003 ft/ft

Calculate the re–aeration coefficient

$$B = \frac{U}{\sqrt{H\,s}} = \frac{3}{[(3)(0.003)]^{0.5}}$$

$$= 31.6$$

Therefore, the flow is isotropic (B is greater than 17), and k_2 is:

$$k_2 = \frac{(D_L H)^{0.5}}{2.3 H^{1.5}} = \frac{[0.00195 \times 24 \times 3 \times 60 \times 60]^{0.5}}{2.3 \times (3)^{1.5}}$$

$$= 1.88 \ d^{-1}.$$

Example 6.113

Organic Loads on Stream

A STP discharges 25 MGD with effluent BOD of 30 mg/L at 25.5°C alongwith DO content of 2.0 mg/L. The stream flow is 300 ft³/s at 1.5 fps and the average depth is 10 ft. The water temperature is 24°C before the sewage is mixed with the stream. The stream is 95% saturated with DO and has a BOD of 4.0 mg/L. The deoxygenation rate constant (k_1) is equal to 0.60 at 20°C. Calculate the DO of the mixture of water and STP effluent, the temperature of the mixture of water and STP effluent, the value of the initial oxygen deficit for the river just before the STP discharge, and the minimum DO in the stream below the STP. Assume a pressure of 1 atm and a chloride concentration of zero.

Solution

1. DO of the mixture of water and STP effluent (DO_m)

    STP flow (Q_s) = (25 MGD × 1.547 ft³/s), (1 MGD = 1.547 ft³/s)

    $$= 38.7 \ ft^3/s$$

    Total stream flow = 300 + 38.7 ft/s

    $$= 338.7 \ ft/s$$

    The solubility of DO in water 24°C is 8.5 mg/L

    The DO of the river water = 0.95 × 8.5 = 8.075 mg/L

    $$= 8.1 \ mg/L$$

    $$DO_m = \frac{Q_r(DO_r) + Q_s(DO_s)}{Q_r + Q_s}$$

    $$= \frac{300 \ ft^3/s \times 8.1 \ mg/L + 38.7 \ ft^3/s \times 2 \ mg/L}{(38.7)} = 7.40 \ mg/L$$

2. Temperature of mixture (T_m)

    $$T_m = \frac{300 \ ft^3/s \times 24°C + 38.7 \ ft^3/s \times 25.5°C}{(38.7)}$$

    $$= 24.13° \ C$$

3. Initial oxygen deficit at 24.13°C

    Pure water DO saturation at 24.13°C = 8.49 mg/L

    Deficit (D_o) = (8.49 − 7.40) = 1.09 mg/L

4. Minimum DO at the down stream

k_2 (Reaeration coefficient) at 20°C

$$k_2 = \frac{(D_L U)^{0.5}}{H^{1.5}}$$

$$= \frac{(81 \times 10^{-6} \text{ ft/h} \times 1.5 \times 3600)^{0.5} \times 24}{(10)^{1.5}}$$

$$= \frac{15.87}{31.623} = 0.502 \text{ d}^{-1}$$

$$k_1 = 0.60 \text{ d}^{-1} \text{ at } 20°C$$

$$k_2 = 0.502 \text{ d}^{-1} \text{ at } 20°C$$

$$k_1(24.13°C) = 0.60 \cdot 1.047^{(24.13 - 20)}$$

$$k_1 = 0.72 \text{ d}^{-1}$$

$$k_2(24.13°C) = 0.502(1.047^{(24.13 - 20)})$$

$$= 0.61 \text{ d}^{-1}$$

5. The volume of initial BOD (L_o) of the mixture of the river and STP effluent

$$L_o = \frac{[(300 \times 4.0) + 38.7 \times 30]}{(338.7)} \text{ mg/L}$$

$$= 3.9 \text{ mg/L}$$

6. The critical time (t_c) at which the DO is minimum

$$t_c = \frac{1}{(k_2 - k_1)} \ ln\left[\frac{k_2}{k_1}\left(1 - \frac{D_o(k_2 - k_1)}{k_1 L_o} \right) \right]$$

$$= \frac{1}{0.01} \ ln\left[\frac{0.61}{0.72}\left(1 - \frac{1.09(0.11)}{(0.72 \times 3.9)} \right) \right]$$

$$= \frac{1}{0.01} \ ln[0.85(0.957)]$$

$$= \frac{1}{0.01} \ ln[0.810] = 0.2 \text{ d}$$

7. The minimum DO (D_c)

$$D_c = \frac{k_1 L_o}{(k_2 - k_1)}(e^{-k_1 t_c} - e^{-k_2 t_c}) + D_o e^{-k_2 t_c}$$

$$= \frac{0.72 \times 3.9}{(-0.11)}(e^{-0.72 \times 0.2} - e^{-0.61 \times 0.2}) + 1.09 e^{-0.61 \times 0.2}$$

$$= -2.55 \times (-0.019) + 0.965$$

$$= 0.0485 + 0.965 = 1.0136 \text{ mg/L}$$

8. Distance down stream at which the critical DO occurs from the value of t_c and velocity of flow

$$= \text{Velocity} \times \text{Time}$$
$$= 1.5 \text{ ft/s} \times 0.2 \text{ day}$$
$$= 1.5 \text{ ft/s} \times 0.2 \times 24 \times 60 \times 60$$
$$= 25,920 \text{ ft}$$

Note :

1. In addition, the following processes may be taking place in any given river stretch :

    BOD removal by adsorption or sedimentation

    BOD addition along the river stretch by tributary inflow or through other STPs

    BOD addition removal and oxygen removal by the menthol layer

    Photosynthetic addition of oxygen through plankton

    Oxygen removal by plankton respiration.

2. Critical point is where rate of change of deficit is zero and the demand equals the re–aeration rate.

Example 6.114

The Streeter–Phelps Equation for BOD in a Stream : Uncertainty Analysis

The Streeter–Pheps equation for BOD in a stream contains parameters L_o and k_d for initial BOD and deoxygenation, respectively. Suppose many measurements of L_o and k_d are obtained for the streams and they have the following mean and standard deviations :

$$L_o = 10.0 \pm 2.0 \text{ mg/L}$$
$$k_d = 0.6 \pm 0.1 \text{ day}^{-1}$$

The stream velocity is exactly 0.305 m/s and the basic equation is:

$$L = L_o \exp(- k_d \, x/u)$$

The parameters are weekly correlated with each other. When L_o is large, there is a tendency for factor deoxygenation ρ_{Lo}, $k_d = + 0.7$.

Solution

1. Required expression for first order analysis (uncertainty analysis)

    Consider a function or a set of equations where y is dependent variable and x is independent variable:

    $$y = f(x)$$

If the function is smooth and well behaved, and provided that the variance of x is not too large, the expected value of y is approximately equal to the value of the function with the expected value of x:

$$E(y) = E\,(f(x)) \approx f(E(x))$$

where E is the expectation operator. The expected value of x is the mean value of x using Taylor series expression of f(x) to approximate the function

$$f(x) = f(x_o) + (x - x_o)\frac{\partial f(x)}{\partial x}\Big|x_o + \frac{(x-x_o)^2}{2}\frac{\partial^2 f(x)}{\partial x^2}\Big|x_o + ----$$

where X_o is any point that satisfies the function. Linearizing the above equation about its mean $\overline{X}$:

$$f(x) \approx f(\overline{x}) + (x - \overline{x})\frac{\partial f(x)}{\partial x}\Big|\overline{x}$$

The variance about the mean $S^2(f(\overline{x}))$ is zero, so only the last term in the equation is considered:

$$\text{Variance }(S^2 f(x)) \approx S^2(x) + \left(\frac{\partial f(x)}{\partial x}\right)^2 \Big|\overline{x}$$

where   S = Sample standard deviation

S^2 = Sample variance of the function about any point from mean

The equation is valid for a bivarate relationship involving only one independent variable.

For multi-variation relationship, we must consider the co–variation among independent variables:

$$S_y^2 \sim \sum_{i=1}^{n}\left(\frac{\partial f}{\partial x_i}\right)^2 S_{x_i}^2 + 2\sum_{j=1}^{n-1}\sum_{i=j+1}^{n}\left(\frac{\partial f}{\partial x_i}\right)\left(\frac{\partial f}{\partial x_j}\right)S_{x_i}.S_{x_j}.\rho_{x_i x_j}$$

where ρ_{xixj} is the correlation coefficient in a linear least square regression between x_i and x_j variables

This equation provides the basis for the first order uncertainty analysis (higher order terms in the Taylor series expansion are dropped). The variance in the independent variable depends on the variance (uncertainty) of the individual variables or parameters (S_{xi}), the sensitivity of the independent variable compared to changes in each parameter df/dx_i, and the correlation between variables ρ_{xixj} .

2. Required variance in BOD concentration at downstream distances

$$L = L_o \exp\left(-k_d\frac{x}{u}\right)$$

$$\frac{dL}{dk_d} = \frac{-L_o x}{u} e^{-k_d \frac{x}{u}}$$

$$\frac{\partial L}{\partial L_o} = e^{-k_d \frac{x}{u}}$$

$$S_L^2 = \left(\frac{\partial L}{\partial k_d}\right)^2 S_{k_d}^2 + \left(\frac{\partial L}{\partial L_o}\right)^2 S_{L_o}^2 + 2\left(\frac{\partial L}{\partial k_d}\right)\left(\frac{\partial L}{\partial L_o}\right)^2 S_{k_d} S_{L_o} \rho_{k_d L_o}$$

For example, at x = 20 km; the variance in BOD in stream due to uncertainty in L_o and k_d is expressed as:

$$S_L^2 = \left(7.2521 \frac{mg.d}{L}\right)^2 (0.1\ d^{-1})^2 + (0.6342)^2 (2.0\ mg/L)^2 + 2\left(7.2521 \frac{mg.d}{L}\right)$$

$$S_L^{\ 2} = (0.6342)(0.1\ d^{-1})(2.0\ mg/L)(0.7)$$

$$S_L^2 = 3.4225\ (mg/L)^2$$

The mean expected value for L is given by the BOD equation assuming average parameter values

$$L_{20} = L_o e^{-kdx/u} = (10\ mg/L)exp(-0.6 \times 0.759)$$

$$= 6.342\ mg/L$$

The mean and standard deviation of BOD at 20 km downstream from discharge point is:

$$= 6.342 \pm 1.850\ mg/L$$

There is a rather large uncertainty due to the variation in L_o and the positive correlation between L_o and k_d.

Distance (x, km)	Travel Time (day)	(mean, mg/L)	SL (mg/L, standard deviation)
0	0	10.00	2.000
10	0.3795	7.9637	1.817
20	0.7590	6.3421	1.850
40	1.5179	4.0222	1.3067
60	2.2769	2.5509	1.0062
100	3.7948	1.0261	0.5528

The error terms (standard deviation) is well behaved-it does not grow, but the coefficient of variation (ratio of standard deviation to mean value) increases from 20% to 53.9% from x = 0 to x = 100 km.

[Careful attention must be paid to units]

Note :

1. Models include the following types of error:

   Model error (incorrect formulation of the model)

   Errors in state variables (dependent variable and initial conditions)

   Errors in the input data used to derive the model

   Parameter errors (rate constants, coefficients, and independent variables).

2. A comparison of methods for uncertainty analysis

Method	Theory	Advantages	Disadvantages
Sensitivity Analysis	Vary one parameter at a time by ± 1% to determine the change in state variable	Easy to use and conceptually simple	Not truly uncertainty analysis. Parameter may not be independent of other parameters.
First order analysis	Solve the variance equation directly for the state variable as a function of the uncertainty in the parameter value. [It is a linearized approximation that is solved]	Preserves the co-variance relationships among the parameters. Use less computer time for complex models	It is an approximation to the real solution. For highly highly non-linear equations, the method is not very accurate. It is parametric and requires the use of mean and standard deviations.
Monte Carlo Analysis	Select parameters randomly from a known or hypothetical error distribution. Run the model many times to determine the expected value and the variance (uncertainty)	Exact solution Does not depend an parametric statistics (mean and standard devia-tion). Less error at the extremes usually	Uses lots of computer time to generate statistically significant number of realiza-tions. May not preserve co-variance structure among parameters and variables.

3. Monte Carlo analysis (uncertainty analysis)

   Parameter distributions may be determined by multiple samples collected in the field, determined in microcosm experiments in the laboratory, or from expert judgement. Typical distribution of parameters are normal distributions, log normal, hat functions (constant probabilities over the range of parameter values), and trapezoid function (constant probabilities in a predominant range and decreasing linearly at both ends of the function). Normal distributions are thought to arise from additive process in the environment:

   $$y = f(x) = \frac{1}{\sigma(2\pi)^{0.5}} \exp\left[-0.5\left(\frac{x-\mu}{\sigma}\right)^2\right]$$

   where μ = Mean volume of the distribution

   σ = Standard deviation

This expression gives the probability density function (pdf) for a normal distribution. The area under the curve at $x = \pm\, 2\sigma$ contains 95.5% of values.

Log normal distribution is the most common distribution of parameters in the environment. It accounts for environmental measurements with skewness in their distribution, which is often the case. Log normal distribution is thought to arise from multiplicative processes. They can be made into a Gaussian normal distribution supply by performing a log transformation on the data.

$$f(y) = \frac{1}{y\,(2\pi)^{0.5}\,\sigma_x}\, \exp\!\left[-0.5\left(\frac{\ln y - \mu_x}{\sigma_x}\right)^2\right]$$

where μ_x = Geometric mean of the parameter values.

σ_x = Geometric deviation

y = Parameter values (must be positive)

$x = \log y$

where $y \geq 0$ and $-\infty \leq x \leq +\infty$. The area under the curve is normalized to 1.0, as the use of pdf's allows a ready comparison to be made in the shape of various parameters.

Steps in the analysis and computer programme include the following:

1. Determine the uncertainty distribution (pdf) for each parameter, input function, and variable that you want to analyse.

2. Sample each of the parameter distributions using a random number generator that takes into account the probability of each volumes.

3. Use the set of parameter values selected as input for the first simulation. Run the model and save the output to tape or file.

4. Repeat steps 2 and 3 a large number of times (100 to 200) or initial the statistical output no longer changes by repeated realization of the model.

5. Sort the stored output data and plot the output as the mean value plans or minus the probability range that is desired.

Example 6.115

Rates of Bacterial Self–purification

Estimate the values of die–away constant (k) and retardation coefficient (n) for coliform organisms in a small stream (X) and a large stream (Y) during summer period. Use the following data :

Time of flow, hr	10	20	100
Stream (X), % coliform remaining	2.5	2.1	1.1
Stream (Y), % coliform remaining	64	41	1.3

Solution

1. Chicks law

   The discharge into receiving water of wastewaters rich in degradable organic matter vastly increases the number and geeneraa of saprophytic bacteria that help to bring self purifications.

   The multiplying organisms are derived in part from wastewater, in part from receiving waters, others enter with run off from agricultural and other soils.

   A balance is reached with the food supply under prevailing environmental conditions do the number and variety of saprophytes began to decline.

   The density of intestinal bacteria also rise appreciably below a sewer outfall during the first 10 to 12 hrs of flow.

   The die–away may be quite different from that observed when pure cultures of representative organisms are suspended in clean water and stored in laboratory (conditions of light and temperature prevailing in receiving water are simulated).

   Principal reasons are: (1) the presence, in polluted natural bodies of water, of pre detors (ciliated protozon) which feed on bacteria and (2) the operation of bio–physical forces (sedimentation and biological flocculation and precipitation, that ally the processes of natural purification to those of a wastewater–treatment plant (which act as a river would up in a small space).

   Bacteria die not merely for lack of food, and their die–away in receiving waters is only approximated by Chick's law. The general purification equation is written as :

   $$\frac{(N_0 - y)}{N_0} = \frac{N}{N_0} = [1 + nkt]^{1/n} \tag{1}$$

   where N_0 is the modal number of bacteria in a pond, y is the number removed during a time of flow–to below the point of modal density to leave a number–N, k is the initial rate of die away for a specific bacterial population, and n is its associated coefficient of non–uniformity (retardation). If $n = 0$ (uniformly of removal), equation (1) can be expressed as :

   $$\frac{(N_0 - y)}{N_0} = \frac{N}{N_0} = \exp[-kt] \tag{2}$$

   Phelps (Stream Sanitation, John Wiley & Sons, N.Y., 211 (1944) suggested the following expression :

   $$\frac{N}{N_0} = p \exp[-k_1, t] + (1 - p) \exp(-k_2 t) \tag{3}$$

where p is the less resistant fraction $(0 < p < 1)$ with a die–away rate of k_1 and $(1 - p)$ is the more resistant fraction, $(1 > (1 - p) > 0)$ with a die–away rate of k_2. If $p = 1$,

$$\frac{N}{N_0} = \exp(-kt) \text{ is similar to equation (2).}$$

k and n vary with the nature of stream, concentration of pollution, and with the temperature.

Only in very clean and quiescent bodies of water k can be expected to be as low as observed in laboratory experiments on die–away of pure cultures in pure waters.

Shallow and turbulent streams provide large interfaces of contact between the following water and attached growths, and stirring of their water promotes contact and transfer opportunities.

For deep, sluggish streams with dilution factor of considerable magnitude, the opposite is true, and the bacterial die–away may be slow. However, rapid purification is accompanied by much non–uniformity (or high n values), because heavy pollution is quickly succeded by a cleaner environment with poorer purification powers (the die–away of the remaining 5% or less coliforms, may endure for much the same length of time in the two types of streams. Decrease in temperature decrease k values and increase n (the initial rate of die–away drops and the non–uniformily of purification rises. Bacterial self purification is thereby delay.

For high values of k and n, equation (1) can be approximately as :

$$n = \frac{\log t_2 / t_1}{\log (p_1 / p_2)}$$

$$n = \frac{\log t_2 - \log t_1}{\log p_1 - \log p_2}, \quad p = \frac{N}{N_0}$$

2. Determination of k and n for streams (X and Y)

2.1 Retardation factor (n) for stream–X at various flow times

$$n_1 = \frac{\log (20) - \log(10)}{\log(0.025) - \log(0.021)} = 3.98$$

$$n_2 = \frac{\log (100) - \log(20)}{\log(0.021) - \log(0.011)} = 2.49$$

$$n_3 = \frac{\log (100) - \log(10)}{\log(0.025) - \log(0.011)} = 2.80$$

$$\text{Average } n = \frac{(n_1 + n_2 + n_3)}{3} = n = \frac{(3.98 + 2.49 + 2.80)}{3} = 3.09$$

2.2   De–away rate (k) for stream–X at various flow times

$$\left(\frac{N}{N_0}\right)^{-n} = [1+nkt]$$

$$\left(\frac{N}{N_0}\right)^{-n} - 1 = nkt \text{ or } k = \frac{\left(\dfrac{N}{N_0}\right)^{-n} - 1}{nt}$$

$$k_1 = \frac{(p_1)^{-n} - 1}{nt_1} = \frac{(0.025)^{-3.09} - 1}{3.09 \times 10} = 2886\,h^{-1}$$

$$k_2 = \frac{(p_2)^{-n} - 1}{nt_2} = \frac{(0.021)^{-3.09} - 1}{3.09 \times 20} = 2474\,h^{-1}$$

$$k_3 = \frac{(p_3)^{-n} - 1}{nt_3} = \frac{(0.011)^{-3.09} - 1}{3.09 \times 100} = 3649\,h^{-1}$$

$$k = \frac{(k_1 + k_2 + k_3)}{3} = 3003\,h^{-1}$$

2.3   Retardation factor (n) for stream (Y) at various flow times

$$n_1 = \frac{\log(20) - \log(10)}{\log(0.64) - \log(0.41)} = 1.551$$

$$n_2 = \frac{\log(100) - \log(20)}{\log(0.41) - \log(0.013)} = 0.4663$$

$$n_3 = \frac{\log(100) - \log(10)}{\log(0.64) - \log(0.013)} = 0.5909$$

$$\text{Average } n = \frac{n_1 + n_2 + n_3}{3} = 0.869$$

[It can be assumed that retardation factor for stream–Y is negligible in comparison to stream–X (n = 0). Semi log plot of stream Y can be approximated by a straight line with a slope of 0.044 h^{-1}. No such simple fit is possible for stream–X]

2.4   Die–away rates in streams (X and Y)

After 100 hour of flow, the % removals are seen to be essentially the same in both streams and the rates of die–away are :

The rate of die–away in stream Y

$$= kN = 0.044\ (100 - 98.7)$$

$$= 0.0572\ \%\ \text{per hour (4.4\% per hour for stream–X)}$$

The rate of die–away in stream–X

$$k\left(\frac{N}{N_0}\right)^n = 3000\left(\frac{100 - 98.9}{100}\right)^{3.09} \times (100 - 98.9)$$

$$= 2.93 \times 10^{-3}\%\ \text{per hour.}$$

Note :

1. The destruction of entric bacteria is more rapid (1) in heavily polluted streams than in clean streams, (2) in warm weather than in cold weather, and (3) in shallow, turbulent than in deep, sluggish bodies of water.

2. The time of flow, rather than distance of flow is the controlling factor is strikingly shown by the improvement in water quality during relatively slow passage of stream through the lake.

3. Self purification of polluted groundwater differs appreciably from that of surface water (the variety of micro–organisms that seize upon the pollutional substances for food is greatly restricted in the confinement and darkness of the pore space of the soil (other processes do occur: physical purification through filtration).

4. The die–away of coliforms inoculated into natural seawater under laboratory conditions is reported to be many times as rapid (about 25 times) as in auto claved seawater (it is not due to sea salt but organise, but labile substances). A dilution factor of 200–250 for municipal wastewaters and seawater normally reduces the coliforms organisms to 10/mL or less.

Example 6.116

The Oxygen Economy of Polluted Waters

The Dissolved Oxygen Sag (Self Purification)

1. Basic differential equation incorporating the rate of oxygen utilization by BOD and rate of oxygen absorption by reaeration in times of DO deficit along the path of water movement :

$$\frac{dD}{dt} = k\,(L_a - y) - r\,D \tag{1}$$

where r is the rate of re–oxygenation, k is the rate of deoxygenation by BOD, y is the oxygen demand exerted in time t_1 and L_a is the initial or first stage BOD of the water.

Integration between the limits D_a at the point of pollution or reference point [$t = 0$, $(L_a - y) = L_a$] and D at any point distant a time of flow t from the reference point yields :

$$D = \frac{kL_a}{(r-k)}[\exp(-kt) - \exp(-rt)] + D_a \exp(-rt) \qquad (2)$$

Incorporating self purification (or oxygen–recovery) factor :

$f = r/k$ in equation (2) :

$$D = \frac{L_a}{f-1}\exp(-kt)\left\{1 - \exp(-(f-1)kt)\left[1-(f-1)\frac{D_a}{L_a}\right]\right\} \qquad (3)$$

2. Sag curve posses two characteristics

    A point of maximum deficit or critical point with co–ordinates (t_c and D_c)

    A point of maximum rate of recovery (a point of inflection with cordinates (t_i and D_i)

2.1 Critical point Maximum deficit ($dD/dt = 0$ and $d^2 D/dt^2 < 0$)

$$t_c = \frac{2.3}{k(f-1)}\log\left\{f\left[1-(f-1)\frac{D_a}{L_a}\right]\right\}$$

$$D_c = \frac{L_a \exp(-kt_c)}{f} = \frac{L_a}{f\left\{f\left[1-(f-1)\frac{D_a}{L_a}\right]\right\}^{1/(f-1)}}$$

2.2 Maximum rate of recovery (inflection) point $\left(\dfrac{d^2D}{dt^2} = 0\right)$

$$t_a = \frac{2.3}{k(f-1)}\log\left\{f^2\left[1-(f-1)\frac{D_a}{L_a}\right]\right\}$$

$$D_i = \frac{f+1}{f^2}L_a\exp(-kt_i)\frac{(f+1)L_a}{f^2\left\{f^2\left[1-(f-1)\left(\frac{D_a}{L_a}\right)\right]\right\}^{1/(f-1)}}$$

2.3 Co–ordinates of these two points are related to each other as :

$$(t_i - t_c) = \frac{2.3(\log f)}{[k(f-1)]}$$

$$\frac{D_i}{D_c} = \frac{f+1}{f}\exp[-k(t_i - t_c)] = \frac{f+1}{f^{(f/f-1)}}$$

2.4 If $k = r$ or $f = 1$, then :

$$D = (k\,t\,L_a + D_a)\exp(-k\,t)$$

$$t_c = \frac{(1-D_a/L_a)}{k}$$

$$t_i = \frac{(2-D_a/L_a)}{k}$$

$$(t_i - t_c) = \frac{1}{k} \quad \text{or} \quad \frac{t_i}{t_c} = \frac{(2-D_a/L_a)}{(1-D_a/L_a)}$$

3. Atmospheric re–aeration (re–oxygenation) coefficient (r)

$$(C_s - C_0) = 0; \ (C_s - C) = D$$

$$\frac{dC}{dt} = k_L a (C_s - C)$$

$$\frac{dD}{dt} = -r\,D$$

3.1 Churchill, Elmore and Buckinghum equation for estimating the value of reaeration coefficient :

$$r(\text{at } 20°C) = \frac{5.026v^{0.969}}{R^{1.673}} = 5v / R^{5/3} \quad \text{(Approx.)}$$

$$= 7.55\ S^{0.5}/(Rn) \qquad \text{(Approx.)}$$

where v is the mean velocity of flow in a given river stretch, R is the mean hydraulic radius, n is the Kutter's coefficient, and S is the loss of head or drop in water surface in a river stretch of known length. Departures from predicted values are caused by photosynthesis, changes in the hydrological regiment of streams, and structured shifts in the river channel.

3.2 Camp's equation for estimation of reaeration coefficient (r) incorporating hydraulic mixing

$$r = \frac{29\,G^{2/3}}{H} \quad \text{(days)}$$

where G is the mean temporal gradient (sec^{-1}), and H is the mean hydraulic depth of the receiving water

$$G^2 = \frac{Q\rho g h_f}{\mu V} = \frac{\rho g h_f}{(\mu t_d)}$$

where Q is the rate of flow (ft^3/s), V is the cubic feet of water in a given stretch of water with a displacement time of t_d (sec), h_f is loss of head in the stretch (ft), ρ and μ are respectively the mass density and absolute viscosity of water, and g is the gravity constant in the ft–lb–sec system

Therefore,

$$G^2 = \frac{62.4 \, h_f}{\mu \, t_d} \quad \text{at temperature T}$$

4. Rates of de–oxygenation by the Benthal Load.

4.1 If part of the pollution load settles in the immediate vicinity of point of pollution, the benthal oxygen demand may be calculated as follows and assumed to be an added DO deficit at this point :

$$y_m = 3.14 \times 10^{-2} y_0 \, C_T w \left(\frac{5+160w}{1+160w} \right) t_a^{0.5}$$

$$= 3.14 \times 10^{-2} y_0 \, C_T w' \left(\frac{5+0.02w'}{1+0.02w'} \right) t_a^{0.5}$$

where y_m is the maximum daily benthal oxygen demand [g/m² (for w) or lb/acre (w')], y_0 is the BOD$_5$ (20°C) (g/kg of volatile matter), $C_T = \dfrac{y}{y_0} = \dfrac{(1-\exp(-5k))}{(1-\exp(-5k_0))}$ or t = t$_0$ = 5 days and T$_a$ = 20°C, and w and w' are respectively the daily rates of deposition of VS in kg/m² or lb/ft², and t$_a$ is the time in days upto 365 days during which the accumulation takes place.

4.2 If the deposited load is dispersed over a long river stretch, the rate of removal of BOD by benthal decomposition (d) may be added to deoxygenation constant (k), and the resultant relationships are expressed as :

$$\frac{dD}{dt} = k L_a \exp[-(k+d)t] - r D$$

$$D = \frac{k L_a}{r-(k+d)} = \{\exp[-(k+d)t] - \exp(-rt)\} + D_a \exp(-rt)$$

The coefficient of deposition d must reflect:

Composition of wastewater and the receiving water

Relative quiescence of the receiving water

In times of considerable turbulence, scour of deposited sludge may render de–negative, and the stream receives an additional load. However, turbulence will increase re–aeration at the same time, and thus may more than offset the influence of the greater load.

4.3 If photosynthesis adds measurable amounts oxygen to the stream, the basic equation may be further expended to include the rate of photosynthetic oxygenation p, r D becoming (r + p) D, and exp(– rt) similarly exp[– (r + p) t].

5. Determine DO deficit at a point one day distant from the point of reference, magnitudes of critical time and critical deficit, and the magnitudes of inflection time and inflection deficit for a large stream. Use the following data :

Rate of self purification (f)	= 2.0
Rate of deoxygenation (k)	= 0.29 d^{-1}
Reference point DO deficit (D$_a$)	= 3.0 mg/L*
First stage BOD$_5$ (La)	= 20 mg/L*

* Mixture of stream water and wastewater.

5.1 DO deficit after travel flow time of one day

$$D = \frac{L_a}{(f-1)} \exp(-k\,t)\left\{1 - \exp[-(f-1)k\,t]\left[1-(f-1)\frac{D_a}{L_a}\right]\right\}$$

$$= \frac{20}{2.0-1}\exp(-0.2\times1)\left\{1-\exp[-(2-1)\times0.2\times1]\left[1-(2.0-1)\frac{3.0}{20}\right]\right\}$$

$$= \frac{20}{1}\times0.82\{1-0.82\times0.85\}$$

$$= 4.97 \text{ mg/L.}$$

5.2 Critical point

$$t_c = \frac{2.3}{k(f-1)}\log\left\{f\left[1-(f-1)\frac{D_a}{L_a}\right]\right\}$$

$$= \frac{2.3}{0.2\times1}\log\left\{2\left[1-(2-1)\frac{3}{20}\right]\right\}$$

$$= \frac{2.3\times0.23}{0.2} = 2.65 \text{ days}$$

$$D_c = \frac{L_a \exp(-k\,t_c)}{f}$$

$$= \frac{20\exp(-0.2\times2.65)}{2.0} = 5.89 \text{ mg/L.}$$

5.3 Inflection point

$$t_i = \frac{2.3}{k(f-1)}\log\left\{f^2\left[1-(f-1)\frac{D_a}{L_a}\right]\right\}$$

$$= \frac{2.3}{0.2(2-1)}\log\left\{4\left[1-(2-1)\frac{3.0}{20}\right]\right\} = 6.11 \text{ days}$$

Alternatively,

$$t_i = t_c + \frac{2.3(\log f)}{[k(f-1)]}$$

$$= 2.65 + \frac{2.3 \times 0.3}{0.2}$$

$$= 2.65 + 3.45 = 6.1 \text{ days} \qquad \text{(OK)}$$

$$D_i = D_c \left[\frac{f+1}{f^{f/(f-1)}} \right] - 5.89 \left[\frac{2+1}{2^2} \right] = \frac{5.85 \times 3}{4}$$

$$= 4.4 \text{ mg/L}.$$

6. Rate of deoxygenation by benthal load at the reference point.

Determine the maximum daily benthal oxygen demand of the accumulating sediment if water temperature remains constant at 20°C. Use the following information :

BOD$_5$ (2) load   :   40 g of volatile settleable solids per capita is 20 g

Deposited volatile solids   :   2 g/m^2 of stream bottom for 100 days

$$y_m = 3.14 \times 10^{-2} \, y_0 C_T \, w \, \frac{5+160w}{1+160w} \, t_a^{0.5}$$

$$C_T = \frac{y}{y_0} = \frac{[1-\exp(-5k)]}{[1-\exp(-5k_0)]} = 1, \text{ as } k_0 = k \text{ at } 20°C$$

19 g BOD is produced by 40 g of volatile settle solids :

$$y_0 = \frac{20 \times 1000}{40} = 500 \text{ g/kg of volatile solids}$$

Therefore,

$$y_m = 3.14 \times 10^{-2} \times 500 \times 12 \times 10^{-3} \frac{(5+160 \times 12 \times 10^{-3})}{(1+160 \times 12 \times 10^{-3})} \times (100)^{0.5}$$

$$= 3.14 \times 10^{-2} \times 500 \times 12 \times 10^{-3} \times \frac{6.92}{2.92} \times 10$$

$$= 4.465 \text{ g/m}^2.\text{d}$$

7. Determine rate of re–aeration, amount of oxygen added during flow (length) = 3000 ft, and maximum rate of reoxygenation of decomposition is sufficiently active to keep the DO content of the stream at 2.0 mg/L for a large stream. Assume the following data :

Rate constant temperature characteristics (Cr) = 0.024 per °C

Large stream width   :   500 ft

Mean hydraulic radius (R)	:	6 ft
Mean velocity (v)	:	3 ft/s
Energy gradient (s)	:	4×10^{-4}
Roughness coefficient (n)	:	0.033
Stream DO	:	2.0 at 12°C

7.1 Oxygen deficit (D_a) at 12°C

$$D_a = (10.8 - 2.0) - 8.8 \text{ mg/L.}$$

7.2 Re–aeration rate (r)

$$r_0 = \frac{5v}{R^{5/3}} = \frac{5 \times 3}{(6)^{1.67}} = 0.75 \text{ day}^{-1} \text{ at 20°C}$$

$$r = r_0 e^{C_r(T-T_0)}$$

$$= 0.75 \exp [0.024(12 - 20)]$$

$$= 0.75 \times 0.8253 = 0.62 \text{ day}^{-1} \text{ at 12°C.}$$

7.3 Time of flow (t)

$$t = \frac{3000}{3} = 1000 \text{ sec} = 0.0116 \text{ day}$$

7.4 Oxygen deficit at time t (D)

$$D = D_a \exp (- r t)$$

$$= 8.8 \exp (- 0.62 \times 0.0116)$$

$$= 8.7 \text{ mg/L}$$

Rate of maximum re–oxygenation (Decomposition is sufficiently active to keep the DO content of the stream at 2.0 mg/L)

$$\frac{dD}{dt} = -r D$$

$$= -0.62 \times 8.8$$

$$= 5.456 \text{ mg/L.d [Also the rate of deoxygenation].}$$

8. Rates of deoxygenation by suspended and dissolved load

Progressive utilization of oxygen by polluting substances is important for at least three reasons :

Generalised measure of the amount of oxidizable matter contained in water or pollutional load placed on it.

Predicting the progress of aerobic decomposition in polluted waters and the degree of self purification achieved in given intervals of time.

Yardstick of the removal of putrecible matter accompanying different treatment processes.

The oxygen demand for a large (deep and sluggish) stream can be estimated as :

$$\frac{dy}{dt} = k(L-y) = kL\exp(-kt) \text{ or}$$

Percent of the first stage demand is expressed by

$$\left(\frac{dy}{dt}\Big/L\right) \times 100 = 100 k \exp(-kt)$$

The oxygen demand of wastewater for a small (shallow and rapid) stream can be estimated from the general purification equation as :

$$y = L[1 - (1 + n k t)^{-1/n}] \text{ and}$$

Deoxygenation rate as :

$$\frac{dy}{dt} = k\left(\frac{L-y}{L}\right)^n (L-y) = kL[1+nkt]^{-1/n}$$

where n is the coefficient of retardation (and varies widely on stream of the kind under considerations).

8.1 Determine BOD–2 day at 30°C, rate of BOD after 2 days, and ultimate BOD for a mixture of domestic wastewater and river water for a large sluggish and deep/river. Assume the following data :

BOD$_5$(20°C) of the mixture   : 20 mg/L (suspended and dissolved)

BOD rate constant   : 0.20 d^{-1} at 20°C

BOD temperature characteristics (C_k) : 0.046 per °C

8.1.1   BOD rate constant at 30°C

$$\frac{k}{k_0} = \exp[C_k(T-T_0)]$$

$$= k(30) = k(20) \exp [0.046(30 - 20)]$$

$$= 0.2 \times 1.584$$

$$= 0.3168 \text{ day}^{-1} \text{ at } 30°C.$$

8.1.2   5 day BOD

The first order BOD equation is:

$$y(5) = L[1 - \exp(-kt)]$$

$$20 = L[1 - \exp(-0.2 \times 5)]$$

$$\text{Ultimate first (L)} = \frac{20}{[1-\exp((-0.2\times5))]} = 31.64 \text{ mg/L}$$

Stage of BOD after 2 days

$$y(2) = 31.64[1 - \exp(-0.3168 \times 2)$$
$$= 16.8 \text{ mg/L.}$$

8.1.3 Rate of BOD after 2 days at 30°C

$$\frac{dy}{dt} = 100\,k\,\exp(-k\,t)$$

$$= 100 \times 0.3168 \exp(-0.3168 \times 2)$$

$$= 16.82 \text{ \% per day of first stage demand compared with an initial rate :}$$

$$\frac{dy}{dt} = 100\,k\,\exp(-k\,t); \quad t = 0$$

$$= 100\,k \times 1$$

$$= 100 \times 0.3168 = 31.68 \text{ \% of the first demand.}$$

9. Allowable BOD loading of receiving

The magnitude of the following parameters determine the allowable pollutional load (L_a) for a given receiving water :

Deoxygenation constant (r)

Self purification constant (f = r/k)

Critic DO deficit (D_c)

Initial deficit (D_a)

For a large stream of normal velocity k as reported to be equal to 0.23 d^{-1}. Departure from this value is expected to be noticeably great in shallow, swift streams filled with boulders and debris. The values of self purification (f = r/k) for some specific situations are as :

Nature of stream	f = r/k at 20°C
Small ponds and back waters	0.5–1.0
Sluggish streams and large lakes or impoundments	1.0–1.5
Large streams of low velocity	1.5–2.0
Large streams of moderate velocity	2.0–3.0
Swift streams	3.0–5.0
Rapids and waterfalls	greater than 5.0

9.1 Determination of self purification (f)

Using maximum oxygen depletion (D_c) and final stage BOD (L_c), f can be estimated

$$L_c = L_a \exp(-k\,t_c)$$

$$D_c = \frac{L_a \exp(-k\,t_c)}{f} = \frac{L_c}{f}$$

or $\qquad f = \dfrac{L_c}{D_c}$

This equation assist in locating the critical point. Field tests for DO not only should confirm the position of the critical point but also account for changes in stream properties. Field measurements of f at different river stages together with determinations of associated stream gradient make it possible to find values of r and k.

9.1.1 Temperature dependence for self purification constant (f)

$$f = \frac{r}{k} = \frac{r_0 \exp[C_r(T - T_0)]}{k_0 \exp[C_k(T - T_0)]}$$

$$= f_0 \exp[(C_r - C_k)(T - T_0)]$$

$$f = f_0 \exp[C_f(T - T_0)]$$

If $C_r = 0.024$ per °C and $C_k = 0.046$ per °C,

$\quad C_f = (C_r - C_k) = -0.020$ per °C, implying that the magnitude of f decreases with increases in temperature.

9.2 Avoidance of septic conditions

Maximum critical deficit (D_c) is equal to the DO saturation value (S). For the support of game fish (trout), the DO content should not fall below 5 mg/L. The allowable critical deficit at 25°C becomes (8.2–5.0) = 3.2 mg/L.

9.2.1 Maximum allowable loading on receiving waters

Two boundary conditions exist for this for given values of k, f, Dc and initial deficit (D_a) :

Full DO deficit $(D_a = 0)$

Initial deficit equal to D_c $(D_a = D_c)$

If $D_a = 0$ and D_c equal or less than S [First boundary value]

$$t_c = \frac{2.3}{k(1-f)} \log\left\{ f\left[1 - (f-1)\frac{D_a}{L_a}\right]\right\}$$

$$t_c^* = \frac{2.3}{k(1-f)}\,(\log f),\ \ [D_a = 0]$$

$$D_c = \frac{L_a \exp(-k\,t_c)}{f} = \frac{L_a}{f\left\{f\left[1 - (f-1)\left(\dfrac{D_a}{L_a}\right)\right]\right\}^{f/f-1}}$$

$$D_c = \frac{L_a}{f \cdot f^{f/f-1}}, \quad [D_a = 0]$$

$$\frac{L_a^*}{D_c} = [f]^{f/f-1} = f \exp(k t_c^*)$$

$$t_i = \frac{2.3}{k(f-1)} \log\left\{ f^2\left[1-(f-1)\frac{D_a}{L_a}\right]\right\}$$

$$t_i^* = \frac{2.3}{k(f-1)} \log f^2; \quad [D_a = 0]$$

$$t_i^* = \frac{4.6}{k(f-1)} \log f; \quad \text{Therefore } t_a^{\cdot} = 2 t_c^{\cdot}$$

$$\frac{D_i^*}{D_c} = \frac{(f+1)}{(f)^{f/(f-1)}} = (f+1)D_c L_a^*$$

If $D_a = D_c$ and D_c is less or equal to S (Second Boundary Condition)

$$t_c^{**} = 0$$

$$\frac{L_a^{**}}{D_c} = f$$

$$t_i^{**} = \frac{2.3(\log f)}{[k(f-1)]} = t_c^*$$

$$D_i^{**} = D_i^*$$

Ratio of the first boundary value to the lower limit of maximum loading is :

$$\frac{L_a^*}{L_a^{**}} = (f)^{1/f-1} = \exp(k\, t_c^*)$$

Therefore, allowable loading and co–ordinators of the characteristic points of the oxygen sag became simple functions of the self purification coefficient (f) when the loading is expressed in terms of the critical deficit and the critical time is expressed in terms of (kt$_c$). If f = 1, than,

$t_c^* = 1/k$, and $\dfrac{L_a^*}{D_c} = e = 2.718.$

9.3   Determine the allowable loading and characteristic coordinates of the oxygen sag of a stream under the following conditions, (1) maximal ($D_a = 0$), (2) minimal ($D_a = D_c$), and (3) intermediate ($D_a = 0.4D_c$). Use the following data :

Water temperature   : 25°C

Minimum DO content   : 4.5 mg/L

Rate of deoxygenation   : 0.2 d⁻¹ at 25°C

Rate of selfpurification (f)   : 2.0 at 25°C

9.3.1 Critical DO deficit of the stream (D_c)

$$= (C_s - C_c) = (8.2 - 4.5) = 3.7 \text{ mg/L at } 25°C$$

9.3.2 Allowable loading

(a) Maximal ($D_a/D_c = 0$ as $D_a = 0$)

$$\frac{L_a^*}{D_c} = f^{[f/(f-1)]}$$

$$= (2)^{(2/1)} = 4$$

$$L_a^* = 4D_c = 4 \times 3.7 = 14.8 \text{ mg/L}$$

(b) Minimal ($D_a/D_c = 1.0$ as $D_a = D_c$)

$$\frac{L_a^{**}}{D_c} = f$$

$$L_a^{**} = f\,D_c = 2.0 \times 3.7 = 7.4 \text{ mg/L}$$

(c) Intermediate ($D_a = 0.4\ D_c$)

$$D_c = \frac{L_a}{f\left\{f\left[1-(f-1)\dfrac{D_a}{L_a}\right]\right\}^{1/(f-1)}}$$

$$= \frac{L_a}{2\left\{2\left[1-(2-1)\dfrac{D_a}{L_a}\right]\right\}}$$

$$D_c = \frac{L_a}{2\left\{2\left[1-\dfrac{D_a}{L_a}\right]\right\}}$$

$$4D_c = \frac{L_a}{\left[1-\dfrac{0.4D_c}{L_a}\right]} \quad \text{or} \quad L_a = 4D_c - \frac{1.6D_c}{L_a}$$

$$L_a^2 - 4D_cL_a + 1.6D_c = 0$$

$$L_a^2 - 4 \times 3.7L_a + 5.92 = 0 \quad (a=1,\ b=-14.8,\ c=5.92)$$

or $\qquad$ $L_a = \dfrac{-b \pm \sqrt{b^2 - 4ac}}{2} = 14.38$ mg/L.

9.3.3 Characteristic co–ordinates

(a) Maximal ($D_a/D_c = 0$)

$$t_c^* = \frac{2.3(\log f)}{[k(f-1)]}$$

$$= \frac{2.3 \log 2}{0.2(2-1)} = 3.46 \text{ days}$$

$$t_i^* = 2t_c^* = 2 \times 3.46 = 6.92 \text{ days}$$

$$\frac{D_i^*}{D_c} = \frac{f+1}{f} \exp[-k(t_i - t_c)] = (f+1)D_c L_a^*$$

$$= \frac{f+1}{f^{f/f-1}} = \frac{2+1}{2^2} = \frac{3}{4} = 0.75$$

$$D_i^* = 0.75 D_c = 0.75 \times 3.7 = 2.775 \text{ mg/L}$$

(b) Minimal ($D_a/D_c = 1.0$)

$$t_c^{**} = 0$$

$$t_i^{**} = t_c^* = 3.46 \text{ days}$$

$$D_i^{**} = D_i^* = 2.775 \text{ mg/L}$$

(c) Intermediate ($D_a = 0.4\, D_c$)

$$D_c = \frac{L_a \exp(-k t_c)}{f} = \frac{L_a}{f\left\{f\left[1-(f-1)\dfrac{D_a}{L_a}\right]^{1/f-1}\right\}}$$

or $\quad \dfrac{L_a \exp(-k t_c)}{f} = \dfrac{L_a}{f\left\{f\left[1-(f-1)\dfrac{D_a}{L_a}\right]^{1/(f-1)}\right\}}$

or $\left\{f\left[1-(f-1)\dfrac{D_a}{L_a}\right]\right\}^{1/(f-1)} = \exp(k t_c)$

$$\left\{2\left[1-\frac{D_a}{L_a}\right]\right\} = \exp(k t_c)$$

$$2 - \frac{2D_a}{L_a} = \exp(k\, t_c)$$

$$2 - \frac{2 \times 0.4 D_c}{L_a} = \exp(k\, t_c)$$

$$2 - \frac{0.8 \times 3.7}{15.2} = \exp(k\, t_c)$$

or
$$e^{k t_c} = 1.8$$

$$k t_c = \ln(1.8) \text{ or } t_c = \frac{\ln 1.8}{k} = 2.94 \text{ day}$$

$$k(t_i - t_c) = \frac{2.3 \log f}{(f-1)} = \frac{2.3 \log 2}{1} = 0.692$$

or
$$k t_i = k t_c + 0.692$$

$$t_i = \frac{0.2 \times 2.94 \times 0.692}{0.2} = 6.4 \text{ days}$$

$$\frac{D_i}{D_c} = \frac{f+1}{f} \exp[-k(t_i - t_c)] = \frac{f+1}{f^{(f/f-1)}}$$

$$= \frac{2+1}{2^2} = 0.75$$

$$D_i = \frac{f+1}{f^{(f/f-1)}} \cdot D_c$$

$$D_i = 0.75 \times 3.7 = 2.775 \text{ mg/L}$$

Alternatively,
$$\frac{D_i}{D_c} = \frac{2+1}{2} \exp[-0.2(6.4 - 2.94)]$$

$$D_i = 0.7508 D_c = 0.7508 \times 3.7 = 2.778 \text{ mg/L}$$

i.e., D_i is constant

10. Dilution requirements in streams

The amount of water into which waste substances can be discharged without creating objectionable conditions, or nuisance, is the converse of the allowable pollutional loading of the receiving water.

If waste with first stage BOD of L lb per capita per day is charged into a stream carrying Q CfS per 1000 persons, the BOD loading is 185.5 L/Q (mg/L). For a permissible loading of L_a, therefore, the requirement stream flow becomes :

$$Q \text{ (CfS)} = 185.5\ L/L_a$$

10.1 Determine the need dilution corresponding to the allowable loadings of 14.8 mg/L (maximal), 7.4 mg/L (minimal), and 14.38 mg/L (intermediate) for a residual DO of the 4.5 mg/L. Therefore, stage BOD (20°C) of combined sewage is assumed to be 0.18 lb per capita per day.

$$Q = \frac{185.5 \times 0.18}{L_a}$$

$$Q\,(\text{maximal},\ D_a = 0) = \frac{185.5 \times 0.18}{14.8} = 2.26\ \text{CfS}/1000\ \text{person}$$

$$Q\,(\text{maximal},\ D_a = D_c = 3.7\ \text{mg/L}) = \frac{185.5 \times 0.18}{7.4} = 4.51\ \text{CfS}/1000\ \text{person}$$

$$Q\,(\text{int ermediate},\ D_a = 0.4\ D_c = 1.48\ \text{mg/L}) = \frac{185.5 \times 0.18}{14.38} = 2.82\ \text{CfS}/1000\ \text{person}.$$

Note :

1. Aquatic organisms obtain their oxygen from the oxygen dissolved in water. Water contains only 0.8% oxygen by volume at normal temperature (about 50°F), whereas the atmosphere holds about 21% by volume, the aquatic environment is inherently and critically sensitive to the demands of the oxygen that populate it.

2. Determination of the amount of DO in water relative to its saturation value and rate of utilization (BOD), furnishes a ready and useful mean for identifying the pollutional status of water, and by indirection, also amount of decomposable (organic matter).

3. Exertion of the BOD results in dioxygenation of receiving waters. Absorption of oxygen from the atmosphere and from green plants during photosynthesis results in reaeration (re–oxygenation) of receiving waters. The interplay between deoxygenation and reaeration produces the DO profile of a stream called the oxygen sag.

4. In absence of actual tests the following values can be used in first estimates of expected BOD behaviour of raw and treated municipal wastewaters :

Table 70 : Values of k and L for various types of wastewater.

Nature of wastewater	$k\ (d^{-1})$	$L\ (mg/L)$
Weak wastewater	0.35	150
Strong wastewater	0.39	250
Primary effluent	0.35	75–150
Secondary effluent	0.12–0.23	15–75
Tap water	Less than 0.12	0–1

5. Hydraulic control of receiving waters

   Regulating water discharge.

   Controlling the flow of the receiving water (low–water regulation).

   Supplementing, strengthening or preserving the self purifying power of the receiving water structure, hydraulically, or air injection.

   Limiting sludge deposition.

Example 6.117

Stream Sanitation

A tributory merges with a river. Calculate the initial concentration of DO deficit, C–BOD, and temperature for use in a DO model. Assume the following data :

Table 71 : Parameters for the Two Types of Flows.

Type of flow	Temp. (°C)	Cs (mg/L)	C (mg/L)	L* (mg/L)	Q (ft³/s)
River	26.3	8.4	7.3	3.0	2000
Tributory	24.9	8.6	6.8	6.0	500

r–refers to river, t–refers to tributory L–C–BOD, C refers to DO.

Solution :

1. Required BOD, DO and temperature at the confluence point

   River segmentation

   A new river segment is necessary when initial conditions or parameters change in the DO model. Initial concentration may change due to :

   – Tributory input or confluence

   – Wastewater discharges

   – Dam or rapids (rapid re–aeration).

   When a wastewater discharge enters the river, a flow weighted average mass balance of the stream is computed for the new initial conditions, assuming instantaneous complete mixing at x = 0 :

   $$L_o = \frac{W + L_s\,Q}{Q + Q_W}$$

   where L_o = Initial C–BOD at x = 0 (M/L³)

   $\quad\quad\ \ W$ = Wastewater discharge (M/T)

   $\quad\quad\ \ L_s$ = Upstream C–BOD entering at x = 0 (M/L³)

Q = River flow (L³/T)

Q_w = Wastewater flow (L³/T)

If wastewater mass of C–BOD is large compared to upstream sources and if the wastewater flow rate is small relative to the river :

$$L_o = \frac{W}{Q}$$

[Similar mass balances are performed for N–BOD to obtain DO at x = 0]

Initial DO (C_o) :

$$C_o = \frac{Q_1\,C_1 + Q_2\,C_2}{Q_1 + Q_2}$$

Initial temperature (T_o) :

$$T_o = \frac{Q_1\,T_1 + Q_2\,T_2}{Q_1 + Q_2}$$

Initial mixed DO deficit (D_o) :

$$D_o = (C_{sat} - C_o) \text{ at } T_o$$

where C_o = Initial mixed DO (M/L³)

Q_1 = Main river flow (L³/T)

C_1 = Main river DO (M/L³) at x = 0

Q_2 = Tributary flow (L³/T)

C_2 = Tributary DO (M/L³) at x = 0

T_o = Initial mixed temperature (°C)

T_1 = Main river temperature (°C)

T_2 = Tributary temperature (°C)

D_o = Initial mixed DO deficit (M/L³)

C_{sat} = Saturated concentration of DO at T_o (M/L³)

$$L_o = \frac{Q_r\,L_r + Q_t\,L_t}{Q_r + Q_t} = \frac{2000(3.0) + 500(6.0)}{2000 + 500}$$

$$= 3.6 \text{ mg/L (C – BOD)}$$

$$C_o = \frac{Q_r\,C_r + Q_t\,C_t}{Q_r + Q_t} = \frac{2000(7.3) + 500(6.8)}{2000 + 500}$$

$$= 7.2 \text{ mg/L (D) concentration)}$$

$$T_o = \frac{Q_r\,T_r + Q_t\,T_t}{Q_r + Q_t} = \frac{2000(26.3) + 500(24.9)}{2000 + 500}$$

$$= 26.02°C$$

DO deficit at $T_o = 26.02°C$, D_o:

$$D_o = C_s - C_o = 8.5 - 7.2 = 1.3 \text{ mg/L}.$$

Note :

1. Qual–ZE model

   This model simulates DO with all terms in normal Streeter–Phelps equation, and also includes eutrophication and nutrients including organic–nitrogen, ammonia–N, NO_3–N, nitrite–N, org–P, phosphate–P, phytoplankton biomass, chlorophyll–a, coliform bacteria, and temperature (heat balance). It allows for dispersion in large rivers and includes a Monte Carlo uncertainty analysis option. It can be run under steady state or dynamic water quality conditions, but the flow regime is steady.

2. Changes in other parameters in the basic equation that are constant would require a new segment :

   Velocity (u, changes in Q or area)

   Sediment oxygen demand (S)

   Net primary production (P – R)

   Background C–BOD due to non–point source pollution (L_b)

   Re–aeration rate constant (Change in velocity or depth).

3. Mass balance equation for DO (DO deficit)

$$\frac{\partial C}{\partial t} = -\frac{Q}{A}\frac{\partial C}{\partial x} + k_a(C_s - C) - k_d L(x) - k_n N(x) + P(x,t) - R(x) - \frac{S(x)}{H} \qquad (1)$$

where   C = DO concentration

            t = Time

            Q = Flow rate

            A = Cross–sectional area

            x = Longitudinal distance

            k_a = Re–aeration rate constant

            C_s = Saturation DO concentration

            k_d = C–BOD deoxygenation rate constant

            k_n = N–BOD deoxygenation rate constant

            L(x) = C–BOD concentration with distance

            N(x) = N–BOD concentration with distance

            P(x, t) = Primary production as a function of x and t ($M/L^3.T$)

R(x)　　= Plant respiration with distance $(M/L^3.T)$

S(x)　　= Sediment oxygen demand with distance $(M/L^2.T)$

H　　　= Mean depth (L)

Field survey data are very sensitive to cloud cover, which reduces primary production, lowering the average DO concentration throughout the day. Also, estimates of P(x, t) and R(x) are difficult to obtain. R(x) includes only respiration due to plants. Yet a light–dark bottle measurement of R(x) includes deoxygenation due to C–BOD and N–BOD. Furthermore, respiration by zooplankton may be stimulated or deterred within floating bottles. Primary production measurements in light bottles do not include rooted aquatic plants (macrophytes); they may include only planktonic algae and are an underestimate of total primary production. For these reasons, it is desirable to estimate primary production [P(x, t)] and respiration [R(x)], by an analysis of diurnal cycle obtained at each sampling station within the stream. This is equivalent to obtaining P(x, t) and R(x) from calibration of the model with field data.

The above equation requires a forcing function for primary production in each segment by the river. The rate of photosynthetic oxygen production, P(t) may be represented as a half cycle sine–wane :

$$P(t) = P_m \, S_{in}\left[\frac{\pi}{f}(t - t_s)\right] \quad \text{for } t_s \le t \le (t_s + f)$$

$P(t) = 0$ when $(t_s + f) \le t \le t_s$

where　P_m = Maximum rate of primary oxygen production (mg/L.d)

　　　　t_s = Time at which the source begins (days)

　　　　f = Fraction of the day over which the source is active (usually 0.5 d)

To extend the periodic function of primary production for more than 1 day, a Fourier series is used :

$$P(t) = P_m\left[\frac{2f}{\pi} + \sum_{n=1}^{\infty} b_n \, Cos\left[2\pi n\left(t - t_s - \frac{f}{2}\right)\right]\right] \tag{2}$$

where　$b_n = Cos(n\,\pi f) \times \dfrac{\dfrac{4\pi}{f}}{\left(\dfrac{\pi}{f}\right)^2 - (2\pi n)^2}$

　　　　n = No. of cycles

There is a diurnal variation of oxygen concentration at the head end of the segment can be handled in a manner similar to equation (2) :

$$C_0(t) = \frac{A_o}{2} + \sum_{n=1}^{\infty} A_n \cos\left(\frac{n\pi}{T}\right) + B_n \sin\left(\frac{n\pi t}{T}\right)$$

where   T = One half period (T = 0.5 day)

Dropping all but the first two terms of the Fourier series :

$$C_0(t) = \frac{A_o}{2} + A_1 \cos\left(\frac{\pi}{12} \times t\right) + B_1 \sin\left(\frac{\pi}{12} \times t\right) + A_2 \cos\left(\frac{\pi}{6} \times t\right)$$

$$+ B_2 \sin\left(\frac{\pi}{6} \times t\right)$$

4. Model includes the following types of errors

    Model errors (Incorrect formulation of model).

    Errors in the state variables (dependent variables and initial conditions omitted or included improperly more variables).

    Errors in input data used to drive model.

    Parameter error (rate constants, coefficients, and independent variables).

Example 6.118

Stream Sanitation

A plug flow stream has a velocity of 0.3048 m/s, a mean depth of 1.056, and a deoxygenation rate constant of 0.6 d^{-1}. The initial ultimate BOD concentration at x = 0 is 10.0 mg/L, and the initial deficit is 0.0 mg/L. Estimate the re–aeration rate constant using the O'Connor–Dobbins re–aeration formula. Where is the re–aeration rate equal to the deoxygenation rate? What is the critical deficit and distance?

Solution

1. Required re–aeration rate constant (k_a)

$$k_a = \frac{12.9\bar{u}^{0.5}}{H^{1.5}} = \frac{12.9(1)^{0.5}}{(3.465)^{1.5}} = 2.0 \ d^{-1}$$

2. Required BOD equation and the Streeter–Phelps DO deficit equation

$$L = L_o \exp\left[-k_d \frac{x}{\bar{u}}\right]$$

$$D = D_o \exp\left[-k_a \frac{x}{\bar{u}}\right] + \frac{k_d L_o}{k_a - k_d}\left[\exp\left(-k_d \frac{x}{\bar{u}}\right) - \exp\left(-k_a \frac{x}{\bar{u}}\right)\right]$$

where k_a = 2.0 d^{-1}, k_d = 0.6 d^{-1}, L_o = 10 mg/L,

D_o = 0 mg/L, and $\bar{u}$ = 16.4 mile/d

Table 72 : Required parameters.

x (miles)	L (mg/L)	k_d L (mg/L.d)	D (mg/L)	k_a D (mg/L.d)
0	10.0	6.00	0	0
5	8.33	5.00	1.24	2.48
10	6.94	4.16	1.71	3.42
15	5.78	3.47	1.79	3.58
20	4.81	2.89	1.69	3.38
30	3.33	2.00	1.32	2.64
40	2.31	1.39	0.96	1.92
50	1.60	0.96	0.68	1.36
75	0.64	0.38	0.27	0.54

3. Required critical deficit and distance

$$x_c = \frac{\bar{u}}{k_a - k_d} \ln\left(\frac{k_a}{k_d}\right) = 14.1 \text{ miles}$$

$$D_c = \frac{k_d L_o}{k_a} \exp\left[-\frac{k_d x_c}{\bar{u}}\right] = 1.8 \text{ mg/L}$$

Rate of re–aeration (k_a D) is exactly equal to the deoxygenation rate (k_dL) where the deficit is a maximum (x_c = 14.1 mile).

Note :

1. Effect of temperature

    DO re–aeration rate constant increases with increasing temperature.

    DO solubility decreases with increasing temperature, so the driving force (D) for re–aeration decreases somewhat.

    Deoxygenation rate constant increases with increasing temperature.

    k_d (T) = k_d (20)(1.048)$^{T-20}$

    k_a (T) = k_a (20)(1.024)$^{T-20}$

    Therefore, the effect of increasing temperature is greater on deoxygenation compared to re–aeration, and the causes the critical deficit (D_c) to increase and to move upstream. Increasing temperature also results in a marked decrease in DO saturation concentration, which decreases the DO concentration (C) by limiting the driving force for re–aeration (C_s – C).

2. Stream flow (Q)

    Increasing stream flow has two effects on DO deficit profiles :

    Initial BOD concentration (L_o) decreases proportionately with flow rate [L_o = W/Q; W = wastewater mass discharge rate (M/T); and

Q = Flow rate of river]

Flow rate influences re–aeration rate constant in complex fashion :

$$k_a \alpha \frac{\bar{u}^m}{H^n}$$

where $\bar{u}$ is the mean stream velocity, H is the mean depth, m is an exponent usually less than 1.0, and n is the exponent usually greater than 1.0

- Increasing the flow rate increases both the mean velocity and the mean depth according to the friction factor energy slope and backwater profile.

- Effect of increasing flow rate on mean depth (to the n–power) is larger than the effect and mean velocity (to the m power), so the reaerator rate constant decreases with increasing stream flow. This is somewhat counter intritive result that is not true in the tail waters of dam releases and rapids.

- Over–riding effect of increased flow rate is on the initial ultimate BOD concentration, L_o at x = 0 and the net effect is to push DO deficit curve further down–stream and to decrease the critical deficit (D_c).

- Effect of flow rate on stream physical parameters are :

  • Mean depth (H) α Q^a

    Mean velocity ($\bar{u}$) α Q^b

    Mean depth (B) α Q^c

  With the condition : a + b + c = 1

  where   Q = Flow rate

$$A = 0.4 - 0.7$$
$$B = 0.3 - 0.5$$
$$C = 0 - 0.25$$

The cross–sectional area of the stream varies with flow rate to the power (1–b) by virtue of continuity equation

$$A \alpha \frac{Q}{\bar{u}} \left(= \frac{Q}{Q^b} = Q^{1-b} \right)$$

3. Stream re–aeration formulae

O'Connor–Dobbins $k_a = \dfrac{12.9\,u^{0.5}}{H^{1.5}}$

Owen–Edwards Gibbs $k_a = \dfrac{23\,u^{0.73}}{H^{1.73}}$ for H = 1 to 2.5, u = 0.1 to 0.5,

Q = 4 – 36

Churchill–Elmove–Buckingham

$$k_a = \frac{11\,u}{H^{1.67}} \quad \text{and}$$

H = 2 to 11

u = 2 to 5

Q = 1600–17,000

USGS

$$k_a = \frac{7.6\,u}{H^{1.33}}$$

Tsivoglou

$$k_a = \frac{0.048\,\Delta S}{t} \quad \text{for a = 5 to 3,000}$$

where k_a = Re–aeration rate constant (base e, d^{-1})

u = Mean stream velocity (ft/s)

H = Mean stream depth (ft)

ΔS = Water surface elevation change (ft)

Q = Flow rate (ft^3/s)

t = Travel time (days)

4. Overall mass balance equation for DO deficit in a plug flow river

$$\frac{\partial D}{\partial t} = -u\,\frac{\partial D}{\partial X} - k_a D + k_d L + k_n N + \frac{S}{H} + (R - P) \tag{1}$$

[Accumulation] = [Transport] – Sinks + Sources of deficit

At steady–state :

$$u\,\frac{\partial D}{\partial x} + k_a D = k_d\,L_o\,e^{-k_r\,x/u} + k_n\,N_o\,e^{-k_n\,x/u} + \frac{S}{H} + (R - P)$$

The solution is :

$$D = D_o\,e^{-k_a x/u} + \frac{k_d\,L_o}{k_a - k_r}(e^{-k_r\,x/u} - e^{-k_a\,x/u}) + \frac{k_n\,N_o}{k_a - k_n}\left(e^{-k_n\,x/u} - e^{-k_a\,x/u}\right)$$

$$+ \frac{S}{k_a H}(1 - e^{-k_a\,x/u}) + \frac{(R - P)}{k_a}(1 - e^{-k_a\,x/u}) + \frac{k_d L_b}{k_a} \tag{2}$$

Table 73 : Mass balance equation (2) with sources and sinks and the solutions.

Source of Deficit/Sinks	Mass Balance Equation	Solution
Initial deficit (D_o)	$u\dfrac{dD}{dx} = -k_a D$	$D = D_o e^{-k_a x/u}$
C–BOD (L)	$u\dfrac{dD}{dx} = -k_a D + k_d L$	$D = \dfrac{k_d L_o}{k_a - k_r}(e^{-k_r x/u} - e^{-k_a x/u})$
N–BOD (N)	$u\dfrac{dD}{dx} = -k_a D + k_n N$	$D = \dfrac{k_n N_o}{k_a - k_n}(e^{-k_n x/u} - e^{-k_a x/u})$
Sediment oxygen demand	$u\dfrac{dD}{dx} = -k_a D + \dfrac{S}{H}$	$D = \dfrac{S}{k_a H}(1 - e^{-k_a x/u})$
Net (Respiration–production)	$u\dfrac{dD}{dx} = -k_a + (R - P)_{avg}$	$D = \dfrac{R - P}{k_a}(1 - e^{-k_a x/u})$
Back ground deficit (D_b) [Carried by non–point source]	$u(0) = -k_a D_b + k_d L_b$	$D_b = \dfrac{k_d L_b}{k_a}$

Summation of sources and sinks	**Overall equation**	**Solid equation (2)**

Schematics of the DO deficit equation :

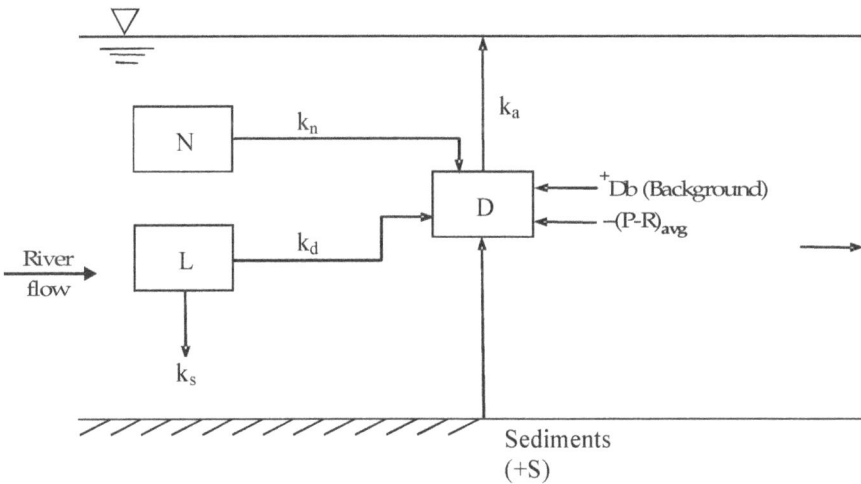

Figure 49 : Schematics of the DO deficit.

If respiration is greater than photosynthesis (night time conditions), then net photosynthesis is –ve, and the term (R–P) contributes to the DO deficit.

If primary production is greater than respiration (day time), and DO is created.

Worst (critical case) condition occur before sunrise where R > P

where   u   = Mean velocity (L/T)

$$L \qquad = \text{Carbonaceous BOD } (M/L^3)$$

$$k_r \qquad = \text{Total C BOD loss rate constants } (T^{-1})$$

$$\qquad = k_s + k_d$$

$$L_o \qquad = \text{Initial C–BOD concentration at } x = 0 \ (M/L^3)$$

$$x \qquad = \text{Longitudinal distance } (L)$$

$$k_d \qquad = \text{C–BOD deoxygenation rate constant } (T^{-1})$$

$$k_a \qquad = \text{Re–aeration constant } (T^{-1})$$

$$D \qquad = \text{DO deficit } [C_s - C], \ M/L^3]$$

$$\text{N–BOD} = \text{Nitrogenous BOD (N) with nitrogenous deoxygenation rate constant } (k_n)$$

$$k_s \qquad = \text{Sedimentation of C–BOD}$$

$$S \qquad = \text{Sediment oxygen demand}$$

$$L_b \qquad = \text{Background BOD caused by non–point sources}$$

$$H \qquad = \text{Mean depth } (L)$$

At peak photosynthesis (12 Noon and plenty of nutrients) gross primary production is in the range of 10–40 mg/L.d of DO produced.

Photosynthesis can be considered as varying sinusoidally during photo period, while respiration occurs continuously.

Peak concentrations of oxygen in a stream are expected to occur at miday, but if one wants to measure the worst case conditions, they would occur just before sun rise when algal blooms have consumed significant concentrations of DO during night respiration.

Respiration amounts to an average of 2.5% of gross photosynthesis, so the net effect of phyto plankton is to supply DO to the stream prior to death and sedimentation.

Background BOD [non–point source pollution to stream] agricultural runoff, storm water discharges, and highway runoff.

$$u\frac{dD}{dx} = -k_a D + k_d L \quad \text{[at steady state]}$$

$$D_b = \frac{k_d L_b}{k_a} \quad \text{[at steady state and a continuous Conc. of C – BOD]}$$

Solutions for carbonaceous BOD (L), nitrogenous BOD (N), and bacterial concentrations (C) in river are expressed as :

$$L = L_o \exp[- k_r x/u]$$

$$N = N_o \exp[- kn \ x/u]$$

$$C = C_o \exp[- k \ x/u]$$

5. Sedimentation of C–BOD

   Streeter–Phelps equations imply that CBOD is entirely soluble and that loss of CBOD in the stream is only by deoxygenation (oxidation of soluble organics). However, particulate CBOD is discharged at wastewater treatment plants as well. Domestic treated wastewater is permitted to contain 30 mg/L of CBOD and 40 mg/L of total suspended solids (TSS). The total suspended solids contain organic matter that may be exerted in a BOD test. If CBOD is in particulate form, it will settle out of the water column. (Perhaps it will form a sludge deposit below the discharge point and require the specification of a sediment oxygen demand.) We may modify the Streeter–Phelps equations slightly to include the possibility of CBOD sedimentation.

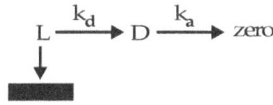

Figure 50 : BOD sedimentations.

$$u\frac{dL}{dx} = -k_r L$$

$$L = L_o \exp(-k_r x/u)$$

$$u\frac{dD}{dx} = -k_d L - k_a D$$

$$u\frac{dD}{dx} = k_d L_o \exp(-k_r x/u) - k_a D$$

6. Method of determination and range of DO model rate constants at 20°C with typical temperature correction factors for rivers

Table 74 : Magnitude for various parameters

Parameters	Value	θ Temperature Correction[a]	Method of Determination
CBOD deoxygenation, k_d	0.05–0.5 day^{-1}	1.048	Lab
CBOD deoxygenation Plus sedimentation, k_r	0.5–5 day^{-1}	1.04	Lab
NBOD deoxygenation, k_n	0.05–0.5 day^{-1}	1.08	Semilog N versus $x/\overline{u}$
Reaeration, k_a			
Slow, deep rivers	0.1–0.4 day^{-1}	1.024	Calibration, field isotopes, or Refer Section 3
Typical conditions	0.4–1.5 day^{-1}	1.024	Calibration, field isotopes Refer Section 3
Swift, deep rivers	1.5–4.0 day^{-1}	1.024	Calibration, field isotopes or Refer Section 3

Contd...

Swift, shallow rivers	4.0–10 day^{-1}	1.024	Calibration, field isotopes or Refer Section 3
Sediment oxygen Demand, S			
Natural to low pollution	0.1 to 1.0 g m^{-2}d^{-1}	1.065	Calibration or *in situ* respirometer
Moderate to heavy Pollution	5–10 g m^{-2} d^{-1}	1.065	Calibration or *in situ* respirometer
Primary production,(P–R)			
Daily average value (P–R)	0.5–10 mg L^{-1} d^{-1}	1.066	*In situ* light and dark bottles or calibration
P_{max} maximum Daily	2–20 mg L^{-1}d^{-1}		*In situ* light and dark bottles or calibration production
R, respiration only	1–10 mg L^{-1}d^{-1}		*In situ* light and dark bottles or calibration
Background DO Deficit,D_b	0.5–2 mg L^{-1}	NA	Model calibration
Coliform bacteria die–away, k			
Freshwater	0.5–5 day^{-1}	1.07	Lab or field semilog plot
Saltwater	2–40 day^{-1}	1.10	Lab or field semilog plot
Virus particles in marine Waters	0.03–0.16day^{-1}	1.10	Lab or field semilog plot

a : $k(T) = k(20)q^{T-20}$.

Example 6.119

Activated Sludge Process [ASP]

1. Determine the biomass concentration ($X_{B,2}$) in an activated sludge process without recycle. Assume the following data :

Influent COD (soluble)	:	450 g COD (S)/m^3
Effluent COD (soluble)	:	60 g COD (S)/m^3
Observed yield coefficient (Y_{obs})	:	0.3 kg COD (B)/kg COD (S)

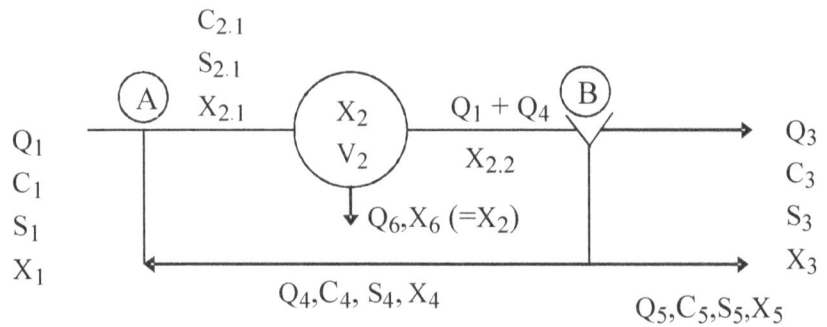

Figure 51 : Schematics of ASP.

1.1 Required equations for the activated sludge process

Water balance is :

$$Q_1 = Q_3 + Q_5$$

General mass balance is for any substance with concentration (C) :

$$Q_1 C_1 - r_{V,S} V_2 = Q_3 C_3 + Q_5 C_5$$

$r_{V,S} V_2$ can be written with respect to volume (V_2), as also with respect to activated sludge concentration ($X_{B.2}$) [$= r_{X,S} = V_2.X_{B.2}$]

where C is a general concentration term (COD, BOD, N, TOC, etc.)

The three processes are responsible for conversion (no nitrification):

Biological growth

Hydrolysis

Growth

The stoichiometric coefficients are presented in Table 75, and the two reaction terms are determined by multiplying the stoichiometric coefficients as given in this table.

$$\underbrace{Q.S_{S,1}}_{\text{Input}} + \underbrace{1.k_n X_S.V_2}_{\text{Hydrolysis}} + \underbrace{\dfrac{\left(-\dfrac{1}{Y_{max.H}}\right)(\mu_{max,H})\left(\dfrac{S_{S,2}}{K_S + S_{S,2}}\right)\left(\dfrac{S_{O_2,2}}{K_{SO_2,2} + S_{O_2,2}}\right)}{X_{B,H2}V_2}}_{\text{Growth}} = Q_3 S_3$$

Oxygen has been expressed as negative COD as there is no oxygen demand when aerating a certain amount of O_2 (but this amount of O_2 can remove an equivalent amount of COD?)

The activated sludge concentration ($X_{B.2}$) can be determined knowing the value of yield coefficient (without recycle)

$$X_{B,2} = Y_{obs}(C_1 - C_3)$$

Mass balance for biomass at point: A is :

$$Q_1.X_{B,1} + Q_4.X_{B,4} = (Q_1 + Q_4) X_{B,2.1}$$

Mass balance for biomass at point: B is :

$$(Q_1 + Q_4) X_{B,2.2} = (Q_1 + Q_4) X_{B,4}$$

$$Q_4 = \dfrac{[Q.X_{B,2.2} - Q_3.X_{B,3}]}{[X_{B,4} - X_{B,2.2}]}$$

Mass balance around the aeration tank is :

$$(Q_1 + Q_4) X_{B,2.1} + (Q_1 + Q_4) (C_{2.1} - C_{2.2}) Y_{obs} = (Q_1 + Q_4) X_{B'2.2}$$

Table 75 : Process Stoichiometry Coefficients for various Components

Parameter	S_S^*	X_S^*	X_I^*	$X_{B,H}^*$	$S_{O_2}^*$	Reaction rate (r_v)
Process						
1. Aerobic heterotrophic growth	$-\dfrac{1}{Y_{max,H}}$			1	$\dfrac{1-Y_{max,H}}{Y_{max,H}}$	$\mu_{max,H}\left[\dfrac{S_S}{S_S+K_S}\right]\left[\dfrac{S_{O_2}}{S_{O_2}+K_{S,O_2,H}}\right]X_{B,H}$
2. Decay of heterotrophs		$1-f_{XB,XI}$	$f_{XB,XI}$	-1		$b_H X_{B,H}$
3. Hydrolysis	1	-1				$k_h\,X_s$
Type of organic matter	Easily degradable organic matter	Slowly degradable organic matter	Inert suspended organic matter	Hetero-trophic biomass	Oxygen	

* kg COD/m³

or alternatively :

$$X_{B,2.2} = X_{B,2.1} + (C_{2.1} - C_{2.2}) \, Y_{obs}$$

The amount of biomass concentration in the aeration tank is $X_{B,2}$:

Ideal well mixed tank : $X_{B,2} = X_{B,2.2}$

Plugflow reactor: It varies from $X_{B,2.1}$ (influent) to $X_{B,2.2}$ (effluent)

For activated process $X_{2.1} = X_{2.2}$

Biomass concentration (X) in terms of COD

$$X_{COD} > X_{B,H} + X_{B,A}$$

where $X_{B,H}$ = Concentration of heterotrophic biomass

$X_{B,A}$ = Concentration of autotrophic biomass (nitrifying micro–organism)

Sludge age (θ_x)

$$\theta_x = \frac{V_2 \, X_2}{[Y_{obs}(C_1 - C_3)Q_1]} = \frac{\theta X_2}{Y_{obs}(C_1 - C_3)}$$

where

$$\theta_x = \frac{M_x}{P_x}$$

$$P_x = Y_{obs}(C_1 - C_3)Q_1$$

$$M_x = V_2 \cdot X_2$$

$$P_x = Q_3 \cdot X_3 + Q_5 \cdot X_5 + Q_1 \cdot X_1$$

Assuming $Q_5 \cdot X_3 + Q_1 \, X_1 \approx 0$

$$P_x = Q_3 \cdot X_3, \text{ and } X_2 = X_3, \text{ (well mixed)}$$

$$\theta_x = \frac{V_2 \cdot X_2}{Q_3 \cdot X_3} = \theta$$

1.2   Required biomass concentration $(X_{B,2})$

$$X_{B,2} = Y_{obs}(C_1 - C_3)$$

$$= 0.3 \, \frac{\text{kg COD(B)}}{\text{kg COD(S)}} \left(0.450 \, \frac{\text{kg COD(S)}}{m^3} - 0.08 \, \frac{\text{kg COD(B)}}{m^3} \right)$$

$$= 0.111 \text{ kg COD (B)}/m^3$$

2. Determine aeration basin volume. Assume the following data

Wastewater flow (Q_1) :   1200 m^3/d

Influent soluble COD   :   450 g COD(S)/m^3

Effluent soluble COD   :   50 g COD(S)/m^3

Reaction rate $(r_{X,S})$:   2.5 kg COD(S)/kg COD(B).d

Yield coefficient (Y_{obs}) :   0.25 kg COD(B)/kg COD(S)

2.1   Required aeration basin volume (V_2)

Aeration biomass concentration $(X_{B,2}) = Y_{obs} (C_1 - C_3)$

$$= 0.25 (0.450 - 0.050)$$

$$= 0.125 \text{ kg COD (B)}/m^3$$

$$V_2 = \frac{(Q_1 C_1 - Q_3 C_3)}{r_{X,S} (X_{B,2})}$$

$$= \frac{1200(450) - 1200(50)}{(2.5)(0.125)}$$

$$= 1{,}536{,}000 \text{ m}^3$$

3.   Determine the effluent biomass concentration $(X_{B,2.2})$ for the ASP.

Assume the following data :

     Wastewater influent flow (Q_1) : 2000 m³/d

$X_{B,2.1}$ $= 4$ kg SS/m³

C_1 $= 0.45$ kg BOD/m³

C_3 $= 10$ mg/L (0.010 kg BOD/m³)

Y_{obs} $= 0.8$ kg SS/kg BOD

     Wastewater effluent flow (Q_4) : 1500 m³/d

3.1   Required effluent biomass concentration $(X_{B,2.2})$

$$X_{B,2.2} = X_{B,2.1} + Y_{obs} (C_{2.1} - C_{2.2})$$

$$(Q_1 + Q_4)(C_{2.1}) = Q_1 C_1 + Q_4 C_4$$

$$C_4 = C_3 = C_{2.2} \text{ [well mixed system]}$$

$$C_{2.1} = \frac{2000(0.45) + 1500(0.010)}{2000 + 1500}$$

$$= 0.25 \text{ kg BOD/m}^3$$

Therefore, $X_{B,2.2} = 4 + 0.8 (0.25 - 0.01)$

$$= 4.19 \text{ kg SS/m}^3$$

4.   Determine the sludge loading (L_X) and sludge production (P_X) for an ASP. Assume the following data :

Q_1 :   10,000 m³/d

C_1 :   450 mg BOD/L (0.45 kg BOD/m³)

C_3 :   50 mg BOD/L (0.05 kg BOD/m³)

Y_{obs} : 0.75 kg SS/kg BOD

Aeration basin volume (V_2) : 6000 m³ with biomass concentration of 4000 mg/L (4 kg SS/m³)

4.1 Required sludge loading (L_x) and sludge production (P_x)

$$L_x = \frac{Q_1 . C_1}{V_2 C_2}$$

$$= \frac{10,000(0.45)}{6000(4)} = 0.1875 \text{ kg BOD/kg SS.d}$$

$$P_x = Y_{obs} (C_1 - C_3) Q_1$$

$$= 0.75 (0.450 - 0.05) (10,000)$$

$$= 3000 \text{ kg SS/d}$$

5. Determine the sludge age (q_x) and hydraulic detention time in the aeration basin. Assume the following data :

Aeration basin volume (V_2)	:	10,000 m³
Aeration biomass Conc. (X_2)	:	4.5 kg COD/m³
Influent wastewater flow (Q_1)	:	1000 m³/h
Recycle flow (Q_4)	:	700 m³/h
Sludge flow from clarifier bottom (Q_5)	:	100 m³/d
Flow rate from the aeration	:	600 m³/d
Tank (withdrawal, Q_6) Effluent biomass Conc. (X_3)	:	50 mg/L
Recycle biomass Conc. (X_5) yield	:	10 kg COD/m³

5.1 Required hydraulic detention time (θ) and sludge age (θ_x)

$$\text{Hydraulic detention time} (\theta) = \frac{V_2}{Q_1} = \frac{10,000 \text{ m}^3}{1000 \text{ m}^3/\text{h}} = 10 \text{ h}$$

$$\left[\theta = \frac{V_2}{Q_1 + Q_4} = \frac{10,000}{1000 + 700} = 5.9 \text{ h} \right.$$

is not the detention time, as it is the time for one flow through of a wastewater particle]

$$\text{Sludge age} (\theta_x) = \frac{M_x}{P_x}$$

$$M_x = V_2 . X_2 + 0 \text{ (in the clarifier)}$$

$$P_x = Q_3 . X_3 + Q_5 . X_5 + Q_6 . [X_6 (= X_2)]$$

$$Q_1 = Q_3 + Q_5 + Q_6$$
$$Q_3 = Q_1 - Q_5 - Q_6 = 1000\,(24) - 100 - 600$$
$$= 23,300 \text{ m}^3/\text{d}$$

$$\theta_x = \frac{10,000(4.5)}{23,300(0.05) + 100(10) + 600(4.5)}$$

$$= 9.22.$$

Note :

1. Biological growth

    The process can be described using the following expression :

    Biological growth rate expression :

    $$r_{V,XB} = m_{max}\, f(s)\, X_B$$

    where $r_{V,XB}$ = Volumetric biological growth rate [kg COD(B)/m^3.d]

    μ_{max} = Max. specific growth rate (d^{-1})

    $f(s)$ = Growth kinetics [zero, first order or Monod kinetics]

    X_B = Biomass conc. [kg COD (B)/m^3 or kg VSS/m^3]

    Substrate consumption rate corresponding to biological growth :

    $$r_{V,S} = \frac{r_{V,XB}}{Y_{max}}$$

    where Y_{max} = Max. yield coefficient [kg COD (B)/kg COD (S)]

    $$r_{V,S} = \frac{\mu_{max}}{Y_{max}}\left[\frac{S}{K_S + S}\right] X_B$$

    Hydrolysis rates

    $$r_{V,XS} = k_h\, X_S \ [\text{Suspended solids } X_S]$$
    $$r_{V,S} = k_h\, S_S \ [\text{Dissolved organic matter } S_S]$$
    $$r_{V,S} = r_{V,XS} . V_{XS} \ [\text{Release of dissolved substrates}]$$

    where n_{XS} = Stoichiometric coefficient converting suspended solids (X) into dissolved solids (S) [If both are in same units n = 1, 1 kg COD (S)/kg COD(B)]

    Monod type hydrolysis ($r_{V,XS}$) for a given biomass (X_B)

    $$r_{V,XS} = K_{h,x} \times \left[\frac{X_S/X_B}{K_X + X_S/X_B}\right] X_B$$

    where $k_{h,x}$ = Hydrolysis rate constant

    K_X = Hydrolysis saturation constant

Table 76 : Hydrolysis rate constant.

Type of electron acceptor	Hydrolysis rate constant			
	Dissolved solids $[k_h,\ d^{-1}]$	Suspended solid $[k_h,\ d^{-1}]$	Monod type $[k_{h,x}$ kg COD (X)/kg COD (B) . d]$	Saturation constant $[K_{X},$ kg COD (X)/ kg COD (B)]$
Oxygen	3–20	0.6–1.14	0.6–1.4	0.02–0.05
Nitrate	1–15	0.15–0.4	0.15–0.4	0.02–0.05
Without O_2 and nitrate	2–20	0.3–0.7	0.3–0.7	0.02–0.05

First order conversion of substrate [substrate removal rate $(r_{V,S})$]

$$r_{V,S} = \frac{\mu_{obs}}{Y_{max}} X_B$$

$$\mu_{obs} = \mu_{max} f(s) f(So_2) f(pH) f(T) f \text{ (toxic substances)}$$

$$\mu_{max}(T) = \mu_{max}(20) \exp[k(T20)].$$

Oxygen :

$$f(S_{O_2}) = \frac{S_{O_2,2}}{K_{S,O_2} + S_{O_2,2}}$$

Oxygen dependency for aerobic process is :

$$\mu_{obs} = \mu_{max} \frac{S_{O_2}}{K_{S,O_2} + S_{O_2}}$$

or $\mu_{obs} = \mu_{max} \left[\frac{S_2}{K_2 + K_{S,2}}\right]\left[\frac{S_{O_2,2}}{K_{S,O_2} + S_{O_2,2}}\right]$

pH :

$$\mu_{max}(pH) = \mu_{max}[opt.pH] \frac{K_{pH}}{K_{pH} + I}$$

where K_{pH} = pH constant

$$I = 10^{|opt\ pH - pH|} - 1$$

$$S_{O2,2} = O_2 \text{ Conc. in the reactor}$$

Toxic substances :

$$\mu'_{max} = \mu_{max} \frac{K_{SI}}{K_{SI} + C_I}$$

$$K'_S = K_S \frac{K_{SI} + C_I}{K_{SI}}$$

where μ'_{max} = Max. specific growth rate with inhibition

$\mu_{max.}$ = Max. specific growth rate without inhibition

C_I = Inhibiting matter Conc.

K_S = Saturation constant without inhibition K'_S

K'_S = Saturation constant with inhibition

K_{SI} = Inhibition constant.

Nitrogen and Phosphorus

Double Monod expression which conceals that the microbial growth is inhibited when the concentrations of N and P are low :

$$\mu_{obs} = \mu_{max}\left[\frac{S_{NH_4}}{K_{S,NH_4} + S_{NH_4}}\right]\left[\frac{S_{PO_4}}{K_{S,PO_4} + S_{PO_4}}\right]$$

where S_{NH_4} = Ammonium concentration

S_{PO_4} = Phosphorus concentration (orthophosphate)

K_{S,NH_4} = Saturation constant (N) [0.1–0.5 g N/m³]

K_{S,PO_4} = Saturation constant (P) [0.1–0.2 g–P/m³]

2. Inorganic and organic materials storage in biomass

Internal storage products are stored as polymers inside the biomass

Biomass can also transform organic substrate into exopolymeric substances (EPS)

Types of compounds stored by the biomass :

*   PHA, polyhydroxy alkonates (lipid polymeric compounds of various fatty acids)

*   PHB (poly β–hydroxy buturate)

*   PHV (poly β–hydroxy valerate)

[Produced as a result of fermentation processes or are produced based on the contents of fatty acids in wastewater]

Phosphorus accumulating organisms (PAO), their metabolism is coupled to storage of glycogen $(C_4 H_6 O_2)_n$

PHA is quickly metabolized within 4–6 hours at 20°C

Summary of storage products found in biological wastewater treatment plants

Table 77 : Summary of storage compounds with Micro-organism type.

S.No.	Storage Compound	Micro–organism type
1.	Exo–polymeric substances (EPS)	Aerobic with carbohydrate load, heterotrophs
2.	Poly–β–hydroxybuturate (PHB)	PAOs, some filamentous organisms, heterotrophs under high acetate load
3.	Poly–β–hydroxyvalerate (PHV)	PAOs, some filamentous organisms, heterotrophs under acetate and propionate load
4.	Polyhydroxy alkanoates (PHA)	Same as PHB
5.	Glycogen (GLV)	Heterotrophs under glucose load PAOs, GAOs
6.	Polyphosphate (PP)	PAOs

3. Substrate removal rate ($r_{V, S}$) and mass balance for dissolved organics

  Substrate removal rate :

$$r_{V,S} = \frac{\mu_{max}}{Y_{max}} \left[\frac{S}{K_S + S} \right] \left[\frac{S_{O_2}}{K_{S,O_2} + S_{O_2}} \right] X_B$$

  where S, S_{O2} and X_B are the substrate, dissolved O_2 and biomass concentrations in the aeration tank

  Mass balance for dissolved oxygenics :

  Input + Hydrolysis – Removal = Output

  $Q_1 S_1 + r_{V, XS} V v_{XS} - r_{V, S} V = Q_3 S_3$

  where Q_1 = Influent flow
  S_1 = Influent dissolved organic conc.
  V = Aeration volume
  Q_3 = Final effluent flow
  S_3 = Final effluent conc.
  v_{XS} = Stoichiometric coefficient for hydrolysis

4. Reaction rate constant for aerobic heterotrophic bacteria (using municipal wastewater)

Table 78 : Rate constants for aerobic heterotrophic bacteria.

Constant	Magnitude
Max. specific growth rate (μ_{max})	4–8 d^{-1} (COD basis)
Decay constant (b)	0.1–0.2
Saturation constant for substrate (K_s)	5–30 g COD/m^3
Saturation constant for oxygen (K_{S,O_2})	0.5–1 g O_2/m^3
Max. yield coefficient (Y_{max})	$0.5 - 0.7 \dfrac{\text{g COD(B)}}{\text{g COD(S)}}$
Temp. coefficient (k) for μ_{max}, k_h, b	0.06–0.1 °C

(Contd...)

Constant	Magnitude
pH constant (K_{pH})	150–250
Hydrolysis constant for suspended solids (k_h)	0.6–1.4 d^{-1}
Hydrolysis constant for dissolved solids (k_h)	3–20 d^{-1}
Hydrolysis constant (k_{hx})	$0.6 - 1.4 \dfrac{\text{kg COD(X)}}{\text{kg COD(B).d}}$
Hydrolysis saturation constant (K_x)	$0.02 - 0.05 \dfrac{\text{kg COD(X)}}{\text{kg COD(B)}}$
Saturation constant for nitrogen (K_{S,NH_4})	0.1–0.5 g-N/m^3
Saturation constant for phosphorus (K_{S,NO_4})	0.1–0.2 g-P/m^3
μ_{max} (T) or b(T) or k_h (T) = $\mu_{max}/b/k_h(20)$ exp[k(T – 20)]	

5. Reaction rate constant for nitrifying bacteria

Table 79 : Reaction rate constants for nitrification.

Constant(s)	Substrate–Oxidation		
	Ammonia–N Oxidation	Nitrite– Oxidation	Total Process
Max. specific growth rate (m_{max}, d^{-1})	0.6–0.8	0.6–1.0	0.6–0.8
Saturation constant (K_{S,NH_4}, g NH_4–N/m^3	0.3–0.7	0.8–1.2	0.3–0.7
Saturation constant (K_{S,O_2}, g-O_2/m^3)	0.5–1.0	0.5–1.5	0.5–1.0
Max. yield constant (Y_{max}, A) g VSS/m^3	0.1–0.12	0.05–0.07	0.15–0.20
Decay constant (b_A, d^{-1})	0.03–0.06	0.03–0.06	0.03–0.6
Temp. coeff. (k) for $m_{max, A}$ and b_A (°C)	0.08–0.12	0.07–0.10	0.08–0.12

6. Reaction rate constants for denitrification

Table 80 : Rate constants for denitrification.

Constant(s)	Magnitude
Max. specific growth rate (μ_{max}, d^{-1})	3–6
Max. specific growth rate with methanol (μ_{max}, d^{-1})	5–10
Decay coefficient (b, d^{-1})	0.05–0.10
Saturation constant (K_{S,NO_3}, g–N/m^3)	0.2–0.5
Inhibition constant (K_{S,O_2} g-O_2/m^3)	0.1–0.5
Saturation constant [methanol, $K_{s,Me}$, g COD/m^3]	5–10
Saturation constant [organic matter (wastewater), $K_{S,COD}$,g COD/m^3	10–20
Hydrolysis constant [suspended solids, k_h, d^{-1}]	0.15–0.4
Hydrolysis constant [dissolved solids, k_h, d^{-1}]	1–15
Hydrolysis constant $\left[k_{hx}, \dfrac{\text{kg COD(X)}}{\text{kg COD(B).d}} \right]$	0.15–0.4

(Contd...)

Constant(s)	Magnitude
Hydrolysis saturation constant $\left[k_x, \dfrac{kg\ COD(X)}{kg\ COD(B)} \right]$	0.02–0.05
Max. yield constant $\left[\text{methanol, } Y_{max}, \dfrac{kg\ COD}{kg\ COD} \right]$	0.5–0.65
Max. yield constant $\left[\text{COD wastewater, } Y_{max}, \dfrac{kg\ COD}{kg\ COD} \right]$	0.5–0.55
Max. yield constant $\left[\text{Wastewater, } Y_{max}, \dfrac{kg\ COD}{kg\ NO_3-N} \right]$	1.6–1.8
Temp. constant (k) for m_{max} and b (°C)	0.06–0.12

Example 6.120

Activated Sludge Process

1. Derive the expressions for SRT (θ_c) and degradable fraction of the biological VSS.

1.1 Expression for sludge retention time (θ_c)

The degradable fraction can be related to the decay coefficient (b) and sludge age (θ_c) by the following expression

$$X_d = \frac{X_d^*}{1 + b\ X_n^*\ \theta_c}$$

In a recycle system such as an ASP, the sludge age is defined as :

$$\theta_c = \frac{X_V \theta}{\Delta\ X_V}$$

The sludge generation from the biological oxidation of soluble substrate ($f_b = 1$) is given by :

$$\Delta\ X_V = a\ S_r - b\ X_d\ X_V\ \theta$$

and $\quad \theta_c = \dfrac{X_V\ \theta}{a\ S_r - b\ X_d\ X_V\ \theta}$

Substituting the value of X_d in θ_c expression, yields :

$$\theta_c = \frac{X_V\ \theta}{a\ S_r - b\left(\dfrac{X_d^*}{1 + b\ X_n^*\ \theta_c}\right) X_V\ \theta}$$

or $\quad \theta_c \left[a\ S_r - b\left(\dfrac{X_d^*}{1 + bX_n^*\ \theta_c}\right) X_V\ \theta \right] = X_V\ \theta$

$$\theta_c [aS_r(1 + bX_n^*\ \theta_c) - bX_d^*\ X_V \theta] = X_V \theta\ (1 + b\ X_n^*\ \theta_c)$$

$$aS_r \theta_c + bX_n^* aS_r \theta_c^2 - b X_d^* X_V \theta \theta_c = X_V \theta + bX_n^* X_V \theta_c \theta$$

$$abS_r X_n^* \theta_c^2 + (aS_r - bX_d^* X_V \theta - b X_V \theta)\theta_c - X_V \theta = 0$$

$$ab S_r X_n^* \theta_c^2 + [aS_r - bX_V \theta (X_d^* + X_n^*)] \theta_c - X_V \theta = 0$$

As, $(X_d^* + X_n^*) = 1$

Therefore,

$$abS_r X_n^* \theta_c^2 + [aS_r - bX_V \theta]\theta_c - X_V \theta = 0$$

$$\theta_c = \frac{-(aS_r - bX_V \theta) + [(aS_r - bX_V \theta)^2 + 4(abX_n^* S_r)X_V \theta]^{0.5}}{2 abX_n^* S_r}$$

1.2 Expression for degradable biological fraction (X_d)

It is possible to estimate X_d as :

$$X_d = \frac{X_d^*}{1 + b X_n^* \theta_c}, \text{ Substituting } \theta_c \text{ in } X_d \text{ expression}$$

$$= \frac{X_d^*}{1 + b X_n^* \left(\dfrac{X_V \theta}{aS_r - bX_d X_V \theta}\right)}$$

$$= \frac{(X_d^*)(aS_r - bX_d X_V \theta)}{aS_r - bX_d X_V \theta + bX_n^* X_V \theta}$$

$$X_d (a S_r - b X_d X_V q + b X_n^* X_V q) = a X_d^* S_r - b X_d X_V X_d^* q$$

$$a S_r X_d - b X_V q X_d^2 + b X_n^* X_V q X_d = a X_d^* S_r - b X_d X_V X_d^* q$$

$$X_d^2 (b X_V q) - X_d (a S_r + b X_V X_d^* q + b X_n^* X_V q) + a S_r X_d^* = 0$$

$$(b X_V q) X_d^2 - [a S_r + b X_V q (X_d^* + X_n^*)] X_d + a S_r X_d^* = 0$$

$$(X_d^* + X_n^*) = 1$$

$$(b X_V q) X_d^2 - (a S_r + b X_V q) X_d + a S_r X_d^* = 0$$

$$X_d = \frac{(aS_r + bX_V \theta) - [a S_r + bX_V \theta^2 - 4(bX_V \theta)(a S_r X_d^*)]^{0.5}}{2 bX_V \theta}$$

$$X_d^* = 0.8$$

Therefore,

$$X_d = \frac{(aS_r + bX_V \theta) - [(aS_r + bX_V \theta)^2 - 4(bX_V \theta)(0.8 \ a S_r)]^{0.5}}{2bX_V \theta}$$

Notation

X_d is the degradable fraction of the biological VSS, X_d^* is the degradable fraction of the biological VSS at generation (usually 0.8) X_n^* is non-biodegradable fraction of the biological VSS at generation [usually 0.2 ($X_d^+ + X_n^*$) = 1.0], b is the decay coefficient, θ_c is sludge age, θ is the hydraulic detention time, X_V is the VSS, ΔX_V is the VSS wasted per day [mg/L based on influent flow], and S_r is the soluble substrate removed.

1.3   Wasted volatile suspended solids (ΔX_v)

If the influent contains bio–resistant VSS and the mass balance is given as:

$$\Delta X_V = a[S_r + f_d\, f_x\, X_i] - b\, [X_d\, f_b\, X_V\, \theta] + (1 - f_d)\, f_x\, X_i + (1 - f_x)X_i$$

where X_i is the influent VSS, f_x is the fraction of influent VSS that is degradable, f_d is the fraction of degradable influent VSS degraded, and f_b is the fraction of the MLVSS that is biomass for soluble substrate (and usually) = 1

$$\text{Total sludge production} = \frac{\Delta X_V}{f_V}$$

where f_V is the volatile fraction of the mixed liquor suspended solids.

The fraction of biomass (f_b) in the overall mixed liquor is :

$$f_b = 1 - [(1 - f_x) + (1 - f_d)f_x]\,\frac{X_i}{X_V}\,\frac{\theta_c}{\theta}$$

[1 mg VSS = 1 mg COD]

Non–biological VSS is given by :

$$X_{VNB} = \frac{X_i\,\theta_c}{\theta}$$

2.   Determine X_d and θ_c for an activated sludge process. Use the following data :

Influent BOD (S_o)	:	400 mg/L
Effluent BOD (S_e)	:	50 mg/L
Effluent SS	:	100 mg/L
Ratio of $\dfrac{\text{mg BOD}}{\text{mg SS}}$	:	0.3
System rate constant (K)	:	4 d^{-1} at 30°C
Decay coefficient (b)	:	0.15 d^{-1} at 30°C
MLVSS (X_v)	:	2500 mg/L
Yield coefficient (a)	:	0.5

2.1   Required detention time (θ)

$$\frac{F}{M} = \frac{S_o - S_e}{X_V\,\theta} = K\,\frac{S_e}{S_o}$$

or

$$\theta = \frac{S_o\,(S_o - S_e)}{K X_V\, S_e}$$

Soluble effluent BOD $(S_e) = (50 - 0.3 \times 50)$

$$= 35 \text{ mg/L}$$

$$\theta = \frac{(400)\,(400 - 35)}{(4)(2500)(35)} = 0.42 \text{ d}$$

$S_r = (400 - 35) = 365 \text{ mg/L}$

2.2 Required X_d and θ_c

$$X_d = \frac{(aS_r + bX_V\,\theta) - [(aS_r + bX_V\,\theta)^2 - (4bX_V\theta)(0.8aS_r)]^{0.5}}{2bX_V\,\theta}$$

$(a\,S_r + b\,X_V\,\theta) = (0.5 \times 365 + 0.15 \times 2500 \times 1.420 = 340$

$(4b\,X_V\,\theta)\,(0.8\,aS_r) = (4 \times 0.15 \times 2500 \times 0.42)(0.8 \times 0.5 \times 365) = 91,980$

$(2b\,X_V\,\theta) = (2 \times 0.15 \times 2500 \times 0.42) = 315$

$$X_d = \frac{340 - 154}{315} = 0.6$$

$$\theta_C = \frac{X_V\,\theta}{aS_\sigma - b}$$

$$\theta_C = \frac{X_V\,\theta}{aS_\sigma - b\,X_d\,X_V\theta}$$

$$= \frac{(2500)(0.42)}{(0.5 \times 365 - 0.15 \times 0.6 \times 2500 \times 0.42)} = 11.93 \text{ days.}$$

2.3 Required θ_c and X_d (check)

$$\theta_c = \frac{-(aS_r - bX_V\,\theta) + [(aS_r - bX_V\theta)^2 + 4(abX_n^* S_r)(X_V\,\theta)]^{0.5}}{2\,ab\,X_n^*\,S_r}$$

$(aS_r - bX_V\,\theta) = (0.5 \times 365 - 0.15 \times 2500 \times 0.42) = 25$

$4(ab\,X_n^*)(X_V\,\theta) = 4\,(0.5 \times 0.15 \times 0.2 \times 365)\,(2500 \times 0.42) = 22,995$

$(2ab\,X_n^*S_r) = (2 \times 0.5 \times 0.15 \times 0.2 \times 365) = 11$

$$\theta_c = \frac{-25 + [625 + 22,995]^{0.5}}{11} = 11.72 \text{ days (OK)}$$

$$X_d = \frac{X_d^*}{1 + b\,X_n^*\,\theta_c}$$

$$= \frac{0.8}{1 + (0.15)(0.2)(11.72)} = 0.59\ (\text{OK}).$$

Example 6.121

Activated Sludge System

1. Determine the amount of each inorganic that must be added to provide for complete removal of ultimate BOD. Assume the following data :

Plateau BOD	:	1500 mg/L
Ultimate BOD	:	2500 mg/L
Total nitrogen	:	25 mg/l
Organic–N	:	20 mg/L
NH_3–N	:	5 mg/L
$NO_2^- - N$	:	0 mg/L
NO_3–N	:	0 mg/L
Total P	:	8 mg/L
Fe	:	2.0 mg/L
Mg	:	3.8 mg/L
Ca	:	30.0 mg/L
S	:	3.0 mg/L

1.1 Required BOD conversion to Cells

$$\text{BOD converted to Cells} = (2500 - 1500)$$

$$= 1000 \text{ mg/L}$$

$$\text{Biomass produced} = \frac{1000 \text{ mg/L} - \text{BOD}}{1.42 \text{ mg/L BOD}_L/\text{mg Cells}}$$

$$= 704 \text{ mg/L.Cells}$$

1.2 Required inorganic nutrient for BOD removal

Inorganic nutrient requirements :

Nitrogen :

$$= 0.14 \ (704) = 98.56 \text{ mg/L}, \qquad [\text{Cell} - \text{N} = 14\%]$$

Additional nitrogen requirement = (98.56 – 25) = 73.56 mg/L

[Assuming all organic – N is available]

Phosphorus

$$= 0.03 \ (704) = 21.12 \text{ mg/L}, \qquad [\text{Cell} - \text{P} = 3\%]$$

Additional P requirement = (21.12 – 8) = 13.12 mg/L

Iron

$$= 0.002 \ (704) = 1.408 \text{ mg/L}, \qquad [\text{Cell} - \text{Fe} = 0.2\%]$$

No additional iron required

Magnesium

$\qquad$ = 0.005 (704) = 3.52 mg/L, $\qquad$ [Cell – Mg = 0.5%]

No additional Magnesium required

Calcium

$\qquad$ = 0.005 (704) = 3.52 mg/L, $\qquad$ [Cell – Sa = 0.5%]

No additional calcium required

Sulphur

$\qquad$ = 0.01 (704) = 7.04 mg/L, $\qquad$ [Cell – S = 1%]

Additional S–requirement = (7.04 – 3.0) = 4.04 mg/L

2. Determine the amount of oxygen and air which must be supplied for aerobic treated, as also the ultimate BOD and the residual COD to be expected after the treatment. Assume the following data :

$\qquad$ Sample volume $\qquad$: 1 L

$\qquad$ Sample COD $\qquad$: 768 mg/L

$\qquad$ Culture volume $\qquad$: 1 L

$\qquad$ Culture COD $\qquad$: 395 mg/L

Table 81 : Time versus COD.

Time (min)	0	5	10	20	30	60	100	130	150
Mixed COD (mg/L)	**583**	565	548	506	469	377	320	319	**317**
Filtered COD (mg/L)	381	346	317	259	206	90	22	21	**20**

2.1. Required ultimate BOD of the diluted sample.

The ultimate BOD of the diluted sample :

$\qquad$ = (Initial filtrate COD – Final filtrate COD)

$\qquad$ BOD_L = 2 (381 – 20) = 722 mg/L

2.2 Required residual COD after treatment

$\qquad$ Residual COD = (768 – 722) = 46 mg/L

2.3 Required oxygen

$\qquad$ O_2 required = (Final mixture COD – Initial value) × Dilution factor

$\qquad$ = 2 (583 – 317) = 532 mg/L

$$\text{Amount of air necessary} = (532)\left(\frac{20}{100}\right)\left(\frac{5}{100}\right)$$

$$= \frac{532 \times 100}{10,000} = 5.32 \ g-air/L$$

Example 6.122

Activated Sludge Process : SVI Effects

Estimate the consequences on process efficiency when SVI changes from 100 to 200 [without any change in SRT (θ_c)], as also the aeration volume, and SRT. Assume the following data :

Wastewater flow	:	1.5 MGD
Influent BOD$_u$ (S$_o$)	:	250 mg/L
Effluent BOD (S$_e$)	:	20 mg/L
Recycle ratio (R = Q$_r$/Q)	:	20 to 50% of wastewater flow
True yield coefficient (Y$_t$)	:	0.5
First order reaction rate constant [k = K/K$_s$]	:	0.1 L/mg.d
Decay coefficient (k$_d$)	:	0.1 d^{-1}
MLVSS	:	2000 mg/L
MLVSS	:	0.8 (MLSS)

Solution

1. Required basic expressions

$$\theta_c^{-1} = \frac{Y_t\, k S_e}{K_s + S_e} - k_d$$

$$S_e = \frac{K_s(1 + k_d\,\theta_c)}{\theta_c[Y_t\, k + k_d] - 1}$$

$$X = \frac{\theta_c\, Y_t\, Q[S_o - S_e]}{V[1 + k_d\,\theta_c]}$$

[If $\dfrac{dS}{dt} = \dfrac{Q(S_o - S_e)}{V} = k X S_e$ instead of $\dfrac{k S_e}{K_s + S_e}$ which presumes first order reaction with $k = K/K_s$ and $K_s \gg S_e$]

$$\theta_c^{-1} = Y_t\, k S_e - k_d$$

And, specific substrate utilization rate (q) = $\dfrac{Q(S_o - S_e)}{V X} = k S_e$,

or $\quad \dfrac{S_e}{S_o} = \dfrac{1}{1 + k X\left(\dfrac{V}{Q}\right)}$

Also, $\quad \theta_c^{-1} = \dfrac{Q}{V}\left[1 + R - R\,\dfrac{X_r}{X}\right]$

The ratio X_r/X is a function of the settling characteristic of the biomass, and of the efficiency of the secondary clarifier. If the secondary clarifiers are operating properly, solids capture should approach 100%, and can be estimated as :

$$[X_r]_{max} = \frac{10^6}{SVI}$$

[SVI refers to the total biomass (suspended solids)]

Minimum SRT (θ_c^m) is calculated when S_e approaches to S_o :

$$(\theta_c)_m^{-1} = Y_t \frac{kS_o}{K_s + S_o} - k_d$$

For first order reaction :

$$(\theta_c)_m^{-1} = Y_t kS_o - k_d; \quad \left[k = \frac{K}{K_s}\right]$$

2. Required aeration basin volume (V) and SRT (θ_c)

$$q = kS_e = \frac{Q(S_o - S_e)}{XV}$$

$$V = \frac{Q(S_o - S_e)}{X(kS_e)} = \frac{1.5(250 - 20)}{2000(0.1 \times 20)} = 0.086 \text{ MG}$$

$$\theta_c^{-1} = Y_t kS_e - k_d$$

$$= 0.5(0.1 \times 20) - 0.1$$

$$= (1.0 - 0.1) = 0.9 \text{ d}^{-1}$$

Or, $\theta_c = 1.1$ d

3. Required biomass (MLVSS) concentration in the aeration basin

$$\theta_c^{-1} = \frac{Q}{V}\left[1 + R - R\frac{X_r}{X}\right]$$

$$[X_r]_{max} = \frac{10^6}{SVI}$$

At SVI = 100 and R = 0.2, the value of X is :

$$0.9 = \frac{1.5}{0.086}\left[1 + 0.2 - \frac{0.2 \times 10^6(0.8)}{100(X)}\right]$$

$$\left[\frac{0.9 \times 0.086}{1.5}\right] - 1.2 = -\frac{0.2 \times 10^4(0.8)}{X}$$

$$X = \frac{0.2 \times 10^4 (0.8)}{1.1484} = 1400 \text{ mg/L}$$

At SVI = 100 and R = 0.5, the value of X is :

$$X = \frac{0.5 \times 10^4 (0.8)}{\left[1.5 - \frac{0.9(0.086)}{1.5}\right]}$$

$$= 2762 \text{ mg/L}$$

At SVI = 200 and R = 0.2, the value of X is :

$$X = \frac{0.1 \times 10^4 (0.8)}{\left[1.2 - \frac{0.9(0.086)}{1.5}\right]} = 697 \text{ mg/L}$$

At SVI = 200 and R = 0.5, the value of X is :

$$X = \frac{0.25 \times 10^4 (0.8)}{\left[1.5 - \frac{0.9(0.086)}{1.5}\right]}$$

$$= 1381 \text{ mg/L}$$

Summary :

Table 83 : Magnitudes of SVI, R and X.

SVI	R	$X_r(mg/L)$	$X(mg/L)$
100	0.2	10000 ×0.8 = 8000	1400
100	0.5	10000× 0.8 = 8000	2762
200	0.2	5000 × 0.8 = 4000	697
200	0.5	5000 × 0.8 = 4000	1381

4. Required effluent concentration (Se)

Specific substrate utilization (q) = $\dfrac{Q(S_o - S_e)}{V\,X}$

$$q = \frac{Q(S_o - q/k)}{X\,V}; \quad q = kS_e$$

$$q = \frac{1.5(250 - q/0.1)}{1400(0.086)}; \text{ at } X = 1400 \text{ mg/L}$$

$$= \frac{1.5(250 - 10q)}{1400(0.086)}$$

$$[80 + 10]q = 250$$

$$q = \frac{250}{90} = 2.8 \text{ d}^{-1}$$

Therefore, $S_e = \dfrac{q}{k} = \dfrac{2.5}{0.1} = 25 \text{ mg/L}$

$$q = \frac{1.5[250 - 10q]}{2762(0.086)}; \text{ at } X = 2762 \text{ mg/L}$$

$168 \text{ q} = 250$

$\qquad q = 1.48 \text{ d}^{-1}$

$$S_e = \frac{q}{k} = \frac{1.48}{0.1} = 14.8 \text{ mg/L}$$

$$q = \frac{1.5}{697} \frac{[250 - 10q]}{(0.086)}$$

$50 \text{ q} = 250$

$$q = 5 \text{ d}^{-1}; \ X_e = \frac{5 \text{ d}^{-1}}{0.1} = 50 \text{ mg/L}$$

$$q = \frac{1.5[250 - 10q]}{1381(0.086)}$$

$90 \text{ q} = 250$

$\qquad q = 2.8 \text{ d}^{-1}$

$$S_e = \frac{q}{k} = 28 \text{ mg/L}$$

Summary [k = 0.1; Q = 1.5 MGD, V = 0.086 MG]

Table 84 : Magnitudes of X, SVI, R and S_e.

X (mg/L)	SVI	R	S_e (mg/L)
1400	100	0.2	25.0
2762	100	0.5	14.8
697	200	0.2	50.0
1381	200	0.5	28.0

(A change in SVI warrants a change in SRT to obtain consistent effluent substrate concentration).

Example 6.123

Some Important Concepts in Activated Sludge Process

1. Under what conditions the true mean solids residence time (θ_m) is equal to sludge age (θ_c)

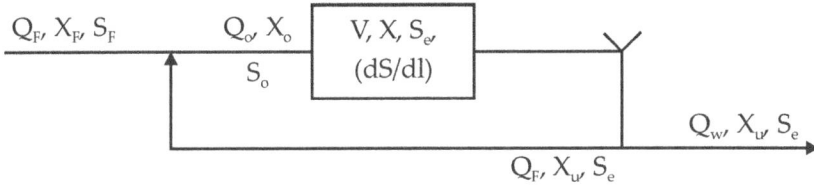

VSS synthesis in the reactor (input) = $Y(S_F - S_e)Q_F$

VSS in fresh feed (input) = $Q_F X_F$

VSS in wastage stream = $Q_w X_u$

VSS lost in clarifier overflow = $Q_e X_e = (Q_F - Q_w) X_e$

VSS lost in reactor by endogenous respiration = $k_d XV$

At steady state conditions :

$Y(S_F - S_e)Q_F + Q_F X_F = Q_w X_u + k_d X V + Q_e X_e$

or, $Q_w X_u = [Y(S_F - S_e)Q_F - k_d X V] + Q_F X_F - Q_e X_e$

Biomass yield $(\Delta X) = [Y(S_F - S_e)Q_F - k_d X V]$

$Q_w X_u = \Delta X + Q_F X_F - Q_e X_e$

True mean solids residence time :

$$\theta_m = \frac{\text{MLVSS in reactor (lb)}}{\text{Total input VSS (lb/d)}}$$

$$= \frac{XV}{Y(S_F - S_e)Q_F + Q_F X_F}$$

At steady–state, the total input VSS = out of VSS from the system:

$$\text{or } \theta_m = \frac{\text{MLVSS in reactor (lb)}}{\text{Total output VSS (lb/d)}} = \frac{XV}{Q_w X_u + k_d XV + Q_e X_e}$$

Sludge age (θ_c) :

$$\theta_c = \frac{\text{Mass of VSS in the system}}{\text{Net biomass yield } (\Delta X)}$$

$$= \frac{XV}{Q_w X_u + Q_e X_e - Q_F X_F}$$

Sludge age (θ_c) becomes equal to true mean solids residence time when

$X_F = 0$ and $k_d = 0$, *i.e.,*

$$\theta_c = \theta_m = \frac{XV}{Q_w X_u + Q_e X_e}$$

2. Prove that Net biomass yield coefficient $(Y_n) = \dfrac{1}{1+k_d\theta_c}$ and $\theta_c = \dfrac{1}{\mu}$

Biomass yield $(\Delta X) = Y(S_F - S_e)Q_F - k_d\, XV$

or $$\frac{\Delta X/V}{X} = \frac{Y(S_F - S_e)}{Xt} - k_d \tag{1}$$

$$\mu = \frac{1}{X}\left(\frac{dX}{dt}\right) = \frac{\Delta X/V}{X} = \frac{mg\ MLVSS/L.d}{mg\ MLVSS} \tag{2}$$

Therefore, using equations (1) and (2) yield :

$$\mu = \frac{Y(S_F - S_e)}{Xt} - k_d$$

$$= Yq - k_d \tag{3}$$

where q = Specific substrate removal rate $\left[\dfrac{(S_F - S_e)}{Xt}\right]$

Under endogenous respiration, biomass synthesized is nearly equal to MLVSS oxidized :

$$Y(S_F - S_e)Q_F = k_d\, XV$$

or $$\frac{S_F - S_e}{X(V/Q_F)} = \frac{S_F - S_e}{Xt} = \frac{k_d}{Y}$$

$$\mu = \frac{1}{X}\left(\frac{dX}{dt}\right) = \frac{Y}{X}\left(\frac{dS}{dt}\right) - k_d$$

or $$\left(\frac{dX}{dt}\right) = Y\left(\frac{dS}{dt}\right) - k_d X$$

$$\theta_c = \frac{XV}{\Delta X}\ \text{or}\ \frac{X}{\Delta X/V} = \frac{1}{\mu}, \text{[equation 1]}$$

Assume that $\Delta X = Y_n(S_F - S_e)Q_F$

Divide by XV on both sides of this equation

$$\frac{\Delta X/X}{V} = \frac{Y_n(S_F - S_e)Q_F}{XV}$$

$$\mu = \frac{Y_n(S_F - S_e)}{Xt}$$

$$= Y_n q$$

where q = Specific substrate utilization rate $\left[\dfrac{(S_F - S_e)}{Xt}\right]$

$$\mu = Yq - k_d$$

$$\mu = Y\left(\frac{\mu}{Y_n}\right) - k_d; \quad \left[q = \frac{\mu}{Y_n}\right]$$

$$\frac{1}{\theta_c} = Y\left(\frac{1}{\theta_c Y_n}\right) - k_d$$

$$1 = \frac{Y\theta_c}{\theta_c Y_n} - k_d \theta_c$$

Therefore,

$$\frac{Y}{Y_n} = (1 + k_d \theta_c)$$

or $\qquad Y_n = \dfrac{Y}{1 + k_d \theta_c} = \dfrac{Y}{1 + k_d/\mu}$

As $\qquad \mu = Y_n\, q :$

Therefore,

$$\mu = \frac{1}{X}\left(\frac{dX}{dt}\right) = Y_n \cdot \frac{1}{X}\left(\frac{dS}{dt}\right)$$

or $\qquad \left(\dfrac{dX}{dt}\right) = Y_n \cdot \left(\dfrac{dS}{dt}\right)$

3. Prove that specific substrate utilization (q) is equal to $k\,S_e$.

Mass balance on substrate at steady–state :

$$0 = Q_F S_F - Q_F S_e - \left(\frac{dS}{dt}\right)V$$

$$\left(\frac{dS}{dt}\right) = \frac{Q_F(S_F - S_e)}{V}$$

Specific substrate removal rate (q) is given as :

$$q = -\frac{1}{X}\left(\frac{dS}{dt}\right) = \frac{1}{X}\left(\frac{dS}{dt}\right)_a$$

or $\qquad q = \dfrac{1}{X}\left(\dfrac{dS}{dt}\right) = \dfrac{Q_F(S_F - S_e)}{VX} = \dfrac{(S_F - S_e)}{Xt}$

It is necessary to decide which kinetic model is to be used. If the substrate removal follows first order kinetics:

$$\left(\dfrac{dS}{dt}\right) = KS_e$$

[only valid at low substrate concentration and is true for biological treatment systems under well mixed conditions]

$$q = \left(\dfrac{1}{X}\right)KS_e = \left(\dfrac{S_F - S_e}{Xt}\right) = kS_e; \left[k = \dfrac{K}{X}\right]$$

Using Michaelis–Menton model :

$$q = \dfrac{1}{X}\left(\dfrac{dS}{dt}\right) = q_m \dfrac{S_e}{K_s + S_e}$$

If $S_e \gg K_s$; $q = q_m$ [High concentration]

If S_e is in the intermediate concentration

$$q = \dfrac{1}{X}\left(\dfrac{dS}{dt}\right) = \dfrac{S_F - S_e}{Xt} = q_m \dfrac{S_e}{K_s + S_e}$$

If $S_e \ll K_s$

$$q = \dfrac{1}{X}\left(\dfrac{dS}{dt}\right) = \dfrac{S_F - S_e}{Xt} = q_{max}\dfrac{S_e}{K_s} = kS_e$$

$S_e = K_s$ for $q = q_{max}$

In the non–biodegradable substrate (residual COD) is S_n, the specific substrate removal rate is

$$q = \dfrac{S_o - S_e}{Xt} = k(S_e - S_n)$$

Using Grau's model :

$$q = \dfrac{1}{X}\left(\dfrac{dS}{dt}\right) = \dfrac{S_F - S_e}{Xt} = k_1\dfrac{S_e}{S_F}$$

4. Derive a relationship between recycle ratio (r) and sludge age (θ_c)

   At steady–state, the mass balance on biomass is :

   $$0 = Q_F X_F + rQ_F X_u + \left[Y\left(\dfrac{dS}{dt}\right) - k_d X\right]V - Q_F(r+1)X$$

   Assume first order kinetics $\left(\dfrac{dS}{dt}\right) = KS_e$; $K = kX$,

Therefore,

$$Y\left(\frac{dS}{dt}\right) = YkXS_e$$

$$\frac{1}{\theta_c} = Y\left(\frac{1}{X}\right)\left(\frac{dS}{dt}\right) - k_d, \text{ and} \tag{1}$$

$$\frac{1}{\theta_c} = YkS_e - k_d \tag{2}$$

From equation (2),

$$S_e = \left(\frac{1 + k_d\theta_c}{Yk\theta_c}\right)$$

Therefore,

$$Y\left(\frac{dS}{dt}\right) = YkX\left[\frac{1 + k_d\theta_c}{Yk\theta_c}\right]$$

$$= \frac{X(1 + k_d\theta_c)}{\theta_c}$$

$$0 = Q_F X_F + r Q_F X_u + \left[\frac{X(1 + k_d\theta_c)}{\theta_c} - k_d X\right]V - Q_F(r+1)X$$

Solving for θ_c yields :

$$\frac{1}{\theta_c} = \frac{Q_F}{V}\left[1 + r - r\frac{X_u}{X} - \frac{X_F}{X}\right]$$

$$\frac{1}{\theta_c} = \frac{1}{t}\left[1 + r - r\frac{X_u}{X} - \frac{X_F}{X}\right]$$

If $\quad X_F = 0$, then :

$$\frac{1}{\theta_c} = \frac{1}{t}\left[1 + r - r\frac{X_u}{X}\right]$$

where X_u/X depends on the design of the secondary clarifier and the settling properties of the biomass.

5. Derive an expression for recycle ratio

    Mass balance for the biomass around the clarifier :

    Input VSS to the clarifier $= Q_F (r + 1) X$

    VSS in clarifier effluent $= Q_e X_e$

    VSS wastage $= \Delta X + Q_F X_F - Q_e X_e$

    VSS in the recycle stream $= Q_r X_u = r Q_F X_u$

Therefore, biomass mass balance is :

$$Q_F(r + 1)X = Q_eX_e + \Delta X + Q_FX_F - Q_eX_e + rQ_FX_u$$

Solving for r yields :

$$r = \frac{Q_F X - (\Delta X + Q_F X_F)}{Q_F(X_u - X)}$$

$$= \frac{X}{X_u - X} \text{ as } \Delta X + Q_F X_F << Q_F X$$

6. Define the relationship between b and k_d

Parameter b is defined as the lb of oxygen utilized per day per lb of MLVSS in the reactor for the endogenous respiration :

$$b = \frac{\text{lb of } O_2/d}{\text{lb of MLVSS in the reactor}}$$

lb of O_2 required for endogenous respiration = b (lb of MLVSS in the reactor)

lb of O_2/d (endogenous respiration) = b X V

Biomass is expressed as $C_5H_7NO_2$

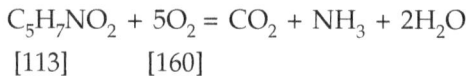

$$C_5H_7NO_2 + 5O_2 = CO_2 + NH_3 + 2H_2O$$

$$[113] \qquad [160]$$

$$k_d = \frac{\text{lb of MLVSS oxidized/d}}{\text{lb of MLVSS in the reactor}}$$

Therefore,

$$\frac{b}{k_d} = \left(\frac{\text{lb } O_2/d}{\text{lb MLVSS}}\right)\left(\frac{\text{lb MLVSS}}{\text{lb of MLVSS oxidized/d}}\right)$$

$$= \frac{160}{113} = 1.42$$

It implies that 1.42 lb of oxygen is required to oxidize 1 lb of MLVSS.

7. Required oxygen to oxidize substrate (without nitrification)

lb O_2/d for substrate oxidation = a $(S_F - S_e) Q_F$

lb O_2/d for endogenous respiration = b X V

Therefore, lb O_2/d = a $(S_F - S_e) Q_F$ + b X V

Considering biomass yield (ΔX), the required O_2 expression can also be expressed as :

$$\text{lb } O_2/d = Q_F (S_F - S_e) - 1.42[Y(S_F - S_e)Q_F - k_d \text{ X V}]$$

$$= Q_F (1 - 1.42Y)(S_F - S_e) + 1.42k_d \text{ X V}$$

where $(S_F - S_e) = BOD_u$ removed

$$Y = lb\ MLVSS\ yield/lb\ BOD_u\ removed$$

Net biomass yield $(\Delta X) = 1.42[Y(S_F - S_e)Q_F - k_d\ X\ V]$

$$a = lb\ O_2\ required/lb\ BOD_u\ removed.$$

Oxygen uptake rate [OUR, M/L^3T] can be utilized to determine the oxygen requirements

$$[OUR][V] = a(S_F - S_e)Q_F + b\ X\ V$$

$$\frac{[OUR][V]}{[XV]} = \frac{a(S_F - S_e)}{XV}Q_F + b$$

Specific oxygen utilization rate (R_{O2})

$$R_{O_2} = \frac{1}{X}\frac{dO_2}{dt} = \frac{OUR}{X} = aq + b; \quad \left[\frac{mg/L.d}{g/L\ of\ MLVSS} = mg/g.d\right]$$

or $\dfrac{dO_2}{dt} = a\left(\dfrac{1}{X}\right)\left(\dfrac{dS}{dt}\right) + bX$

8. Relationship between Y and a

Mechanism of aerobic biological oxidation :

– Cell metabolism is defined by yield coefficient (Y)

– Energy metabolism is defined by kinetic parameter (a)

The schematics of the mechanism of the aerobic biological oxidation is presented as :

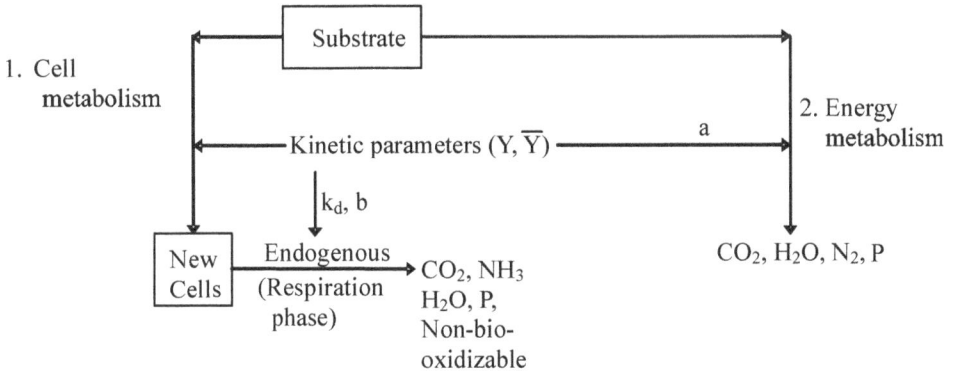

Figure 85 : Schematics of the mechanisms of the aerobic biological oxidation.

Consider lactose (CH_2O) : Influent (CH_2O) Conc. = 550 mg/L

Effluent (CH_2O) Conc. = 50 mg/L

Net (CH_2O) Conc. oxidized = 500 mg/L

It has been found that 65% of lactose is oxidized [= 0.65 × 500 = 325 mg/L] for the energy requirements of the biomass ($C_6H_5NO_2$)

Remaining 35% of lactose is utilized in synthesis of new cells (= 0.35 × 500 = 175 mg/L).

Theoretical oxygen demand (ThOD) of lactose is given by

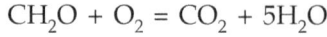

$$CH_2O + O_2 = CO_2 + 5H_2O$$
$$[30] \quad [32]$$

ThOD of biomass reaction is :

$$C_5H_7NO_2 + 5O_2 = 5CO_2 + NH_3 + H_2O$$

Substrate used for this biomass ($C_5H_7NO_2$) is

$$5(CH_2O) \rightarrow C_5H_7NO_2$$
$$[5\times30] \qquad [113]$$

$$Y = \frac{mg\,of\ biomass\ (MLVSS)\ produced}{mg\ of\ total\ substrate\ used}$$

$$= \frac{\dfrac{113}{150}(500)(32/30)(0.35)}{500(32/30)} = 0.2637$$

$$a = \frac{mg\,O_2\ required\ for\ substrate\ oxidation}{mg\ total\ substrate\ ThOD\ removed}$$

$$= \frac{0.65(500)\left(\dfrac{5\times32}{5\times30}\right)}{500(32/30)} = 0.65$$

$$Y = \overline{Y}\left(\frac{113}{150}\right) = 0.35(0.75) = 0.2637$$

General equation is :

$$\overline{Y} + a = 1.0$$
$$0.35 + 0.65 = 1.0\ (OK)$$

$$Y_{ThOD} = \frac{lb\ MLVSS\ produced}{lb\ of\ total\ ThOD\ removed}$$

$$\frac{Y}{Y_{ThOD}} = \frac{lb\ of\ total\ ThOD\ removed}{lb\ of\ total\ substrate\ removed}$$

$$= \frac{32}{30}$$

$$Y = Y_{ThOD} \left(\frac{32}{30} \right)$$

$$\left(\frac{150}{113} \right) Y + a = 1$$

$$\left[\left(\frac{150}{113} \right) \left(\frac{32}{30} \right) \right] Y_{ThOD} + a = 1$$

$$1.42\ Y_{ThOD} + a = 1$$

$$\overline{Y} + a = 1 \text{ is the basic equation.}$$

Average values of oxygen parameter for wastewaters as a fraction of the theoretical oxygen demand (taken as 100)

ThOD	:	100
TOD	:	82
COD (Standard test)	:	83
COD (Rapid test)	:	70
BOD (20)		
– with nitrification	:	65
– Nitrification suppressed	:	55
BOD (5)		
– with nitrification	:	58
– Nitrification suppressed	:	52

ThOD : $1.42 Y_{ThOD} + a_{ThOD} = 1$

$\qquad\qquad 1.42 Y_{ThOD} + a = 1$

COD : $Y_{ThOD} = Y_{COD} \left(\dfrac{COD}{ThOD} \right) = Y_{COD} \left(\dfrac{83}{100} \right)$

$\qquad\qquad a = a_{ThOD} = a_{COD} \left(\dfrac{COD}{ThOD} \right) = a_{COD} \left(\dfrac{83}{100} \right)$

Therefore,

$$1.42 \left(\frac{83}{100} \right) Y_{COD} + \left(\frac{83}{100} \right) a_{COD} = 1.0$$

$$BOD\,(5) : Y_{ThOD} = Y_{BOD} \left(\frac{BOD}{ThOD} \right) = Y_{BOD} \left(\frac{58}{100} \right)$$

$$a = a_{ThOD} = a_{BOD} \left(\frac{BOD}{ThOD} \right) = a_{BOD} \left(\frac{58}{100} \right)$$

Therefore,

$$1.42 \left(\frac{58}{100} \right) Y_{BOD} + \left(\frac{58}{100} \right) a_{BOD} = 1$$

$$0.82 \, Y_{BOD} + 0.58 \, a_{BOD} = 1.0$$

In general for aerobic degradation of organic substances, approx. 2/3 of the substrate removed is oxidized to meet energy demands, and 1/3 of it is converted into biomass.

Biomass consumption: $C_5H_7NO_2$

- N content $= \dfrac{14}{113} \times 100 = 12.4\%$

- P content (normally 2.24%): $C_5H_7NO_2P_n$ and n is given as or n = 0.0833

Therefore, $C_5H_7NO_2P_{0.0833}$ is the empirical formula (or $C_{60}H_{84}N_{12}O_{24}P$) of the biomass.

BOD :

- Carbonaceous oxygen demand $(t < t_c)$

$$y = L_0 (1 - 10^{K_1 t})$$

- Nitrogenous oxygen demand $(t > t_c)$

$$y = L_0 [1 - 10^{-K_1 t}] + L_N [1 - 10^{-K_2(t-t_c)}]$$

$$= L_0 + L_N [1 - 10^{-K_2(t-t_c)}]$$

where L_0 = Ultimate carbonaceous oxygen demand

L_N = Ultimate nitrogenous oxygen demand

- BOD and TOC correlation

$$BOD = 1.87(TOC) - 17$$

TOC represents 95% of the theoretical organic carbon.

9. Material balance for non–volatile

Combined influent stream (Q_0) to the reactor

* Volatile solids

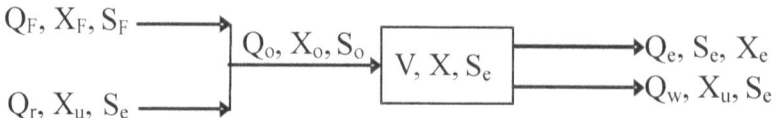

Figure 86 : Schematics of the system.

$$Q_F X_F + r Q_F X_u = Q_o X_o \; ; \; [Q_r = r Q_F]$$

$$Q_F(X_F + r X_u) = Q_o X_o)$$

or $\quad X_u = \dfrac{X_F + r X_u}{1 + r} \; ; \qquad [Q_o = Q_F(1 + r)]$

* Non–volatile solids [n–refers to non–volatile]

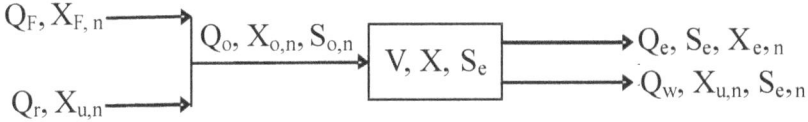

Figure 87 : Schematics of process system.

Fraction of volatile solids (F) in the MLSS $(X + X_n)$ is

$$F = \dfrac{X}{X + X_n} = \dfrac{MLVSS}{MLSS}, \quad \text{[Typically in the range of } 0.76 - 0.9]$$

Overall mass balance for non–volatile suspended solids

* Non–volatile solids wastage

$$Q_w X_{u,n} = Q_F X_{F,n} - Q_e X_{e,n}$$

$$= Q_F X_{F,n} - (Q_F - Q_w) X_{e,n} \; ; \; [Q_e = Q_F - Q_w]$$

$$= Q_F(X_{F,n} - X_{e,n}) + Q_w X_{e,n}$$

* Total sludge wasted (TSS)

(TSS) = (Volatile solids + Non–volatile solids) wasted

$$= \{[Y(S_F - S_e)Q_F - k_d \times V] + [Q_F X_F] - Q_F X_e\}_V +$$

$$\{Q_F(X_{F,n} - X_{e,n}) + Q_w X_{e,n}\}_n$$

10. Detention time in the reactor

Based on influent flow (Q_F)

$$t = \dfrac{V}{Q_F}$$

Based on combined flow $(Q_o = Q_F + r \, Q_F)$

$$t_n = \dfrac{V}{Qo}$$

Based on effluent quality $[S_e \text{ (soluble)}]$

$$t = \dfrac{S_F - S_e}{k(X)S_e}$$

Based on effluent quality containing soluble (S_e) and insoluble substrate (S_n)

$$t = \frac{S_F - S_e}{k(X)(S_e - S_n)}$$

Based on combined flow in terms of recycle ratio (r)

$$t_h = \frac{V}{Q_F + rQ_F} = \frac{V}{Q_F(1+r)} = \frac{t}{(1+r)}$$

11. Food to micro–organism (F/M) ratio

Based on influent feed to the reactor

$$\frac{F}{M} = \frac{QS_F}{XV} = \frac{S_F}{X(t)}$$

Based on combined feed to the reactor

$$\frac{F}{M} = \frac{Q_F S_F + Q_r S_e}{XV}$$

$$= \frac{Q_F(S_F + rS_e)}{XV}; \qquad [Q_r = rQ_F]$$

$$= \frac{S_F + rS_e}{X(t)}$$

If $r\, S_e \ll S_F$

$$\frac{F}{M} = \frac{S_F}{X(t)}$$

Based on substrate removed

$$\frac{F}{M} = \frac{S_F - S_e}{X(t)} = \text{Specific substrate removal rate (q)}.$$

Note :

1. Design equations for activated sludge system

   1.1 General design equations

   Specific substrate removal rate (q) in the reactor

   $$q = -\frac{1}{X}\left(\frac{dS}{dt}\right) = \frac{1}{X}\left(\frac{dS}{dt}\right)_{\text{system}} = \frac{Q_F(S_F - S_e)}{XV} = \frac{(S_F - S_e)}{Xt}$$

   Substrate removal rate in the reactor $\left(\dfrac{dS}{dt}\right) = qX$

   Biomass growth rate in the reactor $\left(\dfrac{dX}{dt}\right) = Y\left(\dfrac{dS}{dt}\right) - k_d X$

Specific biomass growth rate

$$(\mu) \text{ in the reactor} = \frac{1}{X}\left(\frac{dX}{dt}\right) = Yq - k_d$$

$$= Y\left(\frac{1}{X}\right)\left(\frac{dS}{dt}\right) - k_d$$

$$\mu = \left[\frac{\Delta X/V}{X}\right] = \frac{1}{\theta_c}$$

Washout time for sludge $(\theta_c{}^m)$ [where S_e becomes equal to influent substrate concentration (S_F)]

$$\theta_c = \theta_c{}^m$$

Sludge age (θ_c) : $\dfrac{1}{\theta_c} = \dfrac{Y}{X}\left(\dfrac{dS}{dt}\right) - k_d$

Effluent substrate (soluble) concentration : $\dfrac{1}{\theta_c} = Y\left(\dfrac{1}{X}\right)\left(\dfrac{dS}{dt}\right) - k_d$

1.2 Using rectangular hyperbolic relationship

Specific substrate removal rate (q) in the reactor

$$q = \frac{1}{X}\left(\frac{dS}{dt}\right) = q_{max}\frac{S_e}{K_s + S_e}$$

Substrate removal rate in the reactor

$$\frac{dS}{dt} = qX = q_{max} \; X \; \frac{S_e}{K_s + S}$$

Biomass growth rate in the reactor (dX/dt)

$$\left(\frac{dX}{dt}\right) = Y q_{max} \; X \; \frac{S_e}{K_s + S_e} - k_d X$$

Specific biomass growth rate (μ) in the reactor

$$\mu = \frac{1}{\theta_c} = Y q_{max} \frac{S_e}{K_s + S_e} - k_d$$

Washout time for the sludge $(\theta_c{}^m)$

$$\frac{1}{\theta_c{}^m} = Y q_{max} \frac{S_e}{K_s + S_e} - k_d$$

$$\theta_c{}^m = \frac{1}{Y q_{max} \dfrac{S_e}{K_s + S_e} - k_d}$$

Effluent substrate (soluble) concentration (S_e)

$$S_e = \frac{K_s(1+k_d\theta_c)}{\theta_c(Y\,q_{max}-k_d)-1}$$

Using first order substrate kinetics

* Specific substrate removal rate (q) in the reactor

$$q = \frac{1}{X}\left(\frac{dS}{dt}\right) = kS_e; \quad k = \frac{q_{max}}{K_s}$$

* Substrate removal rate in the reactor (dS/dt)

$$\left(\frac{dS}{dt}\right) = qX = kXS_e = KS_e$$

* Biomass growth rate in the reactor (dX/dt)

$$\left(\frac{dX}{dt}\right) = (YkXS_e - k_d\,X) = (YKS_e - k_d\,X)$$

* Specific biomass growth rate (μ) in the reactor

$$\mu = \frac{1}{\theta_c} = Y\,q_{max}\frac{S_e}{K_s+S_e} - k_d$$

* Washout time for the sludge $\left(\theta_c^m\right)$

$$\frac{1}{\theta_c^m} = YkS_F - k_d$$

* Effluent substrate (soluble) concentration (S_e)

$$S_e = \frac{1+k_d\theta_c}{Yk\theta_c}$$

1.3 Using zero–order substrate removal kinetics

* Specific substrate removal rate (q) in the reactor

$$q = \frac{1}{X}\left(\frac{dS}{dt}\right) = q_{max}$$

* Substrate removal rate in the reactor (dS/dt)

$$\frac{dS}{dt} = qX = q_{max}X$$

* Biomass growth rate in the reactor (dX/dt)

$$\left(\frac{dX}{dt}\right) = Y\,q_{max}\,X - k_d X$$

* Specific biomass growth rate (μ) in the reactor

$$\mu = \frac{1}{\theta_c} = q_{max} - k_d$$

* Washout time for the sludge ($\theta_c^{\,m}$)

$$\theta_c^{\,m} = \frac{1}{Y q_{max} - k_d}, \quad [\theta_c^{\,m} \text{ is independent of } S_F]$$

* Effluent substrate (soluble) concentration (S_e)

S_e is independent of θ_c.

1.4 Using Grau's substrate kinetics

* Specific substrate removal rate (q) in the reactor

$$q = k_1 \frac{S_e}{S_F}$$

* Substrate removal rate in the reactor (dS/dt)

$$\left(\frac{dS}{dt}\right) = qX = k_1 X \left(\frac{S_e}{S_F}\right)$$

* Biomass growth rate in the reactor (dX/dt)

$$\frac{dX}{dt} = Y k_1 X \left(\frac{S_e}{S_F}\right) - k_d X = Y K_1 \left(\frac{S_e}{S_F}\right) - k_d X$$

$$[K_1 = k_1 X]$$

* Specific biomass growth rate (μ) in the reactor

$$\mu = \frac{1}{\theta_c} = Y k_1 \left(\frac{S_e}{S_F}\right) - k_d$$

* Washout time for the sludge ($\theta_c^{\,m}$)

$$\theta_c^{\,m} = \frac{1}{Y k_1 - k_d} \quad [\theta_c^{\,m} \text{ is independent of } S_F]$$

* Effluent substrate (soluble) concentration (S_e)

$$S_e = \frac{(1 + k_d \theta_c)}{Y k_d \theta_c}$$

2. Some general expressions for aerobic systems (CSTR)

2.1 Kinetic relationship

* Specific substrate removal rate (q) based on combined flow
$[Q_o = Q_F(1 + r)]$

$$q = -\frac{1}{X}\left(\frac{dS}{dt}\right) = \frac{1}{X}\left(\frac{dS}{dt}\right)_{system} = \frac{Q_o(S_o - S_e)}{XV} = \frac{S_o - S_e}{X t_n}$$

$$t_n = \frac{V}{Q_F(1+r)} = \frac{V}{Q_o}$$

Using first order kinetics $(dS/dt) = K\,S_e$

$$q = \left(\frac{1}{X}\right)(KS_e) = kS_e; \quad \left[k = \frac{K}{X}\right]$$

[useful for determination of k]

Using (COD) :

$q = k(S_e - S_n)$; [useful for determination of S_e and non-biodegradable substrate concentration (S_n)].

2.2 Net biomass yield (ΔX)

* ΔX expressed in terms of Y, and k_d

$\Delta X = Y(S_F - S_e)Q_F - k_d\,XV$ [lb MLVSS/d]

$$\mu\,(\text{Specific biomass growth rate}) = \frac{\text{lb MLVSS yield}}{\text{lb MLVSS.d}}$$

$$\mu = \frac{1}{X}\left(\frac{dX}{dt}\right) = \frac{\Delta X_V/V}{X} = Yq - k_d$$

[useful for determination of Y and k_d]

$$\frac{dX}{dt} = Y\frac{dS}{dt} - k_x X$$

* ΔX expressed in terms of net biomass yield coefficient (Y_n)

$\Delta X = Y_n(S_F - S_e)Q_F$

$\mu = Y_n q$

$$\text{or,} \left(\frac{dX}{dt}\right) = Y_n\left(\frac{dS}{dt}\right), \text{ and}$$

$$Y_n = \frac{Y}{1 + k_d\theta_c}$$

2.3 Oxygen requirements (lb O_2/d)

lb O_2/d $= Q_F(1 - 1.42Y)(S_F - S_e) + 1.42k_d\,X\,V$

$[(S_F - S_e)$ in terms of COD or $BOD_u]$

or, lb O_2/d $= (OUR)V = a(S_F - S_e)Q_F + b\,X\,V$

Oxygen uptake rate [OUR (mg O_2/L.d] $=$ lb O_2/M.lb liquid.d)]

Specific oxygen utilization rate (R_{O2}) is :

$$R_{O_2} = \frac{OUR}{X}, \quad [lb\ O_2/lb\ MLVSS.d]$$

$$= \frac{1}{X}\frac{dO_2}{dt} = \frac{OUR}{X} = aq + b$$

[useful for determination of a and b]

or $\quad \dfrac{dO_2}{dt} = a\left(\dfrac{dS}{dt}\right) + bX$

2.4 General biomass concentration [MLVSS(X)] expression

$$X = \frac{1}{t}\frac{\theta_c\ Y(S_F - S_e)}{(1 + k_d\theta_c)}, \quad S_e \text{ is defined in Note 1 for various models}$$

2.5 **General relationship between recycle ratio (r) and sludge age (θ_c)**

$$\frac{1}{\theta_c} = \frac{1}{t}\left[1 + r - r\frac{X_u}{X} - \frac{X_F}{X}\right]; \quad t = \frac{V}{Q_F}$$

2.6 Total BOD of effluent

Total BOD of effluent = Soluble BOD (S_e) + $\Psi\ X_e$

where Ψ = BOD/VSS in the effluent and is a function of F/M ratio and can vary from 0.15 to 0.65

2.7 Nutrient requirements

N–requirements = $0.12\Delta X$

P–requirements = $0.02\ \Delta X$.

Example 6.124

Dispersion Number (Flow Regime in Aeration Basin)

Determine the dispersion number (D/UL) for a pond system with theoretical detention time of 2.0 days. The tracer test results are presented for a period of 9 hours only. Use both variance and peak–time methods :

Table 85 : Time versus concentration for the determination of dispersion number.

Time (t), hr	Concentration (C), mg/L	t^2	C.t	C.t^2
1	23.0	1	23	23
2	23.5	4	47	94
3	25.0	9	75	225
4	26.5	16	106	424
5	28.0	25	140	700
6	31.0	36	186	1116

(Contd...)

Time (t), hr	Concentration (C), mg/L	t^2	C.t	C.t^2
7	27.0	49	189	1323
8	26.0	64	208	1664
9	25.4	81	228.6	2057.4
Summation (Σ)	258.4	285	1202.6	7626.4

Solution

1. Mean time ($\bar{t}$)

$$\bar{t} = \frac{\Sigma C.t}{\Sigma C} = \frac{1202.6}{258.4} = 4.654 \text{ h}$$

2. Variance (σ^2)

$$\sigma^2 = \frac{\Sigma t^2 C}{\Sigma C} - (\bar{t})^2$$

$$= \frac{7626.4}{258.4} - (4.654)^2 = 7.89$$

3. Variance for a closed vessel of finite length

$$\sigma^2 = 2\left(\frac{D}{UL}\right) - 2\left(\frac{D}{UL}\right)^2 \left[1 - \exp\left(\frac{UL}{D}\right)\right]$$

Trial and error procedure will result in D/UL = 0.24.

4. Peak time procedure

$$\frac{t_p}{t_o} = \frac{\text{Peak time}}{\text{Theoretical detention time}} = \frac{6 \text{ hours}}{2 \times 24 \text{ hours}}$$

$$\frac{D}{UL} = 4.027\left(10^{-2.9(t_p/t_o)}\right), \quad \left[0.3 < \frac{t_p}{t_o} < 0.8\right]$$

or $\quad\dfrac{D}{UL} = 0.2\left(\dfrac{t_p}{t_o}\right)^{-1.34} \quad \left[0.03 < \dfrac{t_p}{t_o} < 0.3\right]$

Therefore,

$$\frac{D}{UL} = 0.2(0.125)^{-1.34} = 3.24$$

Variance method gives lesser values of D/UL because the time–concentration plot is prematurally stopped after 9 hours only without allowing full tail to be developed.

5. Empirical correlations

Empirical correlations between dispersion and certain parameters have been attempted in order to enable greater use to be made of the dispersion number concept in reactor–design without recourse to field tests or model studies everytime.

For diffused air aeration tanks of the type used in activated sludge plants, Murphy has reported the following correlation:

$$D = 3.118 \ (W^2)(q_a) \tag{1}$$

in which W = tank width, ft

q_A = air flow rate, per unit tank volume (standard $ft^3/1000 \ ft^3/min$)

This equation was derived from studies on tanks up to about 30 feet in width, and air supply rates ranging from less than 10 to more than 100 scfm/1000 ft^3.

Arceivala has pooled tracer study data from various sources covering mechanically aerated lagoons, oxidation ponds, rectangular settling tanks and such other units, to arrive at an empirical correlation between dispersion, D, and width, W, of a unit. Essentially, his results show that D varies as W^2 for small–sized units of less than 10 m, whereas the variation is linear for larger widths (W>30 m). Furthernore, in each case he found that units possessing baffles which lengthen the flow path (around–the–end type baffles) invariably gave much higher D values than those without baffles but having same width. This is because bends in flow increase dispersion.

The empirical correlations tentatively obtained can be expressed as follows, where D is in metres per hour and W in metres :

1. For lagoons and ponds of widths larger than 30 m

    (a) With baffles, D = 33 W (2)

    (b) Without baffles, D = 16.7 W (3)

2. For lagoons and ponds of widths less than 10 m

    (a) With baffles, $D = 11 \ W^2$ (4)

    (b) Without baffles, $D = 2 \ W^2$ (5)

    Among the factors affecting dispersion, it may be mentioned that, on three large–scale lagoons studied by Murphy and Wilson, variation in power input from 0.47 to 2.29 kW/1000 m^3 lagoon volume showed no significant effect on D and D/UL values. Thus, the mixing imparted by the aerators apparently did not affect the overall value of D which was already large owing to large lagoon widths (76 and 106 m). However, sufficient data is not in hand at present to be able to state that aerated lagoons of lesser widths will also likewise be unaffected by power input.

    In designing a lagoon or pond to give a desired value of D/UL a trial width first needs to be selected. Generally, the depth is already decided upon other considerations. Thus, for the given flow and detention time, the length is computed, and using one of the equations just given, D is then computed. It is now possible to find D/UL or its equivalent expressed $D.t/L^2$. Inserting a baffle in the unit helps to quickly reduce the D/UL value since the baffle reduces width, and increases both the velocity and length of the flow path.

Polprasert and Bhattarai have developed a dispersion prediction formula for use in waste stabilization pond design relating the pond's dispersion factor to the pond's shape (L, W and depth H), hydraulic detention time, t and kinematic viscosity, v, of pond water, as follows:

$$\frac{D}{UL} = \frac{0.184[t_v(W+2H)]^{0.489}(W)^{1.511}}{(LH)^{1.489}} \qquad (6)$$

The authors state that this formula and the constants proposed should be considered only tentative as they need to be tested with more pond data including variations in flows, inlet and outlet devices and wind conditions.

Presently, a relationship is not available to enable one to predict the overall value of D/UL when two or more cells are placed in series. Preliminary studies show that a sharp decrease in the overall value of D/UL occurs when a group of cells are placed in series although the D/UL value of each individual cell may not be relatively so low. As an approximation it can be said that,

Table 86 : D/UL values for Cell in series.

Cells in series	Likely overall D/UL value
Two	0.2–0.7
Four	0.1 or less

However, it is not essential to know the overall D/UL values. For cases involving cells in series, each cell may be designed individually, using the specific value of D/UL applicable to that cell. In this manner, each cell in the series arrangement can be designed as to promote well mixed or plug flow conditions as desired. The cells may also be unequal in size if beneficial for the process or required to comply with site topography. This flexibility in design is not possible to benefit from when the equal, completely mixed cells–in–series models described below are used.

6. Determination of dispersion number using empirical correlation

A mechanically aerated lagoon provides 5 days detention time to a wastewater flow of 10,000 m^3/day. If its depth is to be restricted to 3 m, estimate the lagoon dimension so that the dispersion number D/UL will be 0.5 or less.

Lagoon volume = $10,000 \times 5 = 50,000$ m^3

$$\text{Lagoon area} = \frac{50,000}{3} = 16,667 \text{ m}^2$$

As a trial, assume L/W = 4/1

This gives L = 258 m and W = 64,5 m

Equation (3) gives D = 16.7 W = 16.7 × 64.5 = 1115 m^2/h

Thus, $\dfrac{D}{UL} = \dfrac{Dt}{L^2} = \dfrac{1115 \times (5 \times 24)}{258^2} = 2.0$

This value is high, and lagoon dimensions need to be readjusted to increase the length/width ratio. This objective can also be achieved by inserting one baffle in the lengthwise direction, so that W = 32.5 m and L = 510 m.

Equation (2) must now be used to estimate D. Thus,

D = 33 W = 33 × 32.5 = 1073 m^2/h

and $\dfrac{Dt}{L^2} = \dfrac{1073 \times (5 \times 24)}{(510)^2} = 0.495$ (OK)

7. Typical values of D/UL for different waste treatment units.

Table 87 : Magnitudes of D/UL with various types of Waste treatment unit.

Waste treatment unit	D/UL* (Likely range)
1. Rectangular sedimentation tanks	0.2–2.0
2. Activated sludge aeration tanks :	
Long, plug flow type	0.1–1.0
Complete–mixing type	3.0–4.0 and over
3. Waste stabilization ponds :	
Multiple cells–in–series	0.1–1.0
Long, rectangular ponds	0.1–1.0
Single ponds	1.0–4.0 and over
4. Mechanically aerated lagoons :	
Rectangular, long	0.2–4.0
Square–shaped	3.0–4.0 and over
5. Pasveer and Carrousel type oxidation ditches	3.0–4.0 and over

*The dispersion number can also be expressed in terms of the detention time, t, of a unit. Since U = L/t where L = length of travel path in reactor, we can write.

$$\frac{D}{UL} = \frac{D.t}{L^2}$$

Example .6.125

Upgradation of Activated Sludge System

Determine the size of roughing filter (RF) and activated sludge system (AS) : (RF/AS). Presently, the ASP is facing operational problems related to discharge standards because of addition of effluent from food processing units. Use the following data:

Table 88 : Characteristics of effluents from PST and SST.

		PST Effluent (mg/L)		SST Effluent (mg/L)	
Condition	Flow (MLD)	BOD_5	TSS	BOD_5	TSS
Maximum–Monthly	3.0	120	80	25	25
Maximum–Weakly	4.0	250	150	5	
Peak–Hour	6.0	–	–	–	–

Aeration volume : 420 m^3 (with 4 of 20 kW high speed aerators at oxygen transfer efficiency of 0.8 kg O_2/kW–h–field conditions)

Sedimentation basin : Over 55 ft diameter; 10 ft SWD.

Solution

1. Determination of roughing filter size (RF) (plastic media)

    Required applied BOD_5 = (4 MLD)(250 mg/L)

    = 1000 kg/d [weakly BOD_5 is the highest value]

    $$\text{Volume of plastic media} = \frac{BOD_5 \text{ load}}{BOD_5 \text{ loading rate to RF}}$$

    $$= \frac{1000 \text{ kg/d}}{1.60 \text{ kg/m}^3.\text{d*}} = 625 \text{ m}^3$$

    $$\left[* \text{ RF loading rate} = \frac{100 \text{ lb BOD}}{1000 \text{ ft}^3.\text{d}} = (1.6 \text{ kg/m}^3.\text{d}) \right]$$

    $$\text{Surface area of bio-tower} = \frac{625 \text{ m}^3}{5 \text{ m}(= \text{depth})} = 125 \text{ m}^2$$

    $$\text{Bio-tower diameter} = \left[\frac{(125)(4)}{3.14} \right]^{0.5}$$

    $$= 12.6 \text{ m}.$$

2. Required hydraulic retention time (HRT), $\dfrac{F}{M}$ ratio, and mean cell residence time (MCRT) for ASP

2.1 Aeration basin HRT

$$= \frac{(400 \text{ m}^3)(1000 \text{ L/m}^3)(24 \text{ hr/d})}{(40,00000)}$$

= 2.4 hours (OK) [Conservative roughing filter BOD organic loading was considered].

2.2 Required $\dfrac{F}{M}$ ratio

Biomass in aeration basin (MLSS) = $(400 \text{ m}^3)(2.5 \text{ kg/m}^3)$*

$$= 1000 \text{ kg}$$

[*Basin MLSS = 2500 mg/L and MLVSS = (0.8) of MLSS]

MLVSS in aeration basin = (0.8)(1000 kg) = 800 kg

$$\text{Aeration basin } \frac{F}{M} \text{ ratio} = \frac{1000 \text{ kg BOD}_5/\text{d}}{800 \text{ kg MLVSS}}$$

$$= 1.25 \text{ (OK)}.$$

2.3 Required MCRT (θ_c) in aeration basin

Assume 0.8 kg TSS/kg BOD_5 of PST

Solids production rate = (0.8)(1000)

$$= 800 \text{ kg TSS/d}$$

$$\theta_c = \frac{VX}{(Q_w X_r + Q_e X_e)}$$

$$\text{or MCRT} = \frac{(\text{kg MLSS}) \text{ in aeration basin}}{\text{kg of solids production}}$$

$$= \frac{1000 \text{ kg MLSS}}{800 \text{ kg TSS/d}} = 1.25 \text{ d (Not OK)}^*$$

*[Suggested range for MCRT = 2 to 6 days].

2.4 Required aeration

Available oxygen in the field :

= (4 aerators)(0.8 kg O_2/kW hr)(20 kW/aerator)(24 hr/d)

= 1536 kg O_2/d

Average DO requirement in the basin

$$= \frac{1536 \text{ kg O}_2/\text{d}}{1000 \text{ kg BOD}_5/\text{d}} = 1.536 \text{ kg O}_2/\text{kg BOD}_5 \text{ [Not OK]}^*$$

*[Suggested range is 0.6 to 1.2 kg O_2/kg BOD_5].

3. Alternatively, reduce loading to the RF to 1.2 kg BOD_5/m^3.d which will load the combined system in TF/SC; operate the basin at a higher MLSS (3000 mg/L) which will affect secondary clarifier (SC) size or construct at an additional aeration basin.

4. At $\dfrac{F}{M}$ = 1.5 kg/kg.d and a bio–filter organic load of 1.2 kg BOD_5/m^3.d, average oxygen demand of less than 0.4 kg O_2/kg BOD_5 would be expected. Existing aeration is OK.

Note :

Table 88 : Design Criteria for combined processes

Process	Range	Typical
Activated biofilter		
Media type	High rate	High rate
BOD loading (lb $BOD_5/1000$ ft^3.d)	10–75	30
Hydraulic loading (gal/ft^2.min)	0.8–5.0	2.0
Filter MLSS (mg/L)	1500–3000	2000
TF–solids contact		
Media type	Rock or High rate	High rate
BOD_5 loading (lb $BOD_5/1000$ ft^3.d)	20–75	40
Hydraulic loading (gal/ft^2.min)	0.1–2.0	1.0
Channel MLSS (mg/L)	1500–3000	2000
HRT [hours based on influent Q only (no RAS)]	0.5–2.0	0.75
MCRT (days, aeration basin)	0.5–1.5	1.0
Return activated sludge (mg/L)	6000–12000	8000
Process	Range	Typical
Minimum Channel mixing		
Diffused air (5 ft^3/MG.min)	2000–4000	3000
Mechanical (hp/MG)	60–130	100
Roughing filter–AS, TF/AS, and BS/AF		
Media type	High rate	High rate
BOD_5 loading (lb $BOD_5/1000$ ft^3.d)	75–200	150
Hydraulic loading (gal/ft^2.d)	0.8–5.0	1.0
Basin MLSS (mg/L)	1500–4000	2500
HRT [hours based on influent Q, (no RAS)]	2.0–4.0	3.0
MCRT (days in aeration basin)	2.0–6.0	3.0
$\dfrac{F}{M}$ (lb PST BOD_5/lb MLSS.d)	0.5–1.2	0.9
Basin oxygen (lb O_2/lb–PST BOD_5 removed)		
Total available	0.6–1.2	0.9
Typically supplied	0.3–0.9	0.6

2. Combined process options

Activated bio–filter

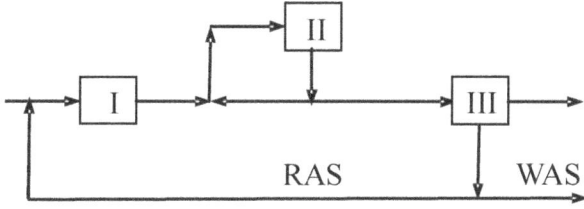

[A]

Trickling filter–solids contact and roughing filter (RF)–activated sludge (AS)

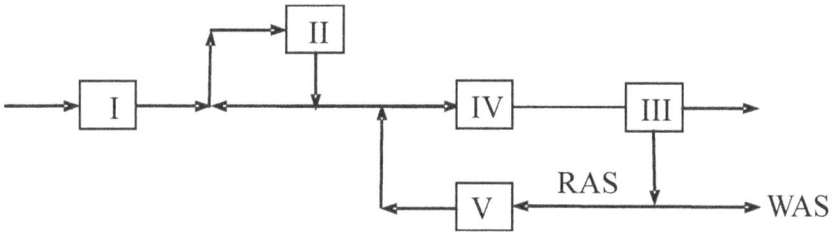

[B]

Bio–filter–activated sludge

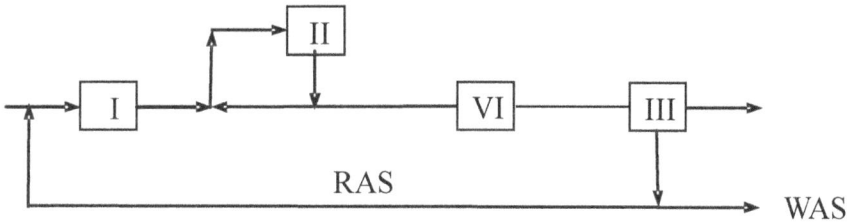

[C]

Trickling filter–activated sludge

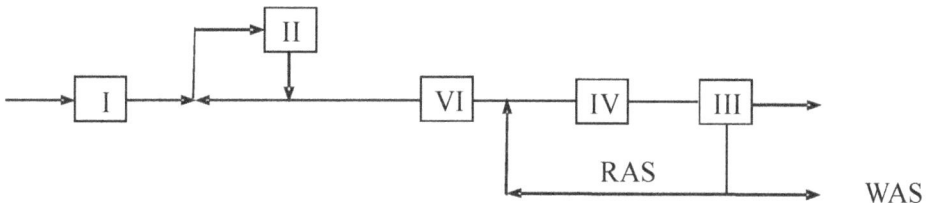

[D]

Figure 88 : Combined process operations, [A – D].

I : PST; II : Filter; III : Secondary clarifier; IV : Aeration system, V : Re-aeration of sludge; VI : Intermediate clarifier RAS : Return activated sludge; WAS–Waste activated sludge.

3. Alternatives for combined biological treatment process (aerobic) commonly used combined processes are typically included in two graphs : (1) Low to moderate organic loading rates to the fixed film reactors, and (2) High (roughing) organic loads.

    3.1 Activated bio–filter (ABF)

    Treats moderate organic loads.

    Red wood or high rate plastic media used as return sludge (RS) is incorporated with primary effluent and recycled over fixed film media [low SVI at high F/M ratio].

    Plug flow of the filter allows heterotrophic bacteria to be more competitive than filamentous bacteria.

    A small aeration basin for RS may be required at low temperatures.

    3.2 Trickling filter–Solids contact (TF/SC)

    Treats low to moderate organic loads with small contact channel.

    Contact channel is 10 to 15% of its sizes than that required for aeration basin for ASP alone.

    Fixed film with solids contact : reactor size is reduced by 10 to 30% of that normally required if the treatment had been accomplished with a trickling filter.

    Low energy requirements for activated sludge portion of the plant because of a relatively high dependence on trickling filter to remove most of the soluble BOD .

    Ability to upgrade existing rock trickling filters through polishing the fixed film effluent by using return activated sludge as a flocculating agent.

    Conventional TF/SC does not include sludge re–aeration.

    3.3 Roughing filter–Activated sludge (RF/AS)

    Upgradation of existing ASP by incorporated RF ahead of ASP.

    Reduces size to the extent of 15–30% of the sizes required if TF alone is used.

    HRT is 35–50% of that required for the ASP alone.

    TF/SC and RF/AS have sane process scheme, but with RF/AS, a much smaller trickling filter is required of that significant amount of DO, BOD removal, and solids digestion is achieved through ASP. This differs from TF/SC where trickling filter is larger and provides all of the S–BOD treatment, allowing the contact channel to provide only enhanced solids flocculation and effluent clarity.

3.4 Bio–filter–Activated sludge (BF/AS)

Similar to RF/AS except return activated sludge (RAS) is recycled over the fixed film reactor similar to the recycle of the ASF.

Recycling (RAS) reduces bulking from filamentous bacteria (especially with food processing wastes).

3.5 Trickling filter–Activated sludge (TF/AS)

Designed for high organic loads similar to those of RF/AS or BF/AS.

Provision of intermediate clarifier to remove the sloughed solids from fixed film before the underflow enters the suspended–growth reactor.

Second–stage is preferred where NH_3–N is to be removed (nitrification).

3.6 Other combined systems

RBC and RBC/SC.

RBC involves RAS from the secondary clarifier to the first stage of the RBC units.

RBC/SC system includes a short term solids contact reactor following a conventional RBC system.

Table 89 : Typical design criteria

System	Fixed Film Total Organic Load[a]	Suspended–Growth System		
		$\dfrac{F^b}{M}$	MCRT[c]	HRT[d]
Conventional ASP	None	0.3–0.5	5–15	4–8
Conventional TF	10–20	None	None	None
RF/AS, BF/AS, TF/AS	75–100	0.7–1.2	2–6	2–4
Tf/SC	20–75	1.5–3.0	0.5–1.5	0.5–2.0

[a] : TOL = Total organic load on filter lb BOD_5 applied/1000 ft^3.d

[b] : $\dfrac{F}{M}$ = lb BOD_5/lb MLVSS.d.

[c] : MCRT–days.

[d] : HRT–hours based on raw wastewater flow rate only.

Example 6.126

Sludge Bulking Controls through Selector Reactor System

Design an influent selector chamber to control sludge bulking in an activated sludge process. Assume the following data :

Wastewater flow	:	1.25 MGD
Influent BOD$_5$	:	350 mg/L
Aeration basin MLVSS	:	2000 mg/L
Hydraulic retention time	:	0.6 day

Solution

1. Required $(F/M)_{Feed}$ ratio

    $$= \frac{(1.25)(350)(8.34)}{(0.6)(1.25)(8.34)(2000)} = 0.29 \ d^{-1}$$

2. Required concentration in the CSTR (ASP)

    $$S_e = \frac{K_s[1 + k_d \ SRT]}{[SRT][Yk - k_d] - 1}$$

    $$[SRT]_{lim}^{-1} = \frac{YkS_o}{K + S_o} - k_d$$

    $$SRT(design) = 20(SRT)_{lim} = 10.36 \ days$$

    $$S_e = 4.85 \ mg/L$$

    [Using $Y = 0.6$; $k_{max}\left(= \dfrac{\mu_{max}}{Y}\right) = 4 \ d^{-1}$; $S_o = 350 \ mg/L$, $K_s = 75 \ mg/L$, $k_d = 0.05 \ d^{-1}$, $X = 2000 \ mg/$]

    As the BOD$_5$ concentration within the reactor is less than 30 mg/L soluble degradable COD, the possibility of sludge filamentous growth is always iminent [other factors are OK.

3. Required selector chamber design

    Assume HRT = 50 min = 0.035 day

    $$\text{Apply a loading of 10 kg/m}^3.d \ \left(= \frac{10 \times 10^6 \ mg}{1000 \ L.d} = 10,000 \ mg/L.d\right)$$

    $$\text{Actual loading} = \frac{350 \ mg/L}{0.035 \ d} = 10,000 \ mg/L.d$$

    Apply a BOD$_5$ feed F/M of at least 3 d^{-1}

$$\text{Actual} \left(\frac{F}{M} \right) = \frac{350 \text{ mg/L}}{(2000 \text{ mg/L})(0.035 \text{ day})} = 5 \text{ d}^{-1}$$

Apply a MLSS loading of 3300 mg/L

$$\text{Actual MLSS} = \frac{2000 \text{ mg/L MLVSS}}{0.8} = 2500 \text{ mg/L MLSS}$$

MLVSS of 2000 mg/L is too low to meet 3300 mg/L.

Therefore, MLVSS of 2600 mg/L = 3250 MLSS ($\approx$ 3300) mg/L (OK)

$$\text{Adjusted actual } \frac{F}{M} = \frac{350 \text{ mg/L}}{(2600 \text{ mg/L})(0.035)} = 3.85 \text{ (OK)}$$

Utilize minimum of 4–chambers

Volume = $(1.25 \times 10^6 \text{ gal/d})(0.035 \text{ d})$

$\qquad$ = 43750 gal (= $43,750 \times 0.13367 = 5848 \text{ ft}^3$)

Assume a depth of 10 ft

$$\text{Area} = \frac{5848 \text{ ft}^3}{10 \text{ ft}} = 584.8 \text{ ft}^2$$

$$\frac{L}{W} = \frac{2}{1} \text{ ; } \qquad W \times L = 584.8 \text{ ft}^2$$

$$W = (292.4)^{0.5} = 17 \text{ ft}$$

$$L = 34 \text{ ft}$$

Selected chamber size = 20 ft $\times$ 36 ft = 720 ft^2

$\qquad$ 4 chambers = 720 ft^2

$\qquad$ One chamber = 180 ft^2 [$\approx$ 10 ft $\times$ 20 ft)

[However, pilot plant studies are required to establish that poor sludge quality is produced because of low substrate concentration in the aeration basin (4.85 mg/L), and also to check the guidelines for the selector chamber design criteria].

Note :

1. Sludge bulking (filamentous growth) control through selector chamber

   Based on the potential problems associated with CSTRS, three reactor augmentation should be considered :

   Wastewater characteristics should be screened for glucose–like saccharide components, high sulphide septic conditions, and toxic compounds. Each stream should be screened and pre–treated to remove (or neutralize the effects) of these components.

Design should assure an environment that suppresses filamentous growth by assuring an adequate oxygen level, a suitable pH range, and sufficient nutrients.

Provision should be made to employ alternative configuration or include a first–stage selector :

- A plug flow reactor to assure contact with the highest level substrate at the process inlet, encouraging the selective growth of flocculating sludge.

- Sequencing batch reactors (SERs) where F/M ratio can be chosen to ensure process stability.

- Intermittent operation of a continuous system at oxygen deficit conditions, encouraging flocculating bacteria growth that can exist under aerobic and anaerobic conditions, and destroying filamentous bacterium that tend to survive under aerobic conditions.

Selector design criteria

- Alternatively, addition of a selector process (a separate initial reactor (selector) may be included to cultivate flocculating and suppress filamentous, bacterium growth. The resulting cultivated population can then be fed to the main aerobic reactor to promote both high process efficiency and a settleable sludge.

- A selector can be a separate reactor or an segregated portion of the biological reactor, designed for high substrate to micro–organism contact. The selector simulates a plug flow reactor, designed as either as a series of CSTRs operating under controlled micro–organism growth conditions or a segregated primary section for near plug flow hydraulic conditions [dispersion number (D/uL) below 0.2].

Selector design criteria can be summarised as :

- Selector chamber should be capable of removing about 80% of the influent degradable COD.

- Selector reactor should contain four compartments.

- Selector should accommodate a volumetric loading (BOD$_5$ based) of 10 kg/m^3.d.

- Selector should operate at a MLSS concentration of 3300 mg/L.

- Selector should accommodate a sludge loading of 3 kg BOD$_5$/kg MLVSS.d.

- Selector should have an oxygenation capacity of 4 kg O$_2$/m^3.d.

These parameters should assure a first compartment BOD$_5$ volumetric loading of 40 kg/m^3.d and a sludge loading of 12 kg/kg.d. The whole aeration system design should fall within the following parameters:

- Volumetric BOD_5 loading of 1 $kg/m^3.d$
- MLSS concentration of 3300 mg/L
- Sludge loading of 0.3 kg/kg.d (BOD_5 based)
- Oxygenation capacity of 2 $kg/m^3.d$.

Plug flow configuration performance may be improved by step or tapered aeration to optimise oxygen utilization.

Selector configurations.

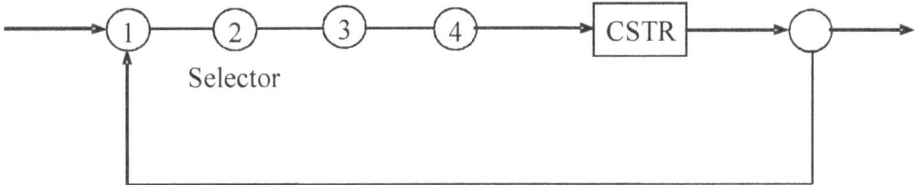

(*a*) Complete mix reactor selector

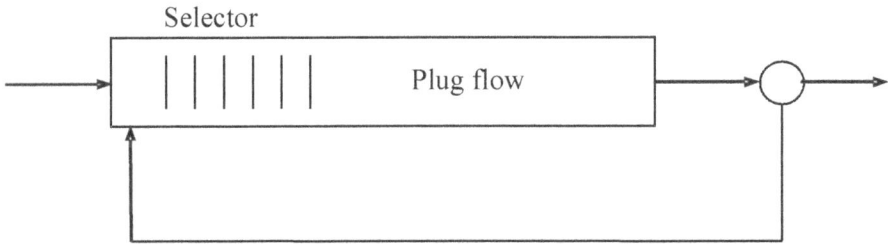

(*b*) Plug flow reactor selector

Figure 89 : Selector configurations.

Example 6.127

1. Nutrient Requirements for an Activated Sludge Process

Determine the nutrient (N and P) requirements for an activated sludge process treating an industrial wastewater. Use the following data :

Wastewater flow (Q)	:	1.0 MGD
Influent BOD (S_o)	:	330 mg/L
Effluent BOD (S_e)	:	30 mg/L
MLVSS (X_V)	:	2500 mg/L
Yield coefficient (a)	:	0.5
Decay coefficient (b)	:	0.15 d^{-1}

Influent Ammonia–N (NH_3–N)	:	4.5 mg/L
Influent Phosphorus	:	1.0 mg/L
Food to micro–organism ratio (F/M)	:	0.3 d^{-1}
Sludge retention time (θ_c)	:	10 d

Solution

1. Required hydraulic detention period [θ]

$$\frac{F}{M} = \frac{S_o}{(f_b\,X_V)(\theta)}$$

or $\qquad \theta = \dfrac{S_o}{(f_b\,X_V)(F/M)}$

$$= \frac{330}{(2500)(0.3)} \quad [f_b = 1,\ \text{the fraction which is biological}]$$

$$= 0.44\ d$$

2. Required degradable fraction of the biological VSS (X_d)

$$X_d = \frac{X_d^*}{1 + bX_n^*\theta_c}$$

where X_d^* is the degradable fraction of the biological VSS at generation (usually 0.8), and X_n^* is the non–degradable fraction of the biological VSS at generation (usually 0.2)

$$X_d = \frac{0.8}{1 + (0.15)(0.2)(10)}$$

$$= 0.61$$

3. Required sludge production (ΔX_V)

$$\Delta X_V = [a\,(S_o - S_e)] - [(b\,X_d\,X_V)(\theta)]$$

$$= [0.5(330 - 30)] - [(0.15)(0.61)(2500)(0.44)]$$

$$= [150 - 100.6] = 49.35\ mg/L$$

$$\Delta X_V = (49.35)(1)(8.34)$$

$$= 412\ lb\ VSS/d$$

4. Required nutrient for the activated sludge

    4.1 Required nitrogen [W] : N

$$N = 0.123 \left(\frac{X_d}{0.8}\right)(\Delta X_V) + (0.07)\left(\frac{0.8 - X_d}{0.8}\right)(\Delta X_V)$$

$$= 0.123 \left(\frac{0.61}{0.8}\right)(412) + (0.07)\left(\frac{0.19}{0.8}\right)(412)$$

$$= 38.6 + 6.8 = 45.4 \text{ lb N}/d$$

$$\text{N–input} = (4.5)(1)(8.34) = 37.53 \text{ lb N}/d$$

Additional N–required $= 45.4 - 37.53 = 7.87$ lb N/d

4.2   Required phosphorus [P] :

$$P = 0.026\left(\frac{X_d}{0.8}\right)(\Delta X_V) + (0.01)\left(\frac{0.8 - X_d}{0.8}\right)\Delta X_V$$

$$= \left[0.026\left(\frac{X_d}{0.8}\right) + 0.01\left(\frac{0.8 - X_d}{0.8}\right)\right](\Delta X_V)$$

$$= \left[0.026\left(\frac{0.61}{0.8}\right) + (0.01)\left(\frac{0.19}{0.8}\right)\right](412)$$

$$= [0.0198 + 0.002375]\,(412)$$

$$= 9.14 \text{ lb} - \text{N}/d$$

$$\text{P–input} = (1.0 \text{ mg/L})(1.0 \text{ mgd})(8.34)$$

$$= 8.34 \text{ lb–P}/d$$

Additional phosphorus (P) required $= (9.14 - 8.34)$ lb–P/d

$$= 0.80 \text{ lb–P}/d.$$

Note :

1.   The N–content of the sludge generated in the process is about 12.3% (average basis of the VSS). The nitrogen content of the sludge declines during the endogenous phase. The N–content of the non–biodegradable cellular mass is about 7%.

2.   The phosphorus content of sludge at generation has been found to the around 2.6% with the non–biodegradable cellular mass having a phosphorus content of 1%.

3.   Not all organic nitrogen compounds are available for synthesis. Ammonia is the most readily available form, and other nitrogen compounds must be converted to ammonia. Nitrite, nitrate, and about 75% of organic nitrogen compounds are also available.

4.   Phosphorus may be fed as phosphoric acid in large plants and ammonia as anhydrous or aqueous ammonia. Diammonium phosphate may be fed as a nutrient in small plants.

5.   In many cases, in aerated lagoons treating pulp and paper mill wastewaters, N and P have not been added but rather SRT has been increased. However, reaction rate constant increases with addition of

optimum dose of N and P. In facultative lagoons, benthal decomposition feeds back N and P.

2. Nitrogen Requirements for the ASP

An activated sludge plant (ASP) is to be designed for an industry which will produce 0.5 MGD of waste with a 5 day BOD of 1000 mg/L, and a total available nitrogen concentration of 8 mg/L. Removal 95% of the BOD is necessary to meet receiving stream conditions. It is assumed that approximately 60% of the BOD removed is assimilated to sludge, that one pound of VSS is equivalent 1.42 lb of ultimate BOD and that a critical cell nitrogen concentration of 7% based on the biological volatile solids is necessary to maintain optimum process efficiency. Calculate the quantity of nitrogen which must be added per day t the system.

Solution

1. BOD_5 removed

$$= 0.5 \times 8.34 \times 1000 \times 0.95 = 3962 \text{ lb/d}$$

2. Ultimate BOD removed

$$= 1.42 \times 3962 = 5625 \text{ lb/d}$$

3. VSS produced

$$= \frac{5625 \times 0.6}{1.42} = 2377^* \text{ lb/d} \ (= 2400 \text{ lb/d})$$

4. Total nitrogen required

$$= 0.07 \times 2400^* = 168 \text{ lb/d}$$

5. Total nitrogen available

$$= 0.5 \times 8.34 \times 8 = 33.4 \text{ lb/d}$$

6. Net nitrogen to be added

$$= 168 - 33.4 = 134.6 \text{ lb/d}$$

7. This is approximately to 4.24 lb of nitrogen per 100 lb of 5–day BOD removed in the system.

8. Similar analysis can be carried out for phosphorus requirements.

Example 6.128

Operating Temperature of Well Mixed Biological Reactor

Derive the expression for biological reactor temperature (T_w) in a well mixed system.

Solution

1. Required biological reactor temperature (T_w)

Heat balance over the reactor system :

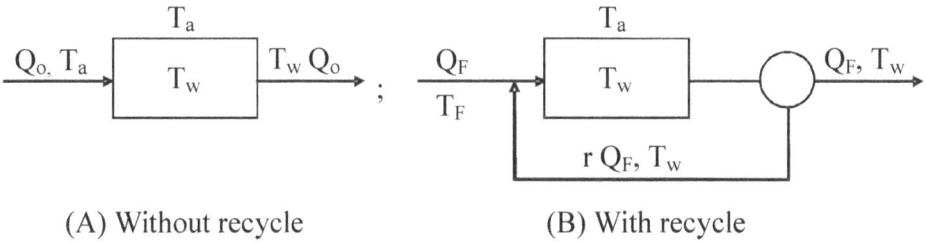

(A) Without recycle (B) With recycle

Figure 90 : Heat balance over the Reactor System

* The recycle stream enters and leaves the reactor systems at the temperature T_w, and this does not add to the heat balance equation.

* Mass flow rate [m_F^*, lb/h] is given by :

$$m_F^* = Q_F(MGD)\left(10^6\ \frac{gal}{MG}\right)\left(8.34\ \frac{lb}{gal}\right)\left(\frac{d}{24\ h}\right)$$

$$= 3.475 \times 10^5\ Q_F$$

Various parameters are :

Q_F = Fresh feed flow (MGD)

T_F = Fresh feed temperature (°F)

T_a = Surrounding air temperature (°F)

h = Overall heat transfer coefficient between reactor liquid and surrounding air (BTU/h.ft^2.°F)

A = Exposed surface area of the surrounding air of biological reactor (ft^2)

C = Specific heat of the liquid (= 1.0 BTU/lb.°F)

Influent enthalpy is :

$$= m_F\ (lb/h)\ (1.0\ BTU/lb,°F)(T_F - T_w);\ [(T_F - T_w)\ in\ °F]$$

$$= 3.475 \times 10^5\ Q_F\ (T_F - T_w)\ ;\ BTU/h$$

Heat lost to the surrounding air is:

$$= h(A)(T_w - T_a)\ ;\ BTU/h$$

Heat balance (considering the heat reaction due to substrate oxidation is small in comparison to influent enthalpy, and can be ignored) is:

$$3.475 \times 10^5\ Q_F\ (T_F - T_a) = hA\ (T_w - T_a)$$

or $Q_F\ (T_F - T_w) = FA(T_w - T_a)$

$$\left[F = \frac{h}{3.475 \times 10^5} \quad \text{and} \quad F = 12 \times 10^{-6} \times h = \frac{100 \text{ BTU}}{\text{ft}^2.\text{d}} \right]$$

$$\text{or } T_w = \frac{(AFT_a + Q_F T_F)}{(AF + Q_F)}$$

Heat transfer coefficient (h) is a function of aerator horsepower, wind speech, solar radiation, relative humidity of the air, and reactor configuration. Horsepower of the aerators is a significant variable in comparison to other variables, and for activated sludge it has been found to be :

$$hA = 2500(hp)$$

where 2500 BTU/hp.h.°F is the quantum of heat liberated from one **horsepower in one hour**, for a 1 °F temperature change

Therefore, 3.475×10^5 °F $(T_F - T_w) = 2500 \times hp \times (T_w - T_a)$

$$\text{or, } T_w = \frac{(3.475 \times 10^5 \, Q_F T_F + 2500 \times hp \times T_a)}{(3.475 \times 10^5 \, Q_F + 2500 \times hp)}$$

A required estimation (initial assessment of hp required) of hp is expressed as :

$$\frac{\text{lb BOD removed/d}}{(40 \text{ to } 50)\left(\dfrac{\text{lb BOD removed}}{\text{hp.day}}\right)} = hp$$

lb BOD removed/d $= 8.34Q_F \, (S_F - S_e)$

Therefore,

$$hp = \frac{8.34Q_F(S_F - S_e)}{(40 \text{ to } 50)}$$

where Q_F = wastewater flow (MGD)

  S_F = Influent BOD

  S_e = Effluent BOD

2. Required kinetic parameters as a function of reactor temperature (T_w)

   Reaction rate constant (k)

   $$k(T_w) = k(20) \, \theta^{(Tw - 20)}$$

   Decay rate constant (k_d)

   $$k_d(T_w) = k_d(20) \, \theta^{(Tw - 20)}$$

   Endogenous respiration rate constant (b)

   $$b(T_w) = b(20) \, \theta^{(Tw - 20)}$$

Biomass yield coefficient (Y), substrate concentration constant (K_s), and lb oxygen required/lb of substrate removed (a) are not sensitive to temperature change.

Note :

1. Heat balance for two CSTRs in series

$$T_w = \frac{A K(7.48 \times 10^{-6}) T_e + Q_F T_F}{A K(7.48 \times 10^{-6}) + Q_F}$$

where overall heat transfer coefficient (ft/d) is given by :

$$K = 0.374 + [0.60 \times 10^{-4} \exp(0.0330 T_a) + 0.003] \times \left(646 \times 10^{-0.1} V_w + 1440 \frac{Q_A}{A} \right)$$

where   X   = Characteristic dimension of the basin [ft, defined as X = $A^{0.5}$]

  V_w = Average wind spread (miles/h)

  Q_A = Airflow through aeration spray (ft³/min),

     = 44 NE V_W

  N   = Number of aerators in the basin

  E   = Aerator spray area [ft², characteristic for the types of aerator selected]

  T_e = Equilibrium temperature between reactor water and surrounding air.

  T_e is calculated from :

$$T_e = T_a + 3.28/K[0.117 H_s + 12.5(\beta - 0.874)] - (0.55 \times 10^{-3}) \exp(0.0335 T_a)$$
$$(1 - f_a/100) \times \left(646 \times 10^{-0.1} V_w + 1440 \frac{Q_A}{A} \right)$$

where H_s = Solar radiation (Btu/ft².h)

  β = Raphael long wave radiation (dimensionless)

  f_a = Relative humidity in air (%).

Example 6.129

Cost Optimization for Combination of Aeration Basin (Activated Sludge) and Secondary Sedimentation Basin

1. Required schematics of the conventional activated sludge process.

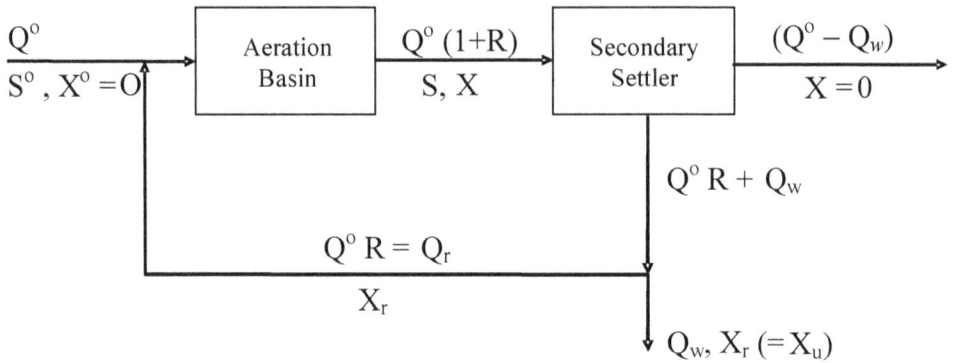

Figure 91 : Schematics of the conventional activated sludge process.

2. Required aeration basin design.

   Given process data = $[Q°, S°, S]$

   Given kinetic data = $[Y$ and $k_d)$ and sludge retention time (θ_c)

   Therefore, the reactor effluent solids concentration (X) going to the settler is :

$$X = \frac{Y(S° - S)}{1 + k_d \, \theta_c}\left(\frac{\theta_c}{\theta}\right) \tag{1}$$

   In this equation, θ_c is the only unknown, and assuming θ, X can be calculated.

3. Required clarifier design

   The clarifier (thickner) can be designed using the batch flux method (Figure 92). Assume that all solids leave in the underflow (X_u or X_r), the solids mass flow rate to the settler is given by :

$$W_t = Q°X + Q°RX = Q° \, (1+R)X \tag{2}$$

   Mass flux of solids in the settler for a reactor with recycle is given as:

$$\frac{W_t}{A} = \left[\frac{Q°}{A}\right][1+R]X, \text{ (Figure 92)} \tag{3}$$

   and is shown as straight line from the origin with slope of $\left[\dfrac{Q°}{A}\right][1+R]$.

   The volumetric flow rates of liquid leaving the top of the clarifier is = $(Q° - Q_w)$

   As Q_w is small compared to $Q°$, the overflow rate = $Q°$ and, therefore, $Q°/A$ is essentially equal to the upward velocity of clear liquid in the clarification segment and must not exceed the settling velocity of solids corresponding to the clarifier feed concentration X.

Draw a vertical line at X intersecting the batch flux plot at :

$$\frac{W_t}{A} = \left[\frac{Q^\circ}{A}\right]_{max} X, \qquad [Flux = v_i\, x_i] \tag{4}$$

Therefore, $[Q^\circ/A]_{max}$ is the initial settling velocity for a solids concentration (X), and represents the maximum possible upward liquid velocity in the clarification zone.

$[Q^\circ/A]_{design} = [Q^\circ/A]_d$ should be less than $[Q^\circ/A]_{max}$

For an assumed solids recycle ratio (R), the total solids flux is found at the intersection of the vertical line at X with the line from the origin of the slope $[Q^\circ/A]$ $[1 + R]$ and is represented as G_1.

Assumed value of R must be checked against the value calculated from a solids balance on the clarifier. The overall solids balance on the settler is :

$$G_1 = \left[\frac{Q^\circ}{A}\right]_d [1+R]X$$

$$= \frac{Q_r\, X_r}{A} + \frac{Q_w X_r}{A} \tag{5}$$

$$= \frac{Q_r\, X_r}{A} + \frac{P_x}{A} \tag{6}$$

It is known that :

$$\theta_c = \frac{\theta}{\left[1+R-R\,\dfrac{X_r}{X}\right]}$$

$$P_x = \frac{X\,V}{\theta_c} = Q_e X_e + Q_w X_w$$

$$= YQ^\circ\,(S^\circ - S) - k_d\, X\, V$$

Therefore,

$$G_1 = \left[\frac{Q^\circ}{A_d}\right]_d R X_r + \left[\frac{Q^\circ}{A}\right]_d \left[\frac{\theta}{\theta_c}\right]X \tag{7}$$

$$= \left[\frac{Q^\circ}{A}\right]_d R \left[\frac{1+R}{1+R-\theta/\theta_c}\right]X_r$$

Bulk underflow velocity (U_b) is the slope of the batch settling curve:

$$U_b = \frac{Q_u}{A} = \frac{W_t}{X_u A} = \frac{G_1}{X_u} = \frac{Q_1}{X_r}$$

Therefore,

$$U_b = \left[\frac{Q^\circ}{A}\right]_d R \left[\frac{1+R}{1+R-\dfrac{\theta}{\theta_c}}\right] \qquad (8)$$

If correct recycle ratio (R) was assumed in the calculation of G_1, the bulk underflow velocity (U_b) calculated by equation (8) should equal that calculated by taking the slope of the tangent to the bulk settling curve passing through the point (0, G_1). If the right value of R was assumed, then X_r is determined by the interaction of the tangent curve with the abscissa.

Procedure for determination of R, A and X_r

* Select a value of θ, calculate X using equation (1).

* G_1 is determined by the intersection of a vertical line at X and a line from origin with a slope of $[Q^\circ/A]_d$ [1+R] using previously assumed values of $[Q^\circ/A]d$ and R.

* From the point (0, G_1), a line is drawn with a negative slope tangent to the batch settling curve.

* The correct recycle ratio was assumed when the slope of the tangent to the batch flux curve measured graphically is equal to that calculated by equation (8).

* The area of the clarifier was fixed by the clarification zone assumption on $[Q^\circ/A]d$, since A = $Q^\circ/[Q^\circ/A]_d$.

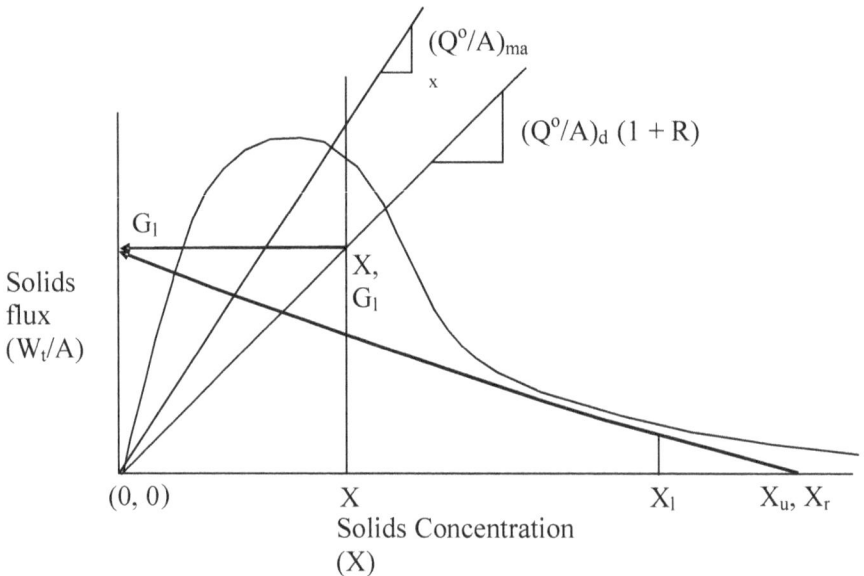

Figure 92 : Reactor-clarifier design procedure.

$$A = \frac{Q^o}{\left[Q^o/A\right]_d}$$

Intersection of the tangent curve with abscissa gives X_r. For each assumed value of q, the corresponding values of R, A, and X_r are determined.

4. Determine the hydraulic residence time (θ), clarifier area (A) and solids recycle flow rate (R) such that the total annual costs are a minimum. Assume the data provided by Keinath, *et. al.* :

   Process data : $S^o = 300$ mg/L

   $S = 8$ mg BOD_L/L

   $Q^o = 0.438$ m^3/s (10 MGD)

   Kinetic data : $Y = 0.625$ mg MLSS/mg BOD_L

   $k_d = 0.06$ d^{-1}

   $\theta_c = 6$ days

   Cost data : Depreciation period = 25 yrs

   Power costs   = 1.5 \$/kWhr

   Labour costs   = \$ 6.90/man–hour

   Inflation rate compounded at 7%

   Constraints :

   * $(Q^o/A)_d$: 32.6 m/d (800 gal/ft^2.d)

   * Min. aeration   : 62 m^3/kg BOD_L consumed (1000 ft^3/lb)

   * Actual aeration capacity   : 150% of minimum

   * Actual recycle pumping   : 150% minimum capacity

   * Batch settling curve for solids : Figure 93.

4.1 Required values of A, maximum recycle pumping capacity, clarifier and pumping costs, aerator basin and aerator costs, and total costs at an assumed hydraulic detention time (q) in aeration basin.

• Assume $\theta = 8$ hours, therefore, MLSS (X) is :

$$X = \frac{Y(S^o - S)}{1 + k_d\theta_c}\left[\frac{\theta_c}{\theta}\right]$$

$$= \frac{0.625(304 - 8)}{1 + 0.06(6)}\left[\frac{6}{8/24}\right]$$

$$= 2415 \text{ mg MLSS/L}$$

Draw a verticle line at solids concentration $(X) = 2415$ mg MLSS/L. [Figure 93]. Assume $R = 0$, 0.3, 0.4 and 0.5, the value of G_1 can be calculated

$$\text{Area} = \frac{Q_o}{[Q_o/A]_d} = \frac{0.438 \text{ m}^3/\text{s} \times 60 \times 60 \times 24}{32.6 \text{ m/d}}$$

$$= 1161 \text{ m}^2 \ (12500 \text{ ft}^2)$$

Mass flux at $R = 0$, 0.3, 0.4 and 0.5

$$G_1 = \frac{W_t}{A} = \frac{Q^o}{A}(1+R)X$$

$$G_1 (R = 0) = \left(\frac{0.438 \times 60 \times 60 \times 24}{1162}\right)(2415 \times 10^{-3} \text{ kg/m}^3)$$

$$= 78.65 \text{ kg/m}^2.\text{d}$$

$G_1 (R = 0.3) = 78.65 \times 1.3 = 102.24 \text{ kg/m}^2.\text{d}$

$G_1 (R = 0.4) = 78.65 \times 1.4 = 110.13 \text{ kg/m}^2.\text{d}$

$G_1 (R = 0.5) = 78.65 \times 1.5 = 117.97 \text{ kg/m}^2.\text{d}$

Draw tangents from ordinate at $G_1 = 102.24$, 110.13 and 117.97 kg/ m².d to the batch settling curve Figure 93. Determine the slope of these lines. U_b is also calculated from the following equation :

$$U_b = \left[\frac{Q^o}{A}\right]_d R\left[\frac{1+R}{1+R-\theta/\theta_c}\right] \tag{9}$$

Calculated and Graphical values of U_b

Table 90 : Calculating and graphical value of U_b.

Recycle Ratio (R)	U_b (Equation–8)	U_b (Graphical)
0.3	10.22	12.02
0.4	13.58	13.46
0.5	16.93	14.81

The corrected value of $R = 0.39$.

Maximum recycle pumping capacity $= 1.5 \ Q^o \ R$

$= 1.5(0.39)(37840 \text{ m}^3/\text{d})$

$= 22000 \text{ m}^3/\text{d}$ (5.8 MGD, Figure 93)

The recycle solids concentration $= 8570$ mg MLSS/L

Clarifier and pumping costs

Clarifier costs (C) $= 2.8 \times 10^4 \ (A)^{0.88}$

$$= 2.8 \times 10^4 \ (12.5)^{0.88} = \$ \ 25.85 \times 10^4$$

Recirculation pumping costs (C) $= 2.0 \times 10^4 \ (MGD)^{0.70}$

$$= 2.0 \times 10^4 \ (5.8)^{0.7} = \$ \ 6.846 \times 10^4$$

Use straight line depreciation, the annual depreciation are calculated in 2002 using compound interest at 7% for 6 years.

$$\text{Annual depreciation} = \frac{[(25.85 + 6.846) \times 10^4][1.07]^6}{25 \ \text{years}}$$

$$= \$ \ 1.96 \times 10^4/\text{yr}.$$

5. Clarifier costs with R = 0.39

5.1 Area (A) $= 1162 \ m^2 \ (12.500 \times 10^3 \ ft^2)$

Installed cost in 1996 $= \$ \ 25.86 \times 10^4$

Labour required $= 1800 \ hr/yr$

Annual labour cost $= (1800 \ hr/yr) \ (\$ \ 6.9/hr)$

in 2002 $= \$ \ 12,400$

Supply cost in 2002 $= \$ \ 2000/yr$

5.1.1 Recycle cost

Pump capacity $= 22,000 \ m^3 \ (5.8 \ MGD)$

Pump cost in 1996 $= \$ \ 68,400$

Power cost in 2002 $= \$ \ 6200$

5.1.2 Total annual depreciation costs in 2002 : $ 19,600

5.1.3 Total annual cost/yr = $ 19,600 + $ 12,400 + $ 6200 + $ 2000

$$= \$ \ 40,200$$

5.2 Aeration basin

Volume of aeration basin $(Q^\circ . \theta) = 37,840 \ m^3/d \times 8/24$

$$= 12,614 \ m^3 \ (4.4510^5 \ ft^3)$$

Cost of aeration basin volume $= 4.2 \times 10^3 \ (V)^{0.79}$

in 1996 $= 4.210^3 \ [4.4210^2]^{0.79}$

$$= \$ \ 517,000 \ \text{in} \ 1996$$

Minimum air requirement $= 62.0 \ m^3/kg \ BOD_L \ \text{consumed} \ (1000 \ ft^3/lb)$

$$= (62) \ [37,840 \times (300 - 8) \times 10^{-3}]$$

$$= 6.85 \times 10^5 \ m^3/d \ \text{of air}$$

Actual installed capacity of aerator = 10.3×10^5 m^3/d (25,200 ft^3/min) [1.5 × Min. requirements]

Actual blower cost in 1996 (C) = $ 400,000

Manpower costs = 5500 hr/yr × $ 6.90/hr

= $ 37,950/yr

Electric power costs = $ 70,000/yr

Supplies costs = $ 6,000/yr

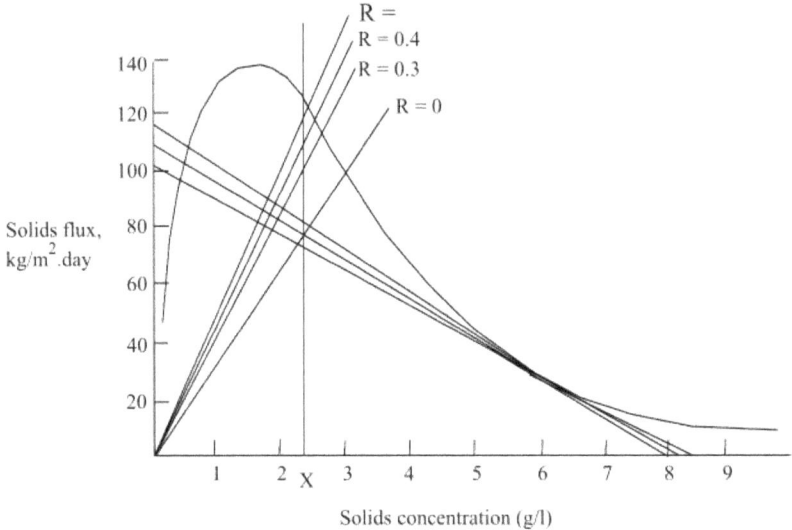

Figure 93 : Batach settling curve

Annual depreciation costs $= \$\dfrac{(517,000 + 400,000)(1.07)^6}{25}$

= $ 55,000/yr

Total aeration system costs = $ 55,000 + $ 37,950 + $ 70,000 + $ 6000

= $ 169,000

5.3 Total annual cost for the aeration plus clarifier system for θ = 8 hours
= $ 169,000 + $ 40,100

= $ 209,100

5.4 A series of residence times is assumed and the above procedure is repeated to arrive at total system cost. The least cost Solution is obtained for θ near 8 hours [Figure 94]. The optimum recycle ratio near 39% of the influent is in the range of present operations.

Note :

1. Cost analysis

$$\text{Total annual costs} = \sum_{i}^{n} (\text{Annual depreciation cost})(\text{DF})$$

$$+ \sum_{i}^{n} (\text{Annual O\&M})(\text{DF}) \qquad (1)$$

where n   = Project life [20 to 30 years]

   DF = Discount factor [(depends upon the type of payment (lump sum or continuous)]

$$= \left[\frac{1}{1+i}\right]^g$$

   (Lump sum payment at the end of gth year from the time of plant start up)

Annual depreciation costs depend upon the total capital cost and the method of depreciation [straight line, double declining balance, or the sum of years digits]. A common method of calculating annual depreciation in municipal wastewaters is to use the sinking fund method, which gives annual depreciation including interest (d_g) on the capital to be :

$$d_g = (V_o - V_n)\left[\frac{(i+1)^n\, i}{(1+i)^n - 1}\right]$$

where V_o = Initial value

   V_n = Value after n years

   n   = Project life

If payments of the capital is to be taken upon completion of the construction then equation (1) becomes.

$$\text{Total annual cost} = \text{Total capital cost} + \sum_{1}^{n} (\text{Annual O\&M costs})\,\text{DF}$$

Capital cost of a plant is commonly taken up to be the sum of installed costs for several unit operations making up the process. Costs are often of the form :

   $C = A$ (unit size)^n

where n is less than or equal to 1.0.

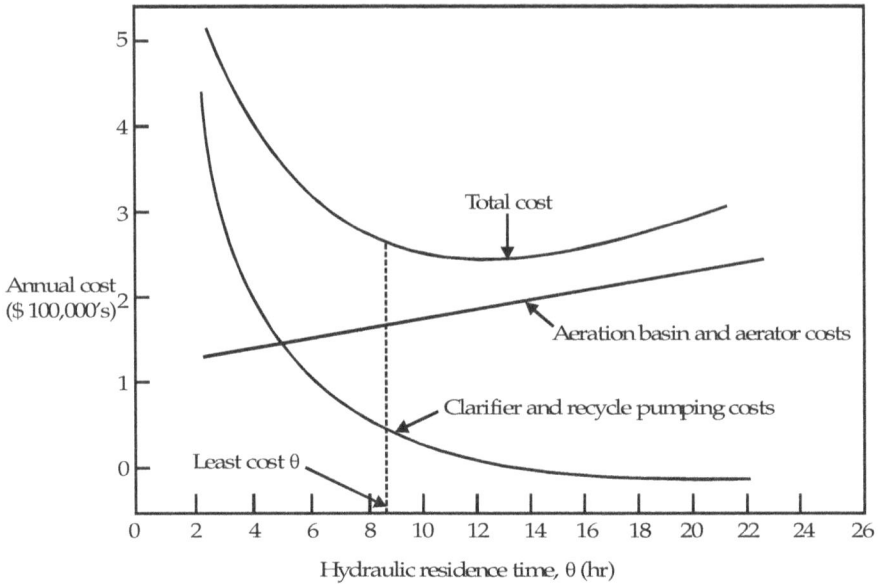

Figure 94 : Costs for activated sludge treatment system.

To update the costs to a current year, a cost index is used where the

$$\text{current cost} = (\text{old cost})\left(\frac{\text{Current index}}{\text{Old index}}\right)$$

Capital cost model for unit operations :

Table 91 : Capital cost model for unit operation.

S.No.	Treatment Unit	Parameter	Model Cost
1.	Raw wastewater pumping	Capacity (MGD)	$C = 2.6 \times 10^3 \, (MGD)^{1.0}$
2.	Screening + grit removal + flow measurement	Capacity (MGD)	$C = 27.6 \times 10^4 \, (MGD)^{0.62}$
3.	Equalization	Volume (MG)	$C = 7.2 \times 10^4 (MG)^{0.52}$
4.	PST or SST	Surface area (in 1000 ft^2)	$C = 2.8 \times 10^4 (A)^{0.88}$
5.	Diffused aeration	Blower capacity (in 1000 ft^3/min)	$C = 9.0 \times 10^4 (Cap)^{0.72}$
6.	Aeration basin	Volume (in 1000 ft^3)	$C = 4.8 \times 10^3 (V)^{0.72}$
7.	Surface aerator	Capacity (hp)	$C = 1.0 \times 10^3 (hp)^{0.89}$
8.	TF	Media volume (in 1000 ft^3)	$C = 3.4 \times 10^3 (V)^{0.84}$
9.	Recirculation pumping	Capacity (MGD)	$C = 2.0 \times 10^4 \, (MGD)^{0.70}$
10.	Sludge digestors + bldgs	Sludge volume (in 1000 ft^3)	$C = 1.4 \times 10^4 (V)^{0.64}$
11.	Lagoons	Volume (MG)	$C = 7.2 \times 10^4 (V)^{0.71}$
12.	Vacuum filtration	Filter area (ft^2)	$C = 5.9 \times 10^3 (A)^{0.67}$
13.	Centrifugation	Capacity (gal/min)	$C = 3.3 \times 10^4 \, (gal/min)^{0.54}$
14.	Incineration	Dry solids capacity (lb/hr)	$C = 1.4 \times 10^4 \, (Cap)^{0.56}$

Example 6.130

Activated Sludge Process : Carbon and Nitrogen Oxidation

Determine the total aeration basin volume and oxygen requirements for the separate and combined processes. Assume the following data:

Wastewater flow (Q)	:	10 MGD
Influent BOD_u [S_o]	:	300 mg/L
Influent total Kjedhal nitrogen $(TKN)_o$	:	30 mg–N/L
Effluent BOD_u [S_e]	:	20 mg/L
First order reaction rate constant [k]	:	0.05 L/mg.d at 20°C
Decay coefficient (k_d)	:	0.05 d^{-1} at 20°C
Critical summer operating temperature	:	30°C
Winter operating temperature	:	15°C
Operating pH [Carbon oxidation stage]	:	7.6
Operating pH [Nitrification stage]	:	7.0
True yield coefficient (Y_t)	:	0.5

Solution

1. Required mathematical expressions for separate–stage process

   Rate limiting step in the formation of NO_2^- from NH_4^+ by *nitrosomonas* bacteria

   Specific growth rate (m_{NS}) for nitrification is

   $$(\mu)_{NS} = [\mu_{max}]_{NS} \times \frac{[NH_4^+ - N]}{K_N + [NH_4^+ - N]}$$

   Influence of residual oxygen concentration

   $$(\mu)_{NS} = (\mu_{max})_{NS} \times \frac{[O_2]}{K_{O_2} + [O_2]}$$

   Influence of operating temperature

   $$(\mu_{max})_{NS} = (\mu_{max})_{NS(20)} 10^{0.033(T - 20)}$$
   $$K_N = 10^{0.051(T) - 1.158}$$

   Influence of operating pH

   $$(\mu_{max})_{NS} = \frac{(\mu_{max})_{NS} \text{ at optimum pH}}{1 + 0.04[10^{(pH)_{opt} - pH} - 1]}$$

Biomass yield

$$\left(\frac{dX}{dt}\right)_{NS} = Y_N \left(\frac{dN}{dt}\right)_{NS}$$

$$(\mu)_{NS} = Y_N(q)_{NS}$$

Bio-kinetic constants for nitrification

Table 92 : Bio-kinetic constants for nitrification.

Constant	Magnitude
$[\mu_{max}]_{NS[20^\circ C,(pH)_{opt}]}$	0.3 to 0.5 d^{-1}
K_N	0.5–2.0 mg/L
Y_N	0.05 mg VSS/mg NH$_4^+$–N
$(pH)_{opt} = (pH)_{optimum}$	8.0–8.4

where $(\mu)_{NS}$ = Specific growth rate of nitrosomonas [T^{-1}]

$(\mu_{max})_{NS}$ = Maximum specific growth rate of nitrosomonas [T^{-1}]

[NH$_4^+$–N] = Ammonia–N concentration [M/L^3]

K_N = Saturation constant [M/L^3]

[O$_2$]   = DO concentration [M/L^3]

KO$_2$ = Saturation constant for oxygen [Typical values range from 0.1 to 2.0 mg/L; commonly used values range from 0.2 to 0.4 mg/L]

T   = Temperature [$^\circ$C]

pH   = Operating pH

$(pH)_{opt}$ = Optimum pH

2. Required aeration basin volume, biomass concentration, oxygen requirements for a separate–stage process

    2.1 Carbon–removal stage

        Correction factor for pH for specific growth rate :

$$pH \text{ factor} = \frac{(\mu_{max})_{NS}}{(\mu_{max})_{NS(pH)_{opt}}} = \frac{1}{1+0.04[10^{(pHopt-pH)} - 1]}$$

$(pH)_{opt} = 8.2$

$$pH \text{ factor} = \frac{1}{1+0.04[10^{8.2-7.6} - 1]}$$

$$= 0.892$$

Temperature correction factor for specific growth rate :

$$\frac{(\mu_{max})_{NS}}{(\mu_{max})_{NS(20^\circ C)}} = 10^{0.033(T-20)}$$

Factor (winter – 15°C) $= 10^{0.033(15\,-\,20)}$

$$= 0.684$$

Factor (summer – 30°C) $= 10^{0.033(30\,-\,20)}$

$$= 2.14$$

Overall $(\mu_{max})_{NS}$ corrected for pH and temperature :

Assume $(\mu_{max})_{NS}$ at 20°C = 0.4 d^{-1}

Therefore, $(\mu_{max})_{NS}$ at T = 15°C and pH = 7.6 for carbon removal stage is :

$$= (0.4)(0.892)(0.684) = 0.244\ d^{-1}$$

$(\mu_{max})_{NS}$ at 30°C and pH = 7.6 for carbon removal stage is :

$$= 0.4(0.82)(2.14) = 0.70\ d^{-1}$$

Determination of SRT$_{min}$ (θ_c^m) below which nitrification will not occur in carbon removal stage is :

$$(\theta_c^m)_N = \frac{1}{(\mu_{max})_{NS}}$$

$$(\theta_c^m)_N = \frac{1}{(\mu_{max})_{NS}} = \frac{1}{0.244\ d^{-1}} = 4.1\ d\ at\ 15^\circ C$$

$$(\theta_c^m)_N = \frac{1}{(\mu_{max})_{NS}} = \frac{1}{0.70\ d^{-1}} = 1.43\ d\ at\ 30^\circ C$$

* SRT of 4.1 d during winter, and 1.43 d during summer should prevent nitrification in the carbon removal stage [First–stage]

* Larger the value of the SRT (θ_c) for the first–stage, the larger will be aeration basin volume.

* It is desirable to have $\theta_c = \dfrac{(1.43+4.1)}{2}$ days 2.765 d for both the summer and winter conditions.

[Normally θ_c = 2.765 d will be too low for effective biomass agglomeration, however, it will not affect process efficiency as long as second stage clarifier is functioning and solids recycle in the first stage is possible].

Specific substrate utilization rate (q) at 15 and 30°C

$$\theta_c^{-1} = Y_t\, q - k_d$$

$$\text{or} \qquad q = \frac{\theta_c^{-1} + k_d}{Y_t}$$

$$k_d(T) = k_d(20)(1.05)^{T-20}$$

$$k_d(15) = 0.05(1.05)^{-5} = 0.0392 \ d^{-1}$$

$$k_d(30) = 0.05(1.05)^{10} = 0.081 \ d^{-1}$$

$$q = \frac{\left(\dfrac{1}{2.765}\right) + 0.0392}{0.5} \qquad \text{at } 15^\circ C$$

$$= 0.8 \ d^{-1} \text{ at } 15^\circ C$$

$$q = \frac{\left(\dfrac{1}{2.765}\right) + 0.081}{0.5} = 0.885 \ d^{-1} \text{ at } 30^\circ C.$$

Soluble effluent concentration (S_e) :

$$q = \frac{Q(S_o - S_e)}{X\,V} = k S_e$$

$$k(T) = k(20)1.03^{(T-20)}$$

$$k(15) = (0.05)(1.03)^{(15-20)} = 0.04 \ L/mg.d$$

$$k(30) = (0.05)(1.03)^{(15-20)} = 0.067 \ L/mg.d$$

$$S_e = \frac{q}{k}$$

$$= \frac{0.8 \ d^{-1}}{0.04 \ L/mg.d} = 20 \ mg/L \text{ at } 15^\circ C$$

$$S_e = \frac{0.885 \ d^{-1}}{0.067 \ L/mg.d} = 13.21 \ mg/L \text{ at } 30^\circ C.$$

Organic (soluble) removal efficiency :

$$E = \frac{S_o - S_e}{S_o}(100)$$

$$= \frac{300 - 20}{300}(100) = 93.3\% \text{ at } 15^\circ C$$

$$E = \frac{300 - 13.21}{300}(100) = 95.6\% \text{ at } 30^\circ C.$$

Aeration basin volume (V) :

$$V = \frac{Q(S_o - S_e)}{X\,q}; \quad \text{Assume MLVSS} = 2000 \ mg/L$$

$$= \frac{10(300-20)}{2000(0.8)} = 1.75 \text{ MG at } 15°C$$

$$V = \frac{10(300-13.21)}{2000(0.885)} = 1.62 \text{ MG at } 30°C.$$

Observed yield coefficient (Y_{ob}) :

$$Y_{ob} = \frac{Y_t}{1+k_d\,\theta_c}$$

$$Y_{ob} = \frac{0.5}{1+0.039(2.765)} = 0.45 \text{ at } 15°C$$

$$Y_{ob} = \frac{0.5}{1+(0.081)(2.765)} = 0.408 \text{ at } 30°C.$$

Biomass production (ΔX) :

$$\Delta X = Y_{ob}Q(S_o - S_e)$$

$$\Delta X = 0.45(10 \times 8.34)(300 - 20)$$

$$= 10508 \text{ Ib/d at } 15°C$$

$$\Delta X = 0.408(10 \times 8.34)(300 - 13.21)$$

$$= 9759 \text{ Ib/d at } 30°C.$$

Sludge wastage from the aeration tank :

$$Q_w = \frac{\Delta X\,(\text{lb/d})}{8.34[X(\text{mg/L})]} = \text{MGD}$$

$$Q_w = \frac{10508 \text{ lb/d}}{8.34(2000 \text{ mg/L})} = 0.63\times10^6 \text{ gal/d at } 15°C$$

$$Q_w = \frac{9759 \text{ lb/d}}{8.34(X)}$$

At T = 15°C, controls the system, the volume for design at T = 15°C will be considered, thereby, the X–concentration will be :

$$X = \frac{Q(S_o - S_e)}{V(15°C)q(30)} = \frac{10(300-13.21)}{(1.75)(0.885)}$$

$$= 1852 \text{ mg/L at } 15°C$$

Therefore, $Q_w = \dfrac{9759}{8.34(1852)} = 0.63\times10^6 \text{ gal/d.}$

N – removal by heterotrophic cells :

N – removal = $0.122\ \Delta X$

$$= 0.122\ [10508 \text{ lb/d}] \text{ at } 15°C$$

$$= 1282 \text{ lb/d at } 15°C$$

$$= 0.122 \text{ [9759 lb/d] at } 30°C$$

$$= 1190 \text{ lb/d at } 30°C.$$

N–concentration in the effluent :

$$= \frac{[\text{Nitrogen loading}] - [\text{Nitrogen removed by biomass}]}{\text{Total flow}}$$

$$= \frac{(30 \text{ mg/L})(10 \text{ MGD})(8.34) - 1282 \text{ lb/d}}{(8.34)(10)} \text{ at } 15°C$$

$$= 14.6 \text{ mg/L at } 15°C \text{ as N}$$

$$= \frac{(30)(10)(8.34) - 1190}{8.34(10)} \text{ at } 30°C$$

$$= 15.7 \text{ mg/L at } 30°C \text{ as N.}$$

Required oxygen :

$$\text{lb O}_2 = 8.34 \text{ Q (S}_o - S_e) - 1.42\Delta X$$

$$= 8.34 (10)(300 - 20) - 1.42(10508)$$

$$= 8431 \text{ lb/d at } 15°C$$

$$\text{lb O}_2 = 8.34(10)(300 - 13.21) - 1.42(9759)$$

$$= 10061 \text{ lb/d at } 30°C.$$

8.2 Required separate nitrification stage

Specific growth rate factor :

$$\frac{(\mu_{max})_{NS}}{(\mu_{max})_{NS,\text{pH opt}}} = \frac{1}{1 + 0.04[10^{(\text{pH}_{opt} - \text{pH})} - 1]}$$

$$= \frac{1}{1 + 0.04[10^{(8.2 - 7.0)} - 1]}$$

$$= 0.63.$$

Temperature factor are same as for carbon removal :

At 15°C = 0.684

At 30°C = 2.14.

Overall $(\mu_{max})_{NS}$ corrected for pH and temperature :

Assume $(\mu_{max})_{NS \text{ at } 20°C} = 0.4 \text{ d}^{-1}$

Therefore, μ_{max} at T = 15°C and pH = 7.0 for nitrification is :

$$= (0.4)(0.684)(0.63) \text{ at } 15°C$$

$$= 0.1724 \text{ d}^{-1} \text{ at } 15°C.$$

μ_{max} at T = 30°C and pH = 7.0 for nitrification is :

$$= (0.4)(2.14)(0.63) = 0.539 \text{ d}^{-1} \text{ at } 30°C.$$

Separation constant (K_N) at 15°C and 30°C

$$K_N = [10]^{[0.05T - 1.158]}; \quad [T = °C]$$

$$= 10^{-0.408} = 0.39 \text{ mg/L as N at } 15°C$$

$$= 10^{0.342} = 2.2 \text{ mg/L as N at } 30°C.$$

Nitrification efficiency at 15°C and 30°C

$$E_N = 1 - \frac{K_N}{[NH_4^+ - N]_o[\theta_c(\mu_{max})_{NS} - 1]}$$

Assume nitrification efficiency (E_N = 0.98 both at 15°C and 30°C)

Therefore, θ_c at 15°C :

$$0.98 = 1 - \frac{0.39}{(14.6)[\theta_c(0.1724) - 1]}$$

$$(0.98)(14.6)[q_c(0.1724) - 1] = 14.6[q_c(0.1724) - 1] - 0.39$$

$$2.47\theta_c - 14.31 = 2.52\theta_c - 14.6 - 0.39$$

$$\theta_c [2.47 - 2.52] = 14.31 - 14.99$$

$$\theta_c = 13.6 \text{ days at } 15°C$$

Also, θ_c at 30°C

$$(0.98)(15.7)[\theta_c(0.539) - 1] = 15.7[\theta_c(0.539) - 1] - 2.2$$

$$8.293\theta_c - 15.39 = 8.46\theta_c - 15.7 - 2.2$$

$$\theta_c = \frac{2.6}{0.167} = 15.7 \text{ days at } 30°C.$$

Assume a sludge detention time (θ_c) of 15 days :

$$E = 1 - \frac{0.39}{14.6[15(0.1724) - 1]} \quad \text{at } T = 15°C$$

$$= [1 - 0.0168] = 0.98 \text{ or } 98\% \text{ [OK] at } 15°C$$

$$E_N = 1 - \frac{2.2}{15.7[15(0.539) - 1]} \quad \text{at } T = 30°C$$

$$= 98\% \text{ [OK] at } 30°C.$$

Volume of nitrification aeration basin (V_N) :

$$V_N = \frac{Y_N(Q - Q_w)[(NH_4^+)_o - (NH_4^+)_e]}{\mu_{NS} X_N}$$

$$\mu_{NS} = \frac{1}{\theta_c} \; ; \; Y_N = 0.05 \text{ mg VSS/mg NH}_4^+ - N$$

Volume (V_N) at 15°C :

Assume that effluent is 15.7 mg/L from the aeration basin (first stage) at 15 and 30°C

$$V_N \text{ (at } 15°C) = \frac{0.05[10 - 0.63][0.98(15.7)]}{\left(\dfrac{1}{15}\right)(100)}$$

$$= 1.08 \text{ MG at } 15°C$$

$$V_N \text{ (at } 30°C) = \frac{0.05[10 - 0.63][0.98(15.7)]}{\left(\dfrac{1}{15}\right)(100)}$$

$$= 1.08 \text{ MG } 30°C.$$

Heterotrophic consideration in second stages :

Specific substrate utilization rate (q) :

$$q = \frac{\dfrac{1}{\theta_c} + k_d}{Y_T}$$

$$= \frac{\dfrac{1}{15} + 0.0392}{0.5} \quad \text{at } 15°C$$

$$= 0.212 \text{ d}^{-1} \text{ at } 15°C$$

$$q = \frac{\dfrac{1}{15} + 0.081}{0.5} \quad \text{at } 30°C$$

$$= 0.295 \text{ d}^{-1} \text{ at } 30°C$$

$$q = k \, S_e.$$

Effluent soluble substrate concentration (S_e) :

$$S_e = \frac{q}{k}$$

$$= \frac{0.212 \text{ d}^{-1}}{0.04 \text{ L/mg.d}} = 5.3 \text{ mg/L at } 15°C \text{ [OK]}$$

$$S_e = \frac{0.295 \text{ d}^{-1}}{0.067 \text{ L/mg.d}} = 4.4 \text{ mg/L at } 30°C.$$

Heterotrophic biomass concentrations at 15°C and 30°C :

$$X_H = \frac{(Q - Q_w)(S_o - S_e)}{q\,V_N}$$

$$= \frac{(10 - 0.63)(20 - 5.3)}{0.212(1.08)} \quad \text{at } 15^\circ C$$

$$= 602 \text{ mg/L}$$

$$X_H = \frac{(10 - 0.63)(13.21 - 4.4)}{0.295(1.08)} \quad \text{at } 30^\circ C$$

$$= 260 \text{ mg/L}.$$

Biomass production (heterotrophic) rate (ΔX) :

$$\Delta X = Y_{ob}[Q - Q_w](8.34)[S_o - S_e]$$

$$Y_{ob} = \frac{Y_t}{1 + k_d\,\theta_c}$$

$$= \frac{0.5}{1 + 0.0392(15)} = 0.315 \text{ at } 15^\circ C$$

$$Y_{ob} = \frac{0.5}{1 + 0.081(15)} = 0.226 \text{ at } 30^\circ C$$

At   T = 15°C; ΔX :

$$\Delta X = 0.315[10 - 0.63](8.34)[20 - 5.3]$$

$$= 362 \text{ lb/d}$$

At   T = 30°C; ΔX :

$$\Delta X = 0.226[10 - 0.63](8.34)[13.21 - 4.4]$$

$$= 155 \text{ lb/d}.$$

Fraction of nitrifier biomass (F_N) is given as :

$$F_N = \frac{Y_N\,E_N[(TKN)_o - 0.122Y_{ob}(E_H)(S_o)]}{Y_{ob}(E_H)(S_o) + Y_N\,E_N[(TKN)_o - 0.122Y_{ob}(E_H)(S_o)]}$$

where F_N = Fraction of total biomass which is composed of nitrifiers

              E_W = Nitrification efficiency in fraction

(TKN)$_o$ = Total Kjeldhal nitrogen concentration (mg/L) in the influent as N

              Y_{ob} = Observed yield coefficient

              E_H = Fractional organic removal efficiency $q = \dfrac{\theta_c^{-1} + k_d}{Y_t}$

              S_o = Influent soluble BOD_u (mg/L)

At T = 15°C, F_N is :

$$F_N = \frac{0.05\,(0.98)\,[14.6 - (0.122)(0.315)(20.1 - 5.3)]}{0.315\,(20.1 - 5.3) + (0.05)(0.98)\,[14.6 - (0.122)(0.315)(20.1 - 5.3)]}$$

$$= \frac{0.6875}{4.6620 + 0.6875} = \frac{0.6875}{5.3495} = 0.1285$$

At T = 30°C, F_N is :

$$F_N = \frac{0.05\,(0.98)\,[15.7 - 0.122(0.226)(13.21 - 4.4)]}{0.226\,[13.21 - 4.4] + 0.05(0.98)\,[15.7 - 0.122(0.226)(13.21 - 4.4)]}$$

$$= \frac{0.7574}{1.9888 + 0.7574} = \frac{0.7574}{2.7462} = 0.276$$

Check : $X_N = \dfrac{602}{1 - 0.1285} - 602$ at $T = 15^\circ C$

= [Total biomass] – [Heterotrophic biomass]

= 691 – 602 = 89 mg/L ≈ 100 mg/L

$$X_N = \frac{260}{1 - 0.276} - 260 \text{ at } T = 30^\circ C$$

= 359 – 260 = 99 mg/L ≈ 100 mg/L.

Required oxygen at T = 15°C and 30°C

Ib O_2/d = [(8.34)(Q – Q_w)(S_o – S_e) – 1.42 ΔX] + (38.1)(Q – Q_w) $E_N [NH_4^+ - N]$

At T = 15°C

lb O_2/d = 8.34(10 – 0.63)(20 – 5.3) – 1.42(362) + 38.1(10 – 0.63)0.98(14.6)

= (1149 – 514 + 5108) = 5743 lb $O_2/2$

At   T = 30°C

lb O_2/d = 8.34(10 – 0.63)(13.21 – 4.4) – 1.42(155) + 38.1(10 – 0.63)(0.98)(15.7)

= (688 – 220 + 5493) = 5961 lb O_2/d.

Schematics of separate–stage nitrification process :

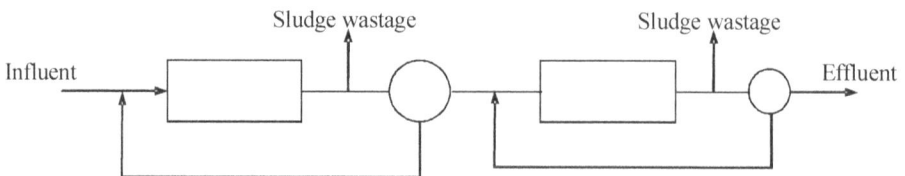

Figure 95 : Schematics of separate stage nitrification process.

Organic yield :

Neglecting the maintenance, the relationship between the absolute growth rate of nitrosomonas and the rate of ammonium degradation is :

$$\left[\frac{dX}{dt}\right]_{NS} = Y_N \left[\frac{dN}{dt}\right]_{NS}$$

$$[\mu]_{NS} = Y_N (q)_{NS}$$

where $\left[\dfrac{dX}{dt}\right]_{NS}$ = Absolute growth rate of nitrosomonas $[M/L^3.T]$

$\left[\dfrac{dN}{dt}\right]_{NS}$ = Ammonia oxidation $[M/L^3.T]$

Y_N = Yield coefficient [mg–nitrosomonas produces per unit of ammonium oxidized]

$[q]_{NS}$ = Specific ammonium oxidation rate $[T^1]$

2. Required mathematical expressions for combined carbon–nitrification process :

    Schematics of the combined process is

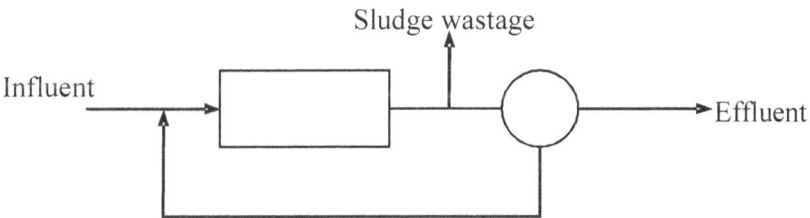

Figure 96 : Schematics of Combined process.

Assumptions are :

Energy of maintenance neglected

Limiting nutrient is ammonium

Steady state.

$$[\mu]_{NS} = [\mu_{max}]_{NS} \frac{[NH_4^+ - N]}{K_N + [NH_4^+ - N]}$$

For completely mixed system at steady state :

$$[\mu]_{NS} = \frac{1}{\theta_c}$$

Therefore, $\theta_c^{-1} = [\mu_{max}]_{NS} \dfrac{[NH_4^+ - N]}{K_N + [NH_4^+ - N]}$

or $\quad [NH_4^+ - N] = \dfrac{K_N}{\theta_c \, [\mu_{max}]_{NS} - 1}$

As, $E_N = 1 - \dfrac{[NH_4^+ - N]_e}{[NH_4^+ - N]_o}$

Therefore, $E_N = 1 - \dfrac{K_N}{(\theta_c \, [\mu_{max}]_{NS} - 1)[NH_4^+ - N]_o}$

$\left[\dfrac{dN}{dt}\right]_{NS} = \dfrac{Q([NH_4^+ - N]_o - [NH_4^+ - N]_e)}{V_N}$

Specific substrate utilization rate (q) is defined as :

$$q = \dfrac{Q(S_o - S_e)}{V\,X}$$

Therefore, $\dfrac{\left[\dfrac{dN}{dt}\right]_{NS}}{X_N} = [q]_{NS} = \dfrac{Q([NH_4^+ - N]_o - [NH_4^+ - N]_e)}{V_N \, X_N}$

or $\qquad V_N = \dfrac{Y_N \, Q([NH_4^+ - N]_o - [NH_4^+ - N]_e)}{(\mu)_{NS} \, X_N}$

$$[\mu_{NS} = Y_N(q)_{NS}].$$

3.1 **Required magnitude of important constants in nitrification**

$[\mu_{max}]_{NS}$ at 20°C $\quad = \qquad$ 0.4 d^{-1}

$[pH]_{opt} \qquad = 8.2$

$pH = \qquad 7.0$

pH correction factor is :

$$f_1 = \dfrac{[\mu_{max}]_{NS}}{[\mu_{max}]_{NS(pH)_{opt}}} = \dfrac{1}{1 + 0.04\,[10^{(pH_{opt} - pH)} - 1]}$$

$$= \dfrac{1}{1 + 0.04\,[10^{-1.2} - 1]} = 0.63$$

Temperature correction factor is :

$$f_2 = \dfrac{[\mu_{max}]_{NS}}{[\mu_{max}]_{NS,\,20°C}} = 10^{0.033(T-20)}$$

At T = 15°C $\qquad f_2 = 0.684$

At T = 30°C $\qquad f_2 = 2.14.$

$[\mu_{max}]_{NS} = (0.4)(f_1)(f_2)$

$\qquad = (0.4)(0.63)(0.684)$ at $T = 15°C$

$\qquad = 0.1724 \ d^{-1}$

$\qquad = (0.4)(0.63)(2.14)$ at $T = 30°C$

$\qquad = 0.539 \ d^{-1}$

Saturation constant for nitrification (K_N) :

$\quad K_N \ (T) \qquad = 10^{0.051(T) - 1.158}$

$\quad K_N \ (15°C) = 0.39 \ mg/L$

$\quad K_N \ (30°C) = 2.2 \ mg/L.$

Specific substrate rate (q) :

$\qquad q = k \ S_e$

$\quad k(T) = k(20)(1.03)^{T- 20}$

$\quad k(15) = 0.05(1.03)^{-5} = 0.04 \ L/mg.d$

$\quad k(30) = 0.05(1.03)^{10} = 0.067 \ L/mg.d$

Therefore, $q = (0.04)(10)$ at $15°C$

$\qquad\qquad = 0.4 \ d^{-1}$ at $15°C$

$\qquad\quad q = (0.067)(10)$ at $30°C$

$\qquad\qquad = 0.67 \ d^{-1}$ at $30°C$

Sludge retention time (θ_c)

$\qquad \theta_c^{-1} = Y_t \ q - k_d$

$\quad k_d(T) = k_d(20)(1.05)^{T- 20}$

$\quad k_d(15) = 0.05(1.05)^{-5} = 0.0392 \ d^{-1}$

$\quad k_d(30) = 0.05(1.05)^{10} = 0.081 \ d^{-1}$

At $\quad T = 15°C$

$\qquad \theta_c^{-1} = 0.5(0.4) - 0.0392 = 0.1608 \ d^{-1}$

$\qquad \theta_c = 6.22 \ d$

At $\quad T = 30°C$

$\qquad \theta_c^{-1} = 0.5(0.67) - 0.081 = 0.254 \ d^{-1}$

$\qquad \theta_c = 3.94 \ d$

Observed yield coefficient (Y_{ob}) :

$$Y_{ob} = \frac{Y_t}{1 + k_d \, \theta_c}$$

At   T = 15°C :

$$Y_{ob} = \frac{0.5}{1 + 0.0392(6.22)} = 0.40$$

At   T = 30°C :

$$Y_{ob} = \frac{0.5}{1 + 0.081(3.94)} = 0.379$$

Biomass production (DX) rate :

$$\Delta X = Y_{ob} \, Q(8.34)[S_o - S_e]$$

At   T = 15°C

$$\Delta X = 0.40(10)(8.34)(300 - 20)$$

$$= 9341 \text{ lb MLVSS/d}$$

At   T = 30°C

$$\Delta X = 0.379(10)(8.34)(280)$$

$$= 8850 \text{ lb MLVSS/d}$$

Nitrogen removal (= 0.122 ΔX) :

At   T = 15°C :

$$N - \text{removal} = 0.122(9341) = 1140 \text{ lb/d}$$

At   T = 30°C

$$N - \text{removal} = 0.122(8850) = 1080 \text{ lb/d}$$

Available nitrogen :

$$= \frac{(N - \text{loading}) - (N - \text{removed})}{\text{Flow rate}}$$

At   T = 15°C

$$\text{N-available} = \frac{10(30)(8.34) - 1140}{10(8.34)}$$

$$= 16.3 \text{ mg/L}$$

At   T = 30°C

$$N - \text{available} = \frac{10(30)(8.34) - 1080}{10(8.34)}$$

$$= 17.0 \text{ mg/L}$$

Assume that N–available is 16.3 mg/L in the combined reactor

Nitrogen removal efficiency (E_N) :

$$E_N = 1 - \frac{K_N}{[NH_4^+ - N]_o\,(\theta_c\,[\mu_{max}]_{NS} - 1)}$$

Assume $E_N = 0.98$, therefore, q_c is :

At   T = 15°C :

$$0.98 = 1 - \frac{0.39}{16.3[\theta_c(0.1724) - 1)]}$$

$$0.98(16.3)[\theta_c(0.1724) - 1)] = 16.3[\theta_c(0.1724) - 1)] - 0.39$$

$$2.75\ \theta_c - 16.0 = 2.81\theta_c - 16.3 - 0.39$$

$$0.056\ \theta_c = 0.69$$

$$\theta_c = 12.32\ \text{days}$$

At T = 30°C :

$$0.98(16.3)[\theta_c(0.539) - 1)] = 16.3[\theta_c(0.539) - 1)] - 2.2$$

$$8.61\theta_c - 16.0 = 8.79 - 16.3 - 2.2$$

$$0.18\ \theta_c = 2.4$$

$$\theta_c = 13.33\ \text{days}$$

Assume $\theta_c = 15.0\ \text{days}$

Specific substrate utilization rate (q)

$$q = \frac{\theta_c^{-1} + k_d}{Y_t}$$

At T = 15°C

$$q(15) = \frac{(15)^{-1} + 0.0392}{0.5} = 0.212\ d^{-1}$$

At   T = 30°C

$$q(30) = \frac{(15)^{-1} + 0.081}{0.5} = 0.295\ d^{-1}$$

Soluble effluent substrate concentration (S_e) :

$$Q = k\,S_e;\ S_e = q/k$$

At   T = 15°C

$$S_e = 5.3\ \text{mg/L [OK]}$$

At   T = 30°C

$$S_e = 4.4\ \text{mg/L [OK]}$$

Aeration basin volume (V)

$$V = \frac{Q(S_o - S_e)}{Xq}; \ X = 2000 \ mg/L$$

At   T = 15°C

$$V = \frac{10(300 - 5.3)}{2000(0.212)} = 6.9 \ MG$$

At   T = 30°C

$$V = \frac{10(300 - 4.4)}{2000(0.295)} = 5.0 \ MG$$

Use V = 6.9 MG

Nitrification fraction (F_N) :

$$F_N = \frac{Y_N \, E_N [(TKN)_o - 0.122 Y \, E_H \, S_o]}{Y_{ob} \, E_H \, S_o + Y_N \, E_N [(TKN)_o - 0.122 Y_{ob} (E_H) S_o]}$$

$$= \frac{0.05(0.98)(16.3)}{0.40(300 - 5.3) + 0.05(0.98)(16.3)}$$

$$= \frac{0.7987}{118 + 0.7987} = 0.0067$$

Nitrifier concentration at T = 15°C

$$= (0.0067)(2000) = 13.5 \ mg/L$$

Aeration basin volume required for nitrification :

$$V_N = \frac{Y_N \, Q([NH_4^+ - N]_o - [NH_4^+ - N]_e)}{(\mu)_{NS} \, X_N}$$

$$(\mu_{NS}) = Y_N (q)_{NS} = \frac{1}{\theta_c}$$

$$V_N = \frac{0.05(10)(0.98)(16.3)}{\left(\dfrac{1}{15}\right)(13.5)} = 8.87 \ MG$$

Use V_N = 8.87 MG as aeration basin volume, as it higher than V.

Biomass production rate (ΔX) :

$$\Delta X = Y_{ob} \ (Q \times 8.34)(S_o - S_e)$$

At   T = 15°C :

$$\Delta X = 0.40(10 \times 8.34)(300 - 5.3)$$

$$= 9831 \ lb \ MLVSS/d$$

At   T = 30°C

$$\Delta X = 0.379 \,(10 \times 8.34)(300 - 4.4)$$

$$= 9344 \text{ lb MLVSS/d}$$

Oxygen requirements :

lb O_2/d = 8.34Q(S_o - S_e) - 1.42 ΔX + 38.1 Q $E_N [NH_4^+ - N]$

At   T = 15°C :

lb O_2/d = [8.34(10)(300 - 5.3) - 1.42(9831) + 38.1(10)(0.98)(16.3)]

$$= (24588 - 13960 + 6086) = 16714 \text{ lb } O_2/\text{d}$$

At   T = 30°C:

lb O_2/d = [8.34(10)(300 - 4.4) - 1.42(9344) + 38.1(10)(0.98)(16.3)]

$$= [24653 - 13268 + 6086] = 17471 \text{ lb } O_2/\text{d}.$$

Example 6.131

Bio–adsorption of Organics in ASP System

1. Bio–adsorption of non–degradable organics on biological solids

    This removal mechanism is called partitioning, and has been related to octanol–water partition coefficient of organic :

    $$k_{SW} = K(k_{OW})^n$$

    where k_{SW} is the bio–solids accumulation factor, ratio of organic sorbed and in solution [(mg/mg) (mg/L)], k_{OW} is the octanol–water partition coefficient [(mg/L)$_o$/(mg/L)$_w$], and K and n are the factors (1.38 × 10^{-5} to 4.3 × 10^{-7} and 0.58 to 1.0 respectively).

    In most industrial wastewaters, partitioning provides negligible S–COD removal but may be a method of bio–accumulation of certain lipid–soluble organic compounds.

2. Adsorption removed equation an biomass

    $$\frac{C_e}{C_i} = \frac{1}{\left[1 + \dfrac{(k_{OW})(Xt)}{\theta_c}\right]}$$

    where C_e is the effluent concentration (mg/L), C_i is the influent concentration (mg/L), X is the MLSS (mg/L), t is the hydraulic detention time (days), and θ_c is the sludge age (d).

    Adsorption is not a significant parameter if log k_{OW} is less than 4.

3. Determine the adsorption of Lindane and tetrachloro–ethane in ASP system. Use the following data :

MLSS (X)   : 3000 mg/L

Hydraulic detention time (t) : 0.20 d

Sludge age (θ_c)   : 10 days

Lindane (k_{OW})   : 12,600

Tetrachloro–ethane (k_{OW})   : 363

Assume k and n : 3.45×10^{-7} (mg/L)$^{-1}$, 1.0 respectively

3.1   Adsorption of lindane

$$\frac{C_e}{C_i} = \frac{1}{\left[1 + \dfrac{(k_{SW})(X)(t)}{\theta_c}\right]}$$

$$= \frac{1}{\left[1 + \dfrac{(3.45 \times 10^{-7})(12,600)(3000)(0.20)}{10}\right]}$$

$$= \frac{1}{(1+0.261)} = 0.79 \text{ (or 21\% adsorbed)}$$

$[k_{SW} = (3.45 \times 10^{-7})(12,600)]$.

3.2   Adsorption of tetrachloro–ethane

$$\frac{C_e}{C_i} = \frac{1}{\left[1 + \dfrac{(3.47 \times 10^{-7})(363)(3000)(0.20)}{10}\right]}$$

$= 0.93$ [0r 7% adsorbed].

Note :

1. Sorption on biomass does not seen to be significant mechanism for removal of toxic organics, sorption on SS in PST may be significant. Toxicity to anaerobic digestion may be significant. Toxicity to anaerobic digestion may result or land disposal alternatives may be restricted.

2. Heavy metals will complex with the cell wall and bio–accumulate. Low concentrations of metals in wastewater are generally not inhibitory to organics removal efficiency, and their accumulation on the sludges markedly affect subsequent sludge treatment and disposal problem.

Example 6.132

Completely Mixed Activated Sludge Process (ASP) [Combined Carbonaceous Oxidation and Nitrification]

Determine the SRT, the aeration basin volume, the food to micro–organism (F/M) ratio, the material requirements, and the nitrification effectiveness for an ASP. Assume the following data:

System constant at 20°C

Decay constant (k_d)	: 0.05 d^{-1}
Maximum specific growth rate (k_{max})	: 4 d^{-1}
Half–substrate saturation constant (k_s)	: 75 mg/L
Yield coefficient (Y)	: 0.6

Process characteristics :

Wastewater flow	: 1.25 MGD
Influent BOD$_5$	: 350 mg/L
Influent COD	: 700 mg/L
Influent ammonia–N	: 40 mg/L
Influent phosphorus–P	: 0.5 mg/L
Effluent BOD$_5$	: 25 mg/L
Effluent–ammonia	: 2 mg/L
Effluent–SS	: 25 mg/L
Ratio of Ultimate BOD solids	: 1.42
Biodegradable solids fraction	: 0.5
Solids BOD$_5$/BOD$_u$	: 0.70
MLVSS/MLSS	: 0.8
BOD : N : P	: 100:5:1
Design temperature	: 20°C
Summer temperature	: 30°C
Winter temperature	: 10°C
Aeration basin depth	: 10 ft
L/W ratio for aeration basin	: 2

Solution

1. Carbon oxidation mass balance

   Required effluent soluble BOD :

   Soluble BOD$_5$ = TBOD$_5$ – BOD solids

   Biodegradable solids = (25 mg/L) (0.5)

   $\qquad\qquad\qquad\qquad$ = 12.5 mg/L

   Inert solids = (25 – 12.5) = 12.5 mg/L

Ultimate BOD of solids :

$$= \left(1.42 \ \frac{BOD_u \text{ of solids}}{\text{Biodegradable solids}} \right) (12.5 \text{ mg/L biodegrable SS})$$

$= 17.75 \text{mg/L}$

BOD_5 of solids :

$$= \left(0.70 \ \frac{BOD_5 \text{ solids}}{BOD_u} \right) (17.75 \text{ mg/L } BOD_u \text{ solids})$$

$= 12.4 \text{ mg/L}$

Allowable soluble effluent BOD_5 :

$= 25 - 12.4 = 12.6 \text{ mg/L (design)}.$

2. Required sludge retention time (SRT)

$$\left[\frac{1}{SRT} \right]_1 = \frac{Y k_{max} S_0}{K_s + S_0} - k_d$$

$$= \frac{(0.6)(4)(350)}{75 + 350} - 0.05$$

$= 1.93 \text{ d}^{-1}$

$SRT_1 = 0.52 \text{ day}$

[If the effluent substrate concentration (S) = Influent substrate concentration (S$_0$); washout occurs].

For activated sludge process $\dfrac{SRT \text{ (design)}}{SRT \text{ (limiting)}} = 20$

SRT (design) = 20 × 0.52 = 10.36 days at 20°C for carbon oxidation.

3. Required aeration basin volume

$$X = \frac{Y[S_0 - S_e][SRT]}{[1 + k_d SRT][HRT]}$$

$$HRT = \frac{V}{Q}$$

$$X = \frac{Y[S_0 - S_e][SRT]}{[1 + k_d \cdot SRT][V/Q]}$$

or $\qquad V = \dfrac{Y[Q(S_0 - S_e)][SRT]}{X[1 + k_d SRT]}$

$$= \frac{(0.5)[1.25(350-12.6)]10.36}{2000[1+0.05(10.36)]} = 0.75 \text{ MG.}$$

4. Required effluent concentration (S_e) at 20°C [design value]

$$S_e = \frac{K_s[1+k_d \cdot SRT]}{SRT[Y k_{max} - k_d]-1}$$

$$= \frac{75[1+0.05(10.36)]}{10.36[0.6(4)-0.05]-1}$$

$$= 4.85 \text{ mg/L} \qquad [OK; S_e < 12.36 \text{ mg/L}].$$

5. Required F/M ratio at 20°C

$$\left(\frac{F}{M}\right)_{\text{Feed based}} = (R) = \left(\frac{QS_o}{X_v \cdot V}\right) = \frac{S_o}{X_v(HRT)}$$

$$= \frac{350 \text{ mg/L}}{(2000 \text{ mg/L})\left(\dfrac{0.75 \text{ MG}}{1.25 \text{ MGD}}\right)}$$

$$= 0.29 \text{ d}^{-1}$$

$$\left(\frac{F}{M}\right)_{\text{Substrate removed}} = \frac{Q(S_o - S_e)}{X_v V} = \frac{1.25(350-4.85)}{(2000)(0.75)}$$

$$= 0.288 \text{ d}^{-1}$$

6. Required effect in summer at 30°C and 15°C on SRT

$$R_S = (F/M)_{\text{substrate utilized}} \text{ (at 30 °C)} = (0.288)(1.02)^{30-20}$$

$$= 0.35 \text{ d}^{-1}$$

$$k_d \text{ at 30°C} = (0.05 \text{ d}^{-1})(1.02)^{30-20}$$

$$= 0.061 \text{ d}^{-1}$$

$$\left[\frac{1}{SRT \text{ at } 30°C}\right]_{\min} = \frac{Y k_{max} S_o}{K_s + S_o} - k_d$$

k_{max} = Maximum substrate utilization rate ($R_s = 0.35 \text{ d}^{-1}$)

As, $K_s \ll S_o$

$$[SRT \text{ at } 30°C]_{\min} = \frac{1}{Y k_m - k_d}$$

$$= \frac{1}{(0.6)(0.35)-0.061}$$

$$= 6.71 \text{ days}$$

$$R_w \left[= \frac{F}{M} \text{ Winter} \right] = [0.288 \text{ d}^{-1}](1.02)^{10-20}$$

$$= 0.288 \times 0.82 = 0.236 \text{ d}^{-1}$$

$$k_d \text{ at } 10°C = (0.05 \text{ d}^{-1})(1.02)^{10-20}$$

$$= 0.04 \text{ d}^{-1}$$

$$[SRT]_{min} \text{ at } 10°C = \frac{1}{Y k_{max} - k_d}$$

$$= \frac{1}{(0.6)(0.236) - 0.04} = 9.84 \text{ days}$$

$$X \text{ at } 30°C = \left[\frac{Y(S_o - S_e)}{1 + (k_d \text{ at } 30°C)(SRT)} \right] \frac{SRT \text{ at } 30°C}{HRT}$$

$$= \left[\frac{0.6(350 - 4.85)}{1 + (0.061)(6.71)} \right] \left(\frac{6.71}{\frac{0.75}{1.25}} \right)$$

$$= 1643 \text{ mg/L}$$

$$X \text{ at } 10°C = \left[\frac{0.6(350 - 4.85)}{1 + (0.04)(9.84)} \right] \left(\frac{9.84}{\frac{0.75}{1.25}} \right)$$

$$= 2436 \text{ mg/L.}$$

7. Required recycle

$$\text{Recycle ratio} = \frac{\text{Recycle flow}}{\text{Feed flow}} = \frac{X}{X_r - X}$$

where X is the biomass concentration in the aeration basin (mg/L), and other clarifier under concentration (mg/L) [All MLVSS].

Design conditions at 20°C :

$$\text{Recycle ratio} = \frac{2000}{(10,000)(0.8) - 2000}$$

$$= 0.33$$

Summer conditions at 30°C :

$$\text{Recycle ratio} = \frac{1643}{(10,000)(0.8) - 1643}$$

$$= 0.26$$

Winter conditions at 10°C :

$$\text{Recycle ratio} = \frac{2436}{(10,000)(0.8) - 2436}$$

$$= 0.44$$

[Design pumping equipment for 100% operating flexibility].

8. Required Nutrients

Influent N = (1.25 MGD)(8.34)(40) = 417 lb/d

BOD_5(lb/d) = (1.25 MGD)(8.34)(350 − 4.85) = 3598 lb/d

$$\text{Nitrogen required (lb/d)} = \frac{(3598 \text{ lb/d})(5 \text{ lb})}{100 \text{ lb } BOD_5 \text{ removed}} = 180 \text{ lb} - N/d$$

Excess nitrogen available = 417 − 180 = 237 lb–N/d

$$NH_3 - N \text{ Conc.} = \frac{237 \text{ lb/d}}{(8.34)(1.25)}$$

$$= 22.7 \text{ mg/L}$$

Influent P = (1.25 MGD)(8.34)(0.5 mg/L) = 5.21 lb/d

$$P - \text{required} = \frac{(3598 \text{ lb/d})(1 \text{ lb})}{100 \ BOD_5 \text{ removed}} = 35.98 \text{ lb/d}$$

Additional phosphorus to be added = 35.98 − 5.21

$$= 30.77 \text{ mg/L.}$$

9. Required nitrification

Maximum growth rate of nitrifiers :

Specific growth rate of nitrifiers

$$\mu_N = 0.047 \ \exp[0.098(T - 15)][1 - 0.833(7.2 - pH)]\left[\frac{DO}{K_o + DO}\right]\left[\frac{N}{N + 10^{[0.051T - 1.158]}}\right]$$

Maximum possible nitrification at specific temperature, pH and DO conditions is possible when $N \gg K_N$:

$$\mu_N(\text{max}) = 0.047 \ \exp[0.098(T - 15)][1 - 0.833(7.2 - pH)]\left[\frac{DO}{K_o + DO}\right]$$

DO = 2 mg/L; K_o = 1.3 mg/L; Design temperature = 20°C; pH = 7.2

$$\mu_N(\text{max}) = 0.047 \ \exp[0.098(5)][1 - 0.833(7.2 - 7.2)]\left[\frac{2}{1.3 + 2.4}\right]$$

$$= 0.46496 \text{ d}^{-1}$$

10. Required mean cell residence (MCRT) for nitrification

$$MCRT = \frac{1}{\mu_{max}} = \frac{1}{0.46496} = 2.15 \text{ days}$$

After applying the safety factor to the carbonaceous design SRT of 10.36 days which is greater than MCRT for nitrification, and minimum required SRT_d of 10 days.

11. Required effluent nitrogen concentration (N)

$$\frac{1}{SRT_d} = \mu_N(max)\frac{N}{K_N + N}$$

$$K_N = 10^{[(0.051\,T - 1.158)]}, \; T = 20°C$$

$$= 0.728$$

$$\frac{1}{10.36} = 0.46496\left[\frac{N}{0.728 + N}\right]$$

$$N = 0.190 \text{ mg/L [OK]}$$

12. Required sludge wastage $[P_x]$

$$P_x = \frac{Y\,Q[S_o - S]}{1 + k_d\,SRT}$$

During Summer, the sludge generation will be more than Winter :

SRT at 30°C = 6.71 days [Refer section 6] which is less than 10 days, choose SRT = 10 days

$$P_x \text{ at } 30°C = \frac{(0.6)(1.25)[350 - 4.85](8.34)}{1 + (0.061)(10)}$$

$$= 1349 \text{ Ib VSS/d}$$

$$\text{Total solids} = \frac{1349}{0.8} = 1687 \text{ lb solids/d}$$

Assume solids content of 1.0%

$$\text{Total solid slurry} = \frac{1687}{0.01} = 168700 \text{ lb/d}$$

$$\text{Sludge volume generation} = \frac{168700 \text{ lb/d}}{(8.34 \text{ lb/gal})} = 20224 \text{ gal/d}$$

$$P_x \text{ at } 20°C = \frac{(0.6)(1.25)(350 - 4.85)(8.34)}{1 + 0.05(10.36)}$$

Sludge generation at 20 °C = 1422 lb VSS/d.

13. Required oxygen demand at

Carbonaceous oxygen demand

$$\text{lb BOD}_5 \text{ removed/d} = (1.25 \text{ MGD})(350 - 4.85)(8.34)$$
$$= 3598 \text{ lb/d}$$

Solids removed $= 1422 \text{ lb/d at } 20°C$

Ultimate BOD of solids $= (1.42)(1422) = 2019 \text{ lb/d}$

Total oxygen requirements :

lb O_2/d = Ultimate BOD removed $- 1.42W_s$

where W_s is the mass of sludge wasted

lb $O_2/d = (5140 - 2019) = 3121 \text{ lb } O_2/d.$

Nitrogenous oxygen demand (NOD)

$$\text{NOD} = (NH_4.N)(4.57), \text{ or}$$
$$= (TKN)(4.57)$$
$$\text{NOD} = (1.25 \text{ MGD})(8.34)(40)(4.57)$$
$$= 1906 \text{ lb } O_2/d.$$

Total oxygen demand (TOD)

$$\text{TOD} = \text{Carbonaceous oxygen demand} + \text{NOD}$$
$$= (3121 + 1906) = 5027 \text{ lb } O_2/d.$$

Note :

1. Food to micro–organism chain

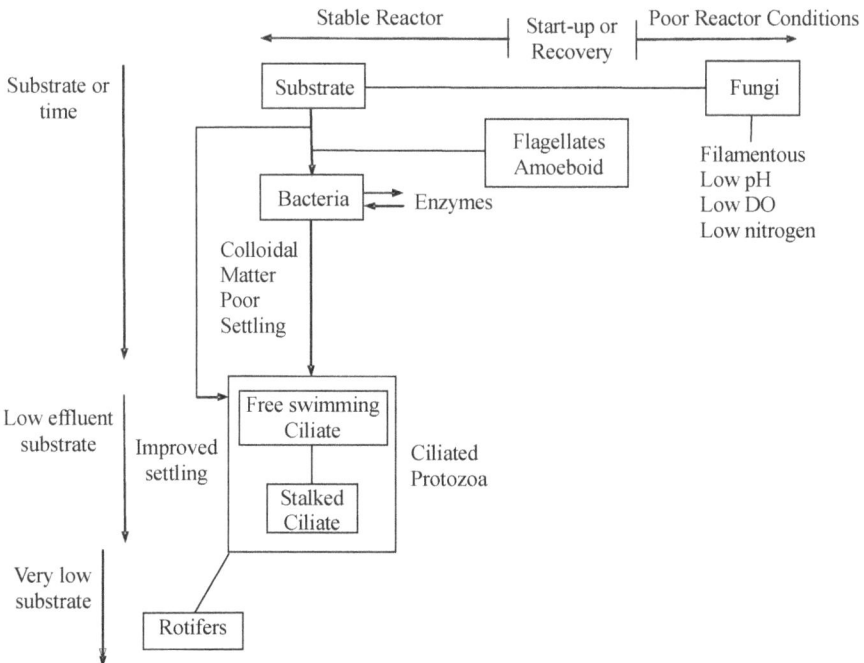

Figure 97 : Food to micro-organism chain.

2. Fate of pollutants

$$Q(S_{in} - S_e) = R_v + R_a + R_b$$

where R_v is the volatile removal rate (lb/d), R_a is the adsorption rate (lb/d), and R_b is the bio–degradation removal rate (lb/d).

Volatilization and stripping : Volatility is measured by Henry's Law and the aerator power output, and is affected by compound diffusivity and wind velocity inducing volatile pollutant transport from the surface.

Ideal gas law can be used to estimate R_V

Assuming that all volatile materials are carried from the system with the escaping gas (air) to its maximum capacity :

$$R_V = -\frac{G(P_e - P_i)}{RT} MW \; [g/d]$$

where G is the total gas volumetric flow rate (m^3/d), P_e is the partial pressure of the component in the exit gas (atm), P_i is the partial pressure of the component in the inlet gas (atm), R is the universal gas constant (0.00008206 m^3 atm/$^\circ$K.mole), T is the reactor temperature in $^\circ$K, and MW is the molecular weight of the component.

Assuming the exit gas concentration is the equilibrium concentration, based on the liquid phase solute concentration :

$$P_e = \frac{HS}{MW}$$

where H is the Henry's constant (atm.m^3/mole), and S is the component liquid concentration (g/m^3).

Entrance P_i is normally = 0. Therefore, R_v becomes :

$$R_v = -\frac{GHS}{RT}$$

[Does not include the effects of aeration devices or process characteristics].

Adsorption rates (R_a)

$$R_a = - W \, X_v \, q$$

where W is the waste sludge flow (m^3/d), X_v is the concentration of the wasted sludge [X_v, g VSS/d], and q is the compound sorption ratio (g/g VSS).

The sorption ratio is expressed as :

$$q = k_p \, S$$

where S is the component concentration in the aerated basin, and k_p is the partition coefficient (m^3/g VSS).

The partitioning ratio is commonly related to the octanol/water coefficient (K_{OW}) and is defined as the ratio of specific compound concentration in the octanol phase to its concentration in water (A large value of K_{OW} indicates high adsorption potential from water to organic (biological solids).

k_p is expressed as :

$$k_p = - (0.00000063)(f_{oc})(K_{ow})$$

where f_{oc} is equal to 0.531 where biological cells are represented by $C_5H_7O_2\,N$

Therefore, final expression for R_a is expressed as :

$$R_a = - (0.0000003345)(W)(X_v)(K_{ow})$$

$$k_p = a(K_{ow})^b$$

where a and b are the constants specific to the sludge type and specific organics adsorbed.

Biodegradation rate (R_b)

$$R_b = - \frac{k\,X_a S}{K_s + S}\,V,\ (g/d)$$

where k is the max specific utilization rate (g/g VSS.d), X_a is the active biomass (g VSS/m³), and S is the aeration basin (effluent) substrate concentration (g/m³).

Most constituents are present in low concentrations, Monod's kinetics is expressed as a first order reaction in the fate model :

$$R_b = - k\,X_a\,S\,V$$

3. Biodegradability (BOD/COD ratio) rules of thumb :

Compounds with a ratio of less than 0.01 are relatively non–biodegradable.

Compounds with a ratio between 0.01 to 0.10 are moderately degradable.

Compounds with ratios greater than 0.1 are degradable.

– Biodegradability decreases with compound branching.

– Biodegradability increases with compound chain length.

– Unsaturated aliphatics are more biodegradable than the corresponding saturated hydrocarbons.

– Commonly, the respiration rate is between 8 to 20 mg MLVSS.hr.

4. Completely mixed system with recycle

$$S = \frac{K_s[1 + k_d\,SRT]}{SRT[(Y\,k - k_d)] - 1}$$

$$X = \frac{Y[S_o - S]}{[1 + k_d \, SRT]} \frac{[HRT]}{[SRT]}$$

$$P_x = \frac{YQ[S_o - S]}{1 + k_d \, SRT}$$

$$(SRT)^{-1} = \frac{YkS}{K_s + S} - k_d \quad \text{or} \quad (SRT)^{-1} = YU_a - k_d$$

$$(SRT)^{-1} = Y.E\left(\frac{F}{M}\right)_{fed} - k_d$$

$$Y_o = \frac{Y}{1 + k_d \, SRT} \quad \text{or} \quad Y_o = \frac{YU_a - k_d}{U_a}$$

If influent concentration and effluent concentration are same, washout conditions develop, the limiting SRT_l is expressed as

$$(SRT)^{-1} = \frac{YkS_o}{K_s + S_o} - k_d$$

where $S_o \gg K_s$,

$$SRT_l = \frac{1}{Yk - k_d}$$

where k is the maximum substrate utilization rate (t^{-1}), K_s is the substrate concentration when $dx/dt = 0.5 \, k$ (mass/volume), X is the biomass concentration (mg/L), k_d is the decay constant of micro–organism (t^{-1}), P_x is the excess biomass generated (mass/time), S is the reactor substrate concentration (mass/volume), S_o is the influent substrate concentration (mass/volume), and U_a is $\dfrac{F}{M}$ ratio $\left[= \dfrac{S_o - S}{(HRT)X} \right]$

SRT design safety factor $\left[\dfrac{SRT \, (design)}{SRT \, (limiting)} \right]$

Table 93 : Safety factor for design purposes [SRT].

Type of oxidation	Safety factor = SRT_d/SRT_l ratio
Carbonaceous Extended aeration	> 70
Other activated sludge configuration	20–70
Short term aeration	4–20
Nitrification	
	Minimum safety factor of 2 based on SRT_d/SRT_l with SRT_d at least 10 days

5. **Plug flow reactor :** Lawrence and Mc Carty suggested the design equations for a plug flow reactor based on biomass in the reactor not significantly changing, the SRT/HRT being greater than 5, and recycle ratio being less than 1; under such situations, the design equations for reactor solids (x), P_x, and HRT are similar to those for a completely mixed system. Equations for effluent substrate concentration (S) are difficult to arrive at mathematically but can be approximated by assuming an infinite number of CSTRs in series.

If recycle ratio is less than 1, the SRT can be expressed as :

$$(SRT)^{-1} = \frac{(Y)(k)(S_o - S)}{(S_o - S) + K_s \ln\left(\dfrac{S_o}{S}\right)} - k_d$$

SRT_{min} (SRT_l) is mathematically difficult to define.

6. Filamentous growth of micro–organisms (sludge bulking).

   Filamentous micro–organisms are present in all activated sludge systems, bulking problems result when their growth rate exceeds that of the floc formers.

   Filamentous micro–organism growth is intensified by some wastewater constituents, low DO, low reactor substrate concentration, low sludge loading (F/M), and high SRT. Glucose, etc. like seccharides favour filamentous growth.

   Filamentous micro–organisms are slow growers, whereas floc formers are fast growers. Low substrate concentration (less than 30 mg/L soluble degradable COD) resulting low loading and high SRT have been identified as major concerns in intensifying filamentous growth.

   Dominant filamentous growth rate will occur in completely mixed system because of inherent low substrate concentration in the reactor, whereas floc farming micro–organisms will dominate in plug flow system because of a substrate gradient allowing suppression of filaments. Settleable sludges with SVI values less than 100 were observed in hydraulic regimes approaching plug flow from multiple tests conducted with reactors operated at near plug, complete mix, and intermediate mix conditions.

   Low pH also results in filamentous growth (pH less than 6). Nutrient deficiency (N and P), and septic wastes also contribute to the development of sludge bulking.

7. **Aerobic or anaerobic treatment :** Theoretically, enough aerator capacity can be included to assure aerobic conditions for a wide range of feed concentration, but from an energy and operating cost aspect, aerobic suspended growth systems are limited to a treatment of wastes at concentration of 3000 mg/L BOD_5 or less, most often in the range of 500

mg/L for large waste volumes. Above, 3000 mg/L BOD_5 an anaerobic system with the possible fuel recovery advantage, because a visible alternative.

8. Aeration basin configuration

A parallel configuration can be employed where production (and waste loads) variation is significant, with one parallel train at optimum conditions and the other adjusted to accommodate varying surplus loads.

A series configuration can be considered where toxic shock may be problem, allowing recovery in the final basins.

A two–stage operation could be more effective where high influent loadings and high effluent quality are a primary criteria. A second stage tends to be unstable, producing a poorer sludge quality because or underloading [can be remedied by including bypass provisions to bleed raw influent into the second stage (provision should be made for independent sludge return and washing to maintain proper loading rates and avoid mixing of biological populations in multicultural systems)].

Multiple units allows repairs to one basin without complete shutdown.

A round or square configuration improves mixing and reduces the above–ground heat loss surface area (because of minimum area for equal value). Heat loss can be minimized by increasing the below ground base depth and thereby reducing the above ground tank surface area.

Deep tanks can improve heat conservation but they are more difficult to aerate and mix.

Complete mix reactors should have L/W ratio of less than 3:1, no greater than 5:1 (ideally 1:1).

Plug flow reactors should have L/W ratio of at least 5:1 and in many cases over 10:1.

CSTR's effectiveness can be improved by increasing mixing energy to the practical economical limits (optimising mixing and turbulence).

Plug flow efficiency is affected by the basin geometry, improving with increasing hydraulic efficiency (HE) :

$$HE = \frac{\text{Hydraulic} - \text{mean detention time (t)}}{\text{Volumetric residence time}\left(\dfrac{V}{Q}\right)}$$

Mixing is minimised to that required to ensure proper oxygen distribution and the reactants contact, providing less than ideal hydraulic efficiency (HE). A correlation relating HE to flow width and length is expressed as :

$$HE = 0.84[1 - \exp[- 0.59 \ L_f/W_f]$$

An alternative, to increasing L/W ratio is to create carousal–type path by adding internal dikes. The flow length to width (L_f/W_f) ratio depends on the number of dikes with a ratio frequency in the range of 0.8 to 1.0, and is expressed as

$$\frac{L_f}{W_f} = \left[\frac{L}{W}\right](r)(N+1)^2$$

For any required residence, flow rate, and depth, the approximate basin area can be evaluated by :

$$LW = \frac{a}{1 - \exp\left[0.59\left(\frac{L}{W}\right)(r)(N+1)^2\right]}$$

where a is $= \left(\dfrac{tQ}{0.84\,D}\right)$ t is the required residence time (T), Q is the influent flow rate [m³(or ft³)time], and D is the basin depth [m(ft)]. The schematics of the rectangular basin configuration is shown hereunder :

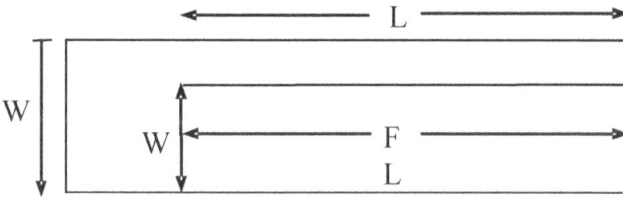

Figure 98 : Schematics of a rectangular a basin configuration.

where r is the ratio of dike length to basin length, L is the physical length [m(ft)], W is the physical width (m³(ft)].

Diffused air system : Width to depth ratio = 1:1 to 2.2:1 (typical 1.5:1) for spiral flow mixing.

Example 6.133

Engineering Considerations of Reoxygenation of Treated Wastewater Effluent through Cascade Aerators

It is necessary to re-oxygenate the treated wastewater effluent. Two principal categories of re-oxygenation systems are cascade and mechanical or diffused air re-oxygenation.

Cascade aeration relies on air and water interfaces at weir overflows, flumes, spillways, and similar hydraulic structures.

Precise measurements of oxygen supplied (mass) or changes in DO are difficult to obtain. Equally, the process is difficult to control. These systems are not energy efficient but they work well and cost little (where hydraulic head is sufficient and the design cannot be modified to conserve head or use it for other efficient purposes).

Re-aeration occurs at the weir's crest during water surface formation and is enhanced by bubble entertainment and splashing in the lowerpool.

Re-aeration over a single weir can be estimated by Nakasone's equation :

Nomenclature of terms in a cascade re–aeration equations

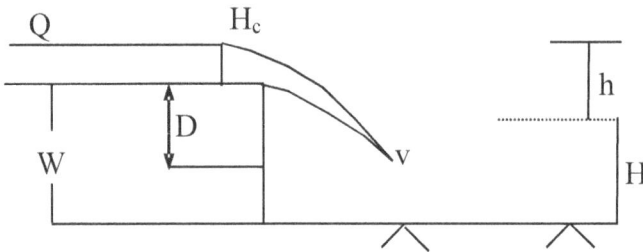

Figure 99 : Cascade reaction system.

The basic equations are :

$$ln\ r\ (20) = 0.0785\ (D + 1.5\ H_c)^{1.31}\ q^{0.428}\ H^{0.310} \tag{1}$$

[For $(D + 1.5\ H_c) > 1.2$ m and $q \le 235$ m^3/m.h]

$$ln\ r\ (20) = 0.0861\ (D + 1.5\ H_c)^{0.861}\ q^{0.428}\ H^{0.310} \tag{2}$$

[For $(D + 1.5\ H_c) \le 1.2$ m and $q > 235$ m^3/m.h]

$$ln\ r\ (20) = 5.39\ (D + 1.5\ H_c)^{1.31}\ q^{-0.363}\ H^{0.310} \tag{3}$$

[For $(D + 1.5\ H_c) > 1.2$ m and $q > 235$ m^3/m.h]

where, D is the drop height (m), H_c is the critical water depth on the weir (m), q is the discharge per width of the weir (m^3/m.h), H is the tailwater depth for downstream channels having horizontal beds (m), and r (20) is the oxygen deficit ratio at 20°C.

Gameson, et al derived the expression for the temperature correction of oxygen deficit ratio (r) as:

$$ln\ r\ (T) = ln\ r\ (20)\ [1 + 0.0168\ (T - 20)]$$

Barret, et al derived another expression for the determination of required cascade height :

$$H = R - 1/0.11\ ab\ (1 + 0.046\ T)\ \text{US units}$$

or   H = R − 1/0.36 ab (1 + 0.046T) SI units

where R is the oxygen deficit [R = $(C_s − C_o)/(C_s − C)$], Cs is the DO saturation concentration of wastewater at temperature T (mg/L), C_o is the DO concentration of pre-aeration (mg/L), C is the final required DO level after post aeration of influent (mg/L), a is the water quality parameter equal to 0.8 for a wastewater treatment plant effluent, b is the weir parameter (for a weir, b = 1.0 ; for steps b = 1.1 ; for step weir, b = 1.3), T is the water temperature (°C), and H is the height through which water falls (ft or m).

If is necessary to determine the critical wastewater temperature that affects the DO saturation concentration.

Example 6.134

Estimation of Sludge Quantities from a Conventional Activated Sludge Plant

Assume the following data for estimation of sludge quantities :

Flow rate	: 5 MGD
Influent BOD	: 250 mg/L
Influent SS	: 150 mg/L
SS removal in PST	: 50%
BOD removal in PST	: 35%
Raw sludge content	: 5%
	: 1601
$\dfrac{F}{M}$ ratio	: 0.33
Waste activated sludge concentration	: 10,000 mg/L
Thickener solids concentration	: 5%
K	: 0.48
Aerobic sludge content	: 1.5%.

Solution

1. Primary sludge (dry)

$$W_p = FSS = 0.5 \times 150 \times 5 \times 8.34 = 3128 \text{ lb/d}$$

$$\text{Volume of sludge } (V_p) = \frac{W}{\left(\dfrac{100-p}{100}\right)} = \frac{3128}{\left(\dfrac{100-95}{100}\right)8.34} = \frac{3128 \times 100}{5 \times 8.34}$$

$$= 750 \text{ gal/d.}$$

2. Sludge from aeration basin to be wasted

$$W_s = K \, BOD_a$$
$$= 0.48 \times 0.65 \times 250 \times 5.0 \times 8.34$$
$$= 3253 \, lb/d$$

$$\text{Volume of sludge } (V_s) = \frac{W_s}{\left(\dfrac{s}{100}\right)8.34} = \frac{3253}{\left(\dfrac{1.5}{100}\right)8.34}$$

$$= 26{,}000 \, gal/d.$$

3. Gravity belt thickener (98% solids capture)

Therefore,

$$\text{Waste solids (activated) volume} = \frac{3253 \times 0.98}{\left(\dfrac{5}{100}\right)8.34}$$

$$= 7645 \, gal/d.$$

4. Total sludge

Total sludge weight (W) = (3128 + 3253) = 6380 lb/d

Total sludge volume (V) = (7500 + 7645) = 15,145 gal/d.

5. Total solids content

$$s = \frac{100}{8.34}\left(\frac{W}{V}\right) = \frac{100}{8.34}\left(\frac{6380}{15{,}145}\right) = 5\%.$$

Example 6.135

Sludge Volume Index (SVI) and Total Sludge Generation

The total sludge generation rate is expressed as P_t:

$$P_t = \frac{X\,V}{\theta_c} = Q_e X_e + Q_w X_r \tag{1}$$

In biomass balance at steady–state:

$$P_t = \frac{\mu_m \, S X V}{K_s + S} - k_d X V \tag{2}$$

SVI is defined as volume in mL occupied by 1 g MLSS in dry weight after 30 min. in a 1000 mL graduated cylinder:

$$SVI = \frac{[\text{Sludge volume (mL)}] \times 10^3}{MLSS \, (mg/L)}$$

[SVI is the volume after settling in 1000 mL cylinder]

SVI can be related to recirculation ratio (R)

$$R = \left[\frac{10^6}{(SVI)(MLSS)} - 1 \right]^{-1}$$

SVI can also be expressed in terms of suspended solids (SS) in the returned sludge :

$$SS \ (mg/L) = \frac{10^6}{SVI}$$

or $\quad SS \ (\%) = \dfrac{10^2}{SVI}.$

Note :

1. *Sludge bulking :* Bulking means that there is an increase of cross–links and a three–dimensional network formation group instead of compact flocs of large particle formations. Sludge bulking is normally viewed through the dominance of filamentous bacteria. Both filamentous bulking and non–filamentous bulking can be explained since the pre–requisite of filamentous bacteria biomass is the type of bio–polymer than can undertake cross–linking. Therefore, the consecutive conditions for both filamentous and non–filamentous bulking are the same :

    Low dissolved oxygen

    Low F/M ratio

    Septic wastewater

    Nutrient deficiency

    Low pH

    Traditional sludge bulking control is based on changing these conditions. New control should be based on colloidal chemistry concepts.

2. Distribution of water in activated sludge

Table 100 : Distribution of water in activated slude.

Free Water	75% by Volume
Floc water	20%
Capillary water	2%
Bound water	2.5%
Solids	0.5%
Volatile solids (S = 1.0)	5% by weight
Fixed solids (S = 2.5)	5%
Water (S = 1)	90%

$$\frac{1}{S} = \sum_{i=1}^{n}\left[\frac{W_i}{S_i}\right] = \frac{0.05}{1.0} + \frac{0.05}{2.5} + \frac{0.9}{1.0} \text{ or } S = 1.03$$

where S = specific gravity of sludge, W_i = Weight fraction of the component in sludge, S_i = Specific gravity of the component.

3. Common water and wastewater sludges

Table 101 : Common water/wastewater sludges.

Sludge	Concentration of Solids (% solids)	Characteristics
Raw primary	4–8	Vile, bad odour, gray–brown, does not well on drying bed, but can be dewatered mechanically
Anaerobic primary digested produce gas	6–10	Dewaters well on drying beds; musty,
Filter humus	3–4	Fluffy; brown
Waste activated	0.5–1.5	Little odour, yellow brown, fluffy; difficult to dewater; very active biologically
Mixed digested (Primary + Waste activities)	2–4	Black–brown; produces gas; musty; not as easy to dewater as digested primary
Aerobic digested	1–3	Yellow–brown; sometimes difficult to dewater; biologically active
Waste alum	0.5–1.5	Gray–yellow; odourless; very difficult to dewater

4. Typical sludge composition of domestic wastewater sludges

Table 102 : Typical sludge composition of various types of sludges.

Components	% of Component		
	Primary	Activated	Digested
Organic matter	60–80	60–75	45–60
Total ash	20–40	25–40	40–55
Protein	20–30	30–40	15–20
Grease and fats	6–35	5–12	3–20
Cellulose	5–15	5–15	5–15
N	2–4	2–6	1.5–6
P_2O_5	1–3	2–7	1.5–4
P_2O_5	0–1	0–2	0–2

5. Most predominant filament types associated with sludge bulking :

Norcardia sp.

Type 1701

Type 021 N

Type 0041

Thiotrix sp.

Sphaerotilus sp

Microthrix parvicella

Type 0092

Haliscomenbacter hydrossis

Type 0675

Type 0803

Nostocolia limicola

Type 1851

Type 0961

Type 0581

Beggiatoa sp.

Fungi

Type 0914

6. **Structure and bulking of sludges :** Chemically, the biomass stays behind since the chemical conversion including biodegradation is largely polymeric in nature. The major components in active sludge are protein, nuclic acid, lipids, and carbohydrates. Yet the components of FSS are largely humus materials, including the more refractor of biopolymers [mucopolysaccharide: polysaccharide containing N and/or S], polyesters [poly Bhydroxyl butyrate], collagen, and resins. In addition to this, surface active polymeric surfactants are present. In the colloidal sense, this sludge is a 3–dimensional cross–linked gel structure where the bridges or links may be a number if polar functional groups. It is possible that different types of transitional metals may be also be parts of the linkages. The skeleton of filamentous bacteria is also created from the cross link process of living biomass. The charges become important in many varieties of poly–saccharides, *e.g.*, glucous, chitin, etc. in particular, the gluconic acid unit carries +ve and –ve charges, and so act as surfactant. Charges in molecules would promote agglomeration into large flocs as seen in the modern concept of fractals.

The filament backbone model which emphasizes the linkages created by filamentous bacteria.

Sludge Surface Model : Metal ions, fibrils, bacterium, capsule/slime.

Swollen network sludge floc: Cross–links, branches (free ends), trapped species, biopolymer web, filamentous net, filamentous net, particles.

7. Microbial composition of an activated sludge floc

   Negatively charged polysaccharide matrix.

   Entrapped bacteria and colloidal particles.

   Rigid filamentous backbone.

   Free swimming and attached protozoa.

Example 6.136

Activated Sludge Process Sludges

Calculate the quantities and solids contents of primary, waste activated and mixed sludges for an operational TF with a flow rate of 1.5 MGD and SS concentration (influent) of 250 mg/L. Also calculate the consolidated sludge volume (95% solids capture) with 5% solids content. Assume the following data :

Primary SS removal	:	65%
BOD removal	:	35%
Water content in raw sludges	:	95%
$\dfrac{F}{M}$ ratio	:	1:3 in aeration basin
$\dfrac{MLVSS}{MLSS}$ ratio	:	0.8
Solids content in waste activated sludge	:	12,000 mg/L
Mixed sludges thickened to (gravity thickened)	:	7.0%
Raw BOD	:	250 mg/L

Solution

1. Primary sludge solids (W_{sp}) and volume (V)

1.1 Primary sludge

$$W_{sp} = f(SS) \times Q(8.34) \ (lb/d)$$
$$= 0.65(250) \times (1.5) \times (8.34) = 2033 \ lb/d$$

1.2 Primary sludge volume

$$V = \frac{W_{sp}}{\left[\left(\dfrac{100-p}{100}\right)\right] \times S} = \frac{2033}{\left[\left(\dfrac{100-95}{100}\right)8.34\right] \times 1}$$

$$= 4875 \ gal$$

2. Biological secondary solids (W_{ss})

$$W_{ss} = k \times BOD \times Q \times 8.34 \ (lb/d)$$

At $\dfrac{M}{F}$ of 3.0, lb VSS produced/lb of BOD applied = 0.27 (Refer Figure 100).

$$k = \frac{0.27}{0.80} = 0.3375$$

$$W_{ss} = 0.3375 \times 250 \times 1.50 \times 0.65 \times 8.34 = 686 \text{ lb/d}$$

$$V = \frac{W_{ss}}{\left(\dfrac{S}{100}\right) \times S} = \frac{686}{0.015 \times 8.34} = 5483 \quad (S = 1)$$

3. Total sludges (Blended sludge volume-V)

$$V = 4875 + 5483 = 10{,}358 \text{ gal/d}$$

3.1 Total sludge weight (W_s)

$$W_s = W_{sp} + W_{ss}$$
$$= 2033 + 686 = 2719 \text{ lb/d}$$

3.2 Solids content in the blended sludge

$$= \frac{2719 \times 100}{10{,}358 \times 8.34 \times 1} = 3.15\%$$

4. Thickened sludge volume

$$= \frac{0.95 \times 2719}{0.05 \times 8.34} = 6194 \text{ gal.}$$

Note :

1. Concentration of organic matter in wastewater is about 200 mg/L (0.02%).

2. Typical raw sludge is 40,000 mg/L (4%).

3. 1.0 MGD wastewater will produce about 5000 gal of sludge.

4. The raw sludge is odorous and putrescible residue must be processed further and reduced in volume for land disposal, incineration or barging to sea. Common methods include : mechanical thickening, biological digestion, and dewatering after chemical conditioning.

5. Dry solids may be calculated 0.20 lb/Capita/day for TF (excluding household garbage grinders when the actual amount realised in treatment process is closer to 0.12 lb/capita/day. The required digester volume may be computed using a conservative figure of 5 ft³/population equivalent.

6. Primary sludge accounting for 50–60% of SS of the SS applied. Screw is usually less than 1% of the settled sludge volume. Typical solids

concentrations in raw primary sludge form settling municipal wastewater are 6–8% and volatile solids vary from 60 to 80%.

7. Trickling filter – humus from secondary filtration : The humus sludge is settled with raw organics in primary clarifier. The combined sludge has solids content of 4–6%, which is slightly thinner than primary residue with raw organics only.

8. Waste activated sludge : The return activated sludge is 0.5–2% SS with a volatile fraction of 0.7 to 0.8. Routing of waste activated sludge to the wet well for settling with raw wastewater is not recommended. CO_2, H_2S and odorous organics are liberated from the settling in the primary basins as a result of anaerobic reactions, and SS is rarely greater than 4%. Waste–activated sludge can be thickened by floatation or centrifugation.

9. Anaerobically digested sludge : Well digested sludge, dewaters rapidly on sand drying beds (releasing in offensive odour resembling that of garden loan). Addition of chemicals necessary to coagulate a digested sludge prior to mechanical dewatering. Dry residue is 30–60% volatile and compactness of digested sludge ranges from 6 to 12%, depending on the mode of digestion.

10. Aerobically digested sludge : Thickness of aerobically digested sludge is less than that of the influent (50 % of VS are converted to gaseous end products). Stabilized sludge, expensive to dewater, is often disposed of by spreading on land for its fertilizer value.

11. Mechanical dewatered sludges : Density of dewatered cake ranges from 15 to 40%.

12. k is the fraction of applied BOD that appears as excess biological growth in waste activated or filter humus, assuming about 30 mg/L of BOD remaining in the secondary treatment in the expression :

$$W_{ss} = k \times BOD \times Q \text{ (gal)} \times 8.34 \text{ (for k, refer Figure 100)}$$

13. Peak loads must be assessed for each ETPs : Maximum weekly dry solids will be about 25% greater than the yearly mean.

14. Compactness of waste activated sludge is difficult to predict. Bulking can easily reduce the solids content to one half say from 15,000 to 7500 mg/L, thus doubling the volume. One guide line is to size mechanical thickeners at 200% of the estimated volume during normal operation in order to ensure consolidation of the diluted slurry and provision should be made for addition of coagulation to and in concentrating waste activated sludge.

www.ingramcontent.com/pod-product-compliance
Lightning Source LLC
Chambersburg PA
CBHW020216290326
41948CB00001B/69